KV-064-685

Recent Developments in Food Analysis

Edited by
W. Baltes
P. B. Czedik-Eysenberg
W. Pfannhauser

Proceedings of the
First European Conference
on Food Chemistry (EURO FOOD CHEM I)
held at Vienna (Austria),
17-20 February 1981

Weinheim · Deerfield Beach, Florida · Basel · 1982

Prof. Dr. Werner Baltes
Institut für Lebensmittelchemie
der Technischen Universität
Müller-Breslau-Str. 10
D-1000 Berlin 12

Dr. Peter B. Czedik-Eysenberg
Verein Österreichischer Chemiker
Arbeitsgruppe Lebensmittelchemie
Eschenbachgasse 9
A-1010 Wien

Dr. Werner Pfannhauser
Forschungsinstitut der Ernährungswirtschaft
Blaasstr. 29
A-1190 Wien

This book contains 192 figures and 94 tables

CIP-Kurztitelaufnahme der Deutschen Bibliothek
Recent developments in food analysis: proceedings
of the 1. Europ. Conference on Food Chemistry
(Euro Food Chem I), held at Vienna (Austria),
17-20 February 1981 / ed. by W. Baltes ... -
Weinheim ; Deerfield Beach, Florida ; Basel :
Verlag Chemie, 1982.
    ISBN 3-527-25942-2

NE: Baltes, Werner [Hrsg.]; Euro Food Chem <01, 1981, Wien>

© Verlag Chemie GmbH, D-6940 Weinheim, 1982
All rights reserved (including those of translation into foreign languages). No part of this book may
be reproduced in any form — by photoprint, microfilm, or any other means — nor transmitted or
translated into a machine language without written permission from the publishers.
Registered names, trademarks, etc. used in this book, even when not specifically marked as such,
are not to be considered unprotected by law.

Printer, Bookbinder: Bitsch KG, D-6943 Birkenau
Printed in the Federal Republic of Germany

Preface

The lectures in this volume were presented during EURO FOOD CHEM I, the First European Congress on Food Chemistry.

This Congress was organised by the Working Party on Food Chemistry of the European Federation of Chemical Societies and served two purposes: to introduce the Working Party to the scientific community and to create contacts among food chemists troughout Europe. To promote contacts seems desirable indeed, as many food chemists are working in specialised areas and lack the opportunity to exchange experiences and to receive first-hand information on new and efficient methods in analytical chemistry.

Looking back we feel that 11 plenary lectures, 57 short communications and 17 posters on the one hand and more than 350 participants from 24 countries on the other proved that these targets were met.

Encouraged by this experience the Working Party is now planning EURO FOOD CHEM II, which is to take place in Rome in 1983.

December 1981

W. Baltes
Chairmann of the Scientific Committee
of the FECS Working Party on
Food Chemistry

P.B. Czedik-Eysenberg
Chairmann of the FECS
Working Party on Food Chemistry

Contents

1 Chromatography and Spectrometry

Practical Implications of the Theory in HPLC 3
J.F.K. Huber

Applications of HPLC 14
R. Battaglia

Analysis of Vitamins in Food by HPLC 27
B. Hedlund

Determination of Fat-soluble Vitamins by HPLC 33
A. Rougerau, A. Guiller, J. Gore, O. Person

Determination of Methyl and Ethyl Caffeate in Vegetables by HPLC with
Electrochemical Detection 41
G. Sontag, F.I. Schäfers, K. Herrmann

High Performance Liquid Chromatographic Analysis of Short-Chain
Carboxylic Acids 47
C. Gonnet, M. Marichy

Separation and Estimation of Galacturonic Acid by HPLC 53
E. Forni, R. Giangiacomo, A. Polesello

The Determination of Food Additives by Ion-Pair Liquid Chromatography 59
G. van de Haar, J. P. Cornet

High Performance Liquid Chromatography of Food Dyes 64
J. Bricout

Analysis of Phenols in Smoke, Smoke Preparations and Smoked Meat
Products 70
L. Toth, R. Wittkowski, W. Baltes

Investigation on Contamination of Food by Polycyclic Aromatic Hydrocarbons 76
K. Tiefenbacher, W. Pfannhauser, H. Woidich

Degradation and Metabolism of Relevant Pesticide Residues 83
R. Engst

Possibilities of the Amino Acid Analyser in the Fruit Juice Analysis 97
W. Ooghe

Determination of Rare Sugars in Fruits After Separation by Gel
Chromatography... 105
H. Scherz

Correlation of Gel Chromatographic and Sensory Profiles used for the
Evaluation of Food Products 108
J. Pokorný, N. N. Lý, J. Karnet, J. Pavlis, A. Marcín, J. Davídek

Separation of Histamine Reverse-Phase by HPTLC and Soap-TLC and its
Determinations in Tuna Fish 114
V. Coas, L. Lepri

Gas Chromatography in Food Analysis 120
L. Boniforti, S. Lorusso

Methodological Problems in Vitamin B_6 Estimation 132
E. Kienzl, P. Riederer, J. Washüttl

Considerations and Remarks about Honey Volatile Components 137
C. Bicchi, C. Frattini, F. Belliardo, G. M. Nano

Use of Glass Capillary Columns for the Determination of Fusarium
Mycotoxins in Food 143
A. Bata, R. Lásztity, J. Galácz

Propylene Carbonate, Superior Solvent for the Extraction of Polycyclic
Aromatic Hydrocarbons 149
K. Potthast

Theory and Special Methods of Mass Spectrometry 155
S. Abrahamsson

Selectivity of Reagent Gases as used in the Analysis of Flavour Mixtures 162
U. Rapp, M. Hoehn, C. Kapitzke, G. Dielmann

Some Examples for the Identification of Artificial Flavours in Food 168
W. Pfannhauser, R. Eberhardt, H. Woidich

Study on the Composition of Cruciferae Oils by Gas Chromatography and
Mass Spectrometry... 174
S. Lorusso, L. Boniforti, A. Selva, E. Chiacchierini

Analysis of Organophosphoric Pesticide Residues in Food by GC/MS Using Positive and Negative Chemical Ionisation 183
H. J. Stan, G. Kellner

Use of Positive and Negative Chemical Ionisation Techniques to Determine and Characterize Free Amino Acids by Mass Spectrometry 190
D. Fraisse, F. Maquin, J. C. Tabet, H. Chaveron

Qualitative and Quantitative Aspects of the Analysis of Trace Contaminants in Foods by Selected Ion Monitoring 196
J. R. Startin, J. Gilbert

2 Bioassay and Enzymatic Methods

Bioassays in Food Analysis 205
J. Lüthy

Immunochemistry in Protein Analysis 215
J. Daussant

A New Method for the Rapid Detection of Microbial Contamination of Fruit Juices 229
J.G.H.M. Vossen, H.D.K.J. Vanstaen

Amines, Food and Brain Function 234
P. Riederer, G.P. Reynolds, K. Jellinger

Determination of Subresidual Proteolytic Activities in Foods 240
P. Rauch, L. Fukal, J. Káš

Applications of Isoelectric Focusing to the Analysis of Food Proteins 247
P.G. Righetti, A.B. Bosisio

Ultrathin-Layer Isoelectric Focusing, Electrophoresis and Protein Mapping of Must and Wine Proteins 264
A. Görg, W. Postel, R. Westermeier, G. Günther

Enzymes in Food Analysis 270
C. Mercier

Fast Method for Enzymatic Determination of Starch Using an Oxygen Probe 286
J.C. Cuber

A Heat Resistant Trypsin Inhibitor in Beans (Phaseolus vulgaris) 292
U.M. Lanfer-Marquez, K. Rubach, W. Baltes

Post Mortem Changes of Enzymatic Activities and Morphology in Meat 300
V. Vána, P. Rauch, J. Kás

Rapid Chromatographic Analysis of Enzymes and Other Proteins 306
O. Mikes

New Chlorine Containing Organic Compounds in Protein Hydrolysates 322
J. Davídek, J. Velíšek, V. Kubelka, J. Janíček

Method for the Qualitative and Quantitative Analysis of Gelling and
Thickening Agents 327
U. Pechanek, W. Pfannhauser, H. Woidich

3 Element Analysis

Assets and Deficiencies in Elemental Analysis of Food-Stuffs 335
G. Tölg

The Analysis of Inorganic Contaminants in Food 356
L.E. Coles

Determination of Arsenic in Muscles and Liver of Some Fresh-Water Fish 367
I. Petrović, F. Mihelić, T. Masina

Quality Assessment of the Edible Part of Mussels: Determination of Lead and
Cadmium by Atomic Absorption Spectroscopy and Voltammetry 372
L.G. Favretto, G.P. Marletta, L. Favretto

A Statistical Approach to the Balance of Lead in Milk and Some Dairy
Products 377
G.P. Marletta, L.G. Favretto, C. Calzolari

Study on the Presence of Heavy Metals in Cereals 383
M. Baldini, M. Centi, C. Micco, A. Stacchini

Trace Elements (Cr, Mn, Cu, Zn, Cd, Hg and Pb) Content in Italian Foods and in
the Meals Consumed in Other European Countries 389
G.C. Santoprete, N. Wolkenstein

4 Chemical Reactions and Interactions

Carbonisation and Caramelisation Measurements by Near Infrared
Reflectance 397
M. Meurens, M. Vanbelle

Importance and Quality of Lipoproteins from Wheat Flour 403
H. Stachelberger, E. Schönwald

Comparison of the Antioxidative Activity of Maillard and Caramelisation
Reaction Products 409
A. Huyghebaert, L. Vandewalle, G. Van Landschoot

Model Studies on the Heating of Food Proteins:
Influence of Water on the Alterations of Proteins Caused by Thermal
Processing 416
J.K.P. Weder, U. Scharf

Fractionation and Sensory Evaluation of the Reaction Products of Reducing
Sugars and their Degradation Products with L-Lysine 422
J. Davídek, J. Pokorný, H. Bulantová, A. Marcín, J. Pavlis, G. Janícek

Interactions of Artificial Sweeteners with Food Additives 428
G. Kroyer, J. Washüttl

5 Poster Session

Determination of Carbamate Pesticides 437
G. Blaicher, W. Pfannhauser, H. Woidich

Evaluation of the Lipid Composition of Some Infant Formula 443
G. Bellomonte, B. Carratu, R. Dommarco, S. Giammarioli, E. Sanzini

Immunological Detection of Meat from Turkey 449
A. Schweiger, K. Hannig, H.O. Günther, S. Baudner

Determination of Glucose, Fructose and Small Amounts of Saccharose in
Honey 453
J. Jarý, M. Marek, J. Bacilek

Gas-Chromatographic Analysis of Amino Acids in Fruit Drinks 458
J.P. Roozen, M.Th. Jannsen

Determination of Pyrethrins in Flour 461
I. Scheidl, W. Pfannhauser, H. Woidich

Correlation of Isotopic Composition and Origin of Foods 464
E.R. Schmid, H. Grundmann, W. Papesch, I. Fogy

Immobilization of Glucose Oxidase for Analytical Purposes 470
O. Valentová, M. Marek

Investigation of Colloid Substances in Must and Wine 476
J.-C. Villettaz, R. Amado, H. Neukom

Investigation of Proteins Regarding their Functionality for Baking 482
H.J.G. Wutzel

Investigation about Lead and Cadmium in Wild Growing Edible Mushrooms
from Differently Polluted Areas 486
J. Dolischka, I. Wagner

Evaluation of Lead by Automated Anodic Stripping Voltammetry in Canned
Juice in Presence of High Tin Concentration 492
A. Carisano, G.P. Cellerino, G.C. Dellatorre

Index 499

1 Chromatography and Spectrometry

Practical Implications of the Theory in HPLC

J.F.K. HUBER

Institute of Analytical Chemistry, University of Vienna,
A-1090 Wien, Waehringer Straße 38

SUMMARY

The article is a report on the theory of high performance liquid chromatography (HPLC) emphasizing results of immediate importance for the practitioner like elution function, resolution, choice of sample volume, sample dilution by the chromatographic process and speed of separation.

INTRODUCTION

The theory of the chromatographic process has achieved such a level that the theory can offer a significant aid in solving practical separation problems. The development of the modern column liquid chromatography was even initiated by theoretical predictions. Since the theory requests small diffusions distances for rapid lateral mass transfer to achieve high separation efficiency, the microparticulate column was developed. This type of column can be operated at higher flow velocity without significant loss in efficiency. Both, the microparticulate column packing and the higher flow velocity, require that the column is operated at higher pressure. Therefore, this modern version of column liquid chromatography is called "High Pressure Liquid Chromatography (HPLC)". Since the separation performance of this modern column liquid chromatography is significantly improved compared to the classical column liquid chromatography it is also called "High Performance Liquid Chromatography (HLPC)".

ELUTION FUNCTION

If the concentration of the sample at the outlet of the chromatographic column is recorded, a sequence of concentration peaks will be observed. In general these peaks will approach a gaussian shape. The theory of chromatography succeeded to derive a mathematical expression for gaussian elution peaks which are good approximations of the empirical elution peaks. The mathematical expression describing the concentration of the sample in the mobile phase at the end of the column is called elution or output function. For the case that the sample enters the column in a very short time, the output function approaches the residence time distribution function, which is determined exclusively by the column process. Otherwise the output function depends on the column process as well as on the input function. By considering the mass transport phenomena in the chromatographic column an equation for the elution function can be derived.

RESIDENCE TIME DISTRIBUTION

If we assume that

- the sample enters the column in a very short time interval,
- the mass distribution ratio between the fixed bed and the moving fluid is constant,
- the diffusion coefficients in the moving fluid and the fixed bed, respectively, are also constant,
- the eluent is fed to the column with constant flow rate and constant composition,
- the temperature of the column is constant,
- the column is sufficiently long,

the following expression [1,2] for the elution peak can be derived from the theoretical model of the mass transport phenomena in the chromatographic column:

$$\langle c_i^{(f)} \rangle = \langle c_i^{(f)} \rangle_{max} \cdot e^{-\frac{1}{2}\frac{(t_{Ri}-t)^2}{\sigma_{ti}^2}} \tag{1}$$

with

$\langle c_i^{(f)} \rangle$ = concentration of the component, i, in the moving fluid, f, at the column outlet averaged over the flow cross section.

$\langle c_i^{(f)} \rangle_{max.}$ = maximum value of $\langle c_i^{(f)} \rangle$ = peak height.
t = time elapsed since entering of the sample into the column.
t_{Ri} = average residence time of component i within the column
= retention time of component i.
σ_{ti}^2 = variance of the elution peak.

Equation 1 describes the concentration in the column effluent as function of time. It contains the three parameters t_{Ri} σ_{ti} and $\langle c_i^{(f)} \rangle_{max.}$. The elution function predicted by the theory is shown in Fig. 1.

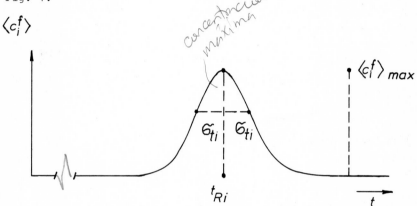

Fig. 1 Theoretical Gaussian elution peak and its parameter.

The three parameters are given by the following expressions:

$$t_{Ri} = \frac{L}{v}(1+\kappa_i) \qquad (1a)$$

$$\sigma_{ti}^2 = H_i \frac{L(1+\kappa_i)^2}{v^2} \qquad (1b)$$

$$\langle c_i^{(f)} \rangle_{max.} = \frac{Q_i}{(2\pi)^{1/2} \epsilon_f A (H_i L)^{1/2} (1+\kappa_i)} \qquad (1c)$$

in which
L = length of the column
v = flow velocity averaged over the flow cross section
κ_i = mass distribution ratio between stationary bed, b, and moving fluid, f.
Q_i = amount of component, i,
A = cross sectional area of the column

ε_f = fraction of cross sectional area in which flow occurs
($\varepsilon_f A$ = flow cross section)

H_i = theoretical plate height = characteristics for the dispersion of the component, i, in the chromatographic column. H_i is a kinetic quantity.

A number of equations have been proposed for the theoretical plate height. The most advanced theoretical plate height equation [3,4,5,6,7] is the following one

$$H = H_{Md} + H_{Mc} + H_{Ef} + H_{Eb} \tag{1d}$$

$$H_{Md} = \varphi_d \frac{D_{mi}}{v}$$

$$H_{Mc} = \frac{\varphi_{c1}}{1 + \varphi_{c2} \left(\frac{D_{mi}}{v d_p}\right)^{1/2}}$$

$$H_{Ef} = \varphi_f \frac{d_p^{3/2} v^{1/2}}{D_{mi}^{1/2}} \left(\frac{\kappa_i}{1 + \kappa_i}\right)^2$$

$$H_{Es} = \varphi_b \frac{d_p^2 v}{D_{pi}} \frac{\kappa_i}{(1 + \kappa_i)^2}$$

in which

H_{Md}, H_{Mc}, H_{Ef} and H_{Eb} = contributions to the theoretical plate height arising from the dispersion due to the mixing by diffusion, the mixing by convective effects, the mass exchange phenomena in the fluid stream, the mass exchange phenomena in the fixed bed.

$\varphi d, \varphi c, \varphi f$ and φb = characteristic numbers describing the influence of the column geometry on mixing by diffusion, mixing by convective effects, dispersion by the mass exchange process in the moving fluid, and dispersion by the mass exchange process in the fixed bed.

D_{mi} = diffusion coefficient of component i in the mobile phase
D_{pi} = diffusion coefficient of component i in the particles of the fixed bed
d_p = particle diameter of the fixed bed material

The flow velocity, v, is the only variable occuring in all four terms of eqn. (1d). It is therefore useful to consider the dependence of the theoretical plate height on the flow velocity. In Fig. 2 such a typical plot is shown together with the splitting into the single contributions. The quality of the packing geometry of the column is mainly reflected by the geometrical factors φ_{c1} and φ_f. Experimental results show, that for a well packed column the minimum value of the theoretical plate height is about $2d_p$. The crucial influence of the particle size on the column efficiency is evident from eqn. (1d). The conclusion from theory was to develop packing materials of very small particle diameter in the order of 10 μm with a narrow particle size range and to learn to prepare stable columns with a regular packing of such fine particles.

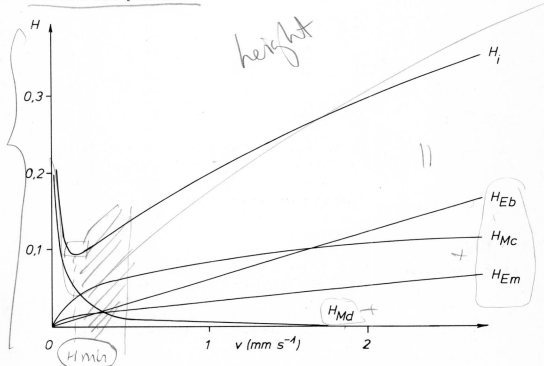

Fig. 2 Plot of the theoretical plate height H and its four contributions (H_{Md}, H_{Mc}, H_{Ef}, H_{Eb}) as function of the flow velocity, v, for d_p = 30 μm.

EFFECT OF SAMPLE VOLUME

With larger sample volumes the input time of the sample is not negligible. Linear system theory allows to describe also such cases. It can be shown that the elution function can be derived from the sample input function and the residence time distribution function. The width as well as the shape of the output function is influenced by the input function. The effect on the width can be described in a very simple manner[3]:

$$\sigma_t^2 = \sigma_{t0}^2 + \sigma_{tc}^2 \qquad (2)$$

with

σ_t^2 = variance of the output peak
σ_{t0}^2 = variance of the input peak
σ_{tc}^2 = variance of the residence time distribution function due to the chromatographic process

Eqn. (2) effects also the peak height according to the following relationship

$$\left\langle c_i^{(f)} \right\rangle_{max.} = \frac{Q_i}{\psi_L \left[(\frac{V_0}{\psi_0})^2 + (\varepsilon_f A)^2 H_i L(1 + \kappa_i)^2 \right]^{1/2}} \qquad (3)$$

where

V_0 = sample volume injected

ψ_L, ψ_0 = geometrical factors depending on the shape of the output peak and input peak, respectively. For a gaussian peak the shape factor is $\sqrt{2\pi}$, for a rectangular peak $\sqrt{12}$.

The shape factor ψ_L approaches the value for a gaussian peak if the first term in equn. (3) becomes negligible. The shape factor ψ_0 depends on the degree of mixing during the sample introduction. In the ideal case its value is $\sqrt{12}$.

RESOLUTION

In order to quantify the degree of separation it is necessary to define an appropriate measure. Such a quantity is the chromatographic resulution which describes the separation of two components. It is defined by the expression

$$R_{ji} = \frac{t_{Rj} - t_{Ri}}{\sigma_{ti}} \qquad t_{Rj} > t_{Ri} \tag{4}$$

where
R_{ji} = chromatographic resolution of the components j and i.

The resolution describes the ratio between the effect of the differential migration, which creates the separation and the effect of the peak dispersion, which works against the separation. The meaning of the resolution is illustrated by Fig. 3.

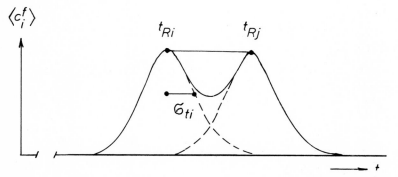

Fig. 3 Illustration of the chromatographic resolution of two components.

The resolution can be expressed by the process variables by inserting eqn. (1a) and (1b) into eqn. (4).
It results the relationship

$$R_{ji} = (r_{ji} - 1) \frac{\kappa_i}{1 + \kappa_i} N_i^{1/2} \tag{5}$$

in which
$r_{ji} = \kappa_j / \kappa_i$ = selectivity coefficient which is the ratio of the distribution coefficients of the components, j and i, for the distribution between the stationary phase, s, and the mobile phase, m.

κ_i = capacity factor for component i. The capacity factor is the mass distribution ratio between the stationary phase and the mobile phase; κ_i must be differentiated from κ_i occuring in the theoretical plate height equation. The latter describes the distribution between the fixed bed, b, and the moving fluid, f. For practical reasons κ_i is used instead of K_i, which is more difficult to measure.

N_i = L/H_i = theoretical plate number of the column for component i.

It characterizes the efficiency of the chromatographic column. Eqn. (5) is the fundamental equation of chromatography. The solution of each separation problem can be discussed on the basis of this equation which has to be applied for all pair of successively eluting components. If the mixture to be separated consists of n components the separation problem is described by n-1 resolution values: R_{21}, R_{32}, ..., $R_{n(n-1)}$. In practice the demand for resolution is combined with other demands like speed of separation or precision and low limit of detection.

SPEED OF SEPARATION

In order to discuss the speed of separation the eqn. (5) for the resolution is combined with eqn. (1a) for the retention time. Further the following substitution is made: $u_i = v/(1 + \kappa_i) = u_0/(1 + \kappa_i)$, where u_i is the migration velocity of the component, i, and u_0 the migration velocity of the mobile phase. The equation for the separation time, t_{Rn}, writes as follows[3,8]

$$t_{Rn} = \left[\frac{R_{ji}(1+\kappa_i)^2}{(r_{ji}-1)\kappa^2} \frac{H_i}{u_0} \right] (1 + \kappa_n) \qquad (6)$$

where n is the last eluting component of the sample and j and i is the most difficult to separate pair of components. From eqns. (6) and (1d) we can conclude that the ratio H_i/u_0, and therefore the separation time, t_{Rn}, decreases with increasing flow velocity and approaches a constant value which depends on d_p^2. This prediction is experimentally verified by the results shown in Fig. 4. In order to understand eqn. (6) it must be emphasized that the theoretical plate number must be kept constant to keep the resolution constant. Since the theoretical plate height increases with the flow velocity, the column length must be adequately increased if the flow velocity is increased in order to remain a constant theoretical plate number. As long as the slope of the $H_i(u_0)$-curve is less than the ratio H_i/u_0 it is possible to increase the speed of separation by increasing the flow velocity and in a smaller ratio the column length.

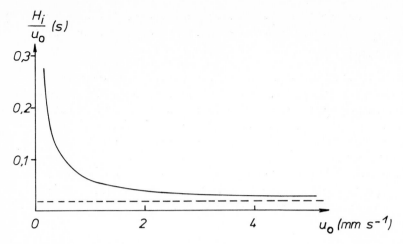

Fig. 4 Plot of the characteristic factor, H/u_0, for the separation time as function of the flow velocity, v, for d_p = 12 μm.

SAMPLE DILUTION

The peak height decreases during the chromatographic process because of the peak dispersion. The height of the elution peak is given by eqn. (3). The maximum resolution obtainable by the column is described by eqn. (5). In order to achieve a high precision and low limit of detection the concentration in the detector cell must be sufficiently high. If we are not limited in sample size the concentration can be increased by increasing the sample volume. With increasing sample volume the resolution will decrease, however. The influence of the sample volume on the resolution can be discussed on the basis of eqns. (2) and (4). Combining both equations results in the expression

$$R_{ji} = \frac{R_{jimax}}{\left[1 + \frac{\sigma^2_{t0}}{\sigma^2_{tci}}\right]^{1/2}} \tag{7}$$

in which
$R_{jimax} = \dfrac{t_{Rj} - t_{Ri}}{\sigma_{tci}}$ = maximum value of resolution which is obtained if the peak width is determined only by the peak broadening effect of the column process.

In order to optimize the chromatographic process with respect to resolution and sample dilution both characteristics, resolution and peak height, have to be considered together. The aim is to achieve maximum resolution and maximum peak height.

In Fig. 5 a plot[9] is shown which can assist in the choice of the proper sample volume. The plot is based on eqns. (3) and (7) taking into account that $Q_i = c_{io} V_0$, c_{io} being the concentration of component i in the sample. It can be seen from Fig. 5 that the sample dilution can be restricted to a factor three without to reduce seriously the resolution. This result can be achieved if a sample volume, V_0, equal to the volume standard deviation, σ_{vc}, created by the column process is chosen.

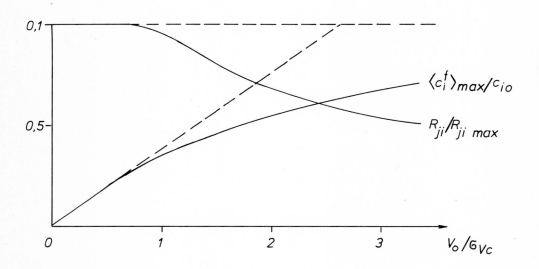

Fig. 5 Plot for the choice of the appropriate sample volume, V_0; $c_{i\ max}^{f} / c_{i0}$ = dilution factor, $R_{ji} / R_{ji\ max}$ = reduction factor of resolution, $\sigma_{vc} = \sigma_{tc} v \varepsilon_f A$.

REFERENCES

1) L. Lapidus and N.R. Amundson, J. Physic. Chem. $\underline{56}$ (1952), 984.
2) J.J. van Deemter, F.J. Zuiderweg and A. Klingenberg, Chem. Eng. Sci. $\underline{5}$ (1956), 271.
3) J.F.K. Huber and J.A.R.J. Hulsman, Anal.Chim. Acta $\underline{38}$ (1967), 305.
4) J.F.K. Huber, J. Chromatogr. Sci. $\underline{7}$ (1969), 85.
5) J.F.K. Huber, Chimia Supplementum 1970, 24.
6) J.F,K. Huber, Ber. Bunsenges. Phys. Chem. $\underline{77}$ (1973), 179.
7) J.F.K. Huber, Z. Anal. Chem. $\underline{277}$ (1975), 341.
8) Sj. van der Wal and J.F.K. Huber, J. Chromatogr. 149 (1978), 431.
9) J.F.K. Huber, J.A.R.J. Hulsman and C.A.M. Meijers, J. Chromatogr. $\underline{62}$ (1971), 79

A version of this lecture has also been published in Ernährung $\underline{5}$ (1981) by agreement of the publishers.

Applications of HPLC

R.BATTAGLIA

Kantonales Laboratorium Zürich, P.O.Box 8o3o Zurich, Switzerland

SUMMARY

HPLC analyses in food are discussed with respect to the various influences stemming from the food matrix. A series of examples is given.

The pure HPLC-part of food-analysis in many cases becomes almost negligible in the face of the overwhelming problems stemming from the food-matrix. This is illustrated in Fig.1, which shows the difference of chromatograms of a standard mixture and an extract. Sobering experiences of this kind force one to have a closer look at the analytical problem as a whole. Fig.2 illustrates the general structure of an HPLC application, which always consists - apart from the chromatography - in a work-up-procedure of some kind, which leads to a solution, that may be injected into the chromatograph. Very often, the interpretation of the resulting chromatograms forces the analyst to modify the work-up. This work-up procedure and the ultimate chromatographic determination are intrinsically interrelated. This leads to an early, important realization. Not every analytical problem is suitable for HPLC ! In other words: the fact alone, that one takes delight in doing LC, is a bad excuse to try and solve every possible problem with this technique. First, some basic thougths on the usefulness of HPLC in food analysis are taken into consideration.

Applications of HPLC   15

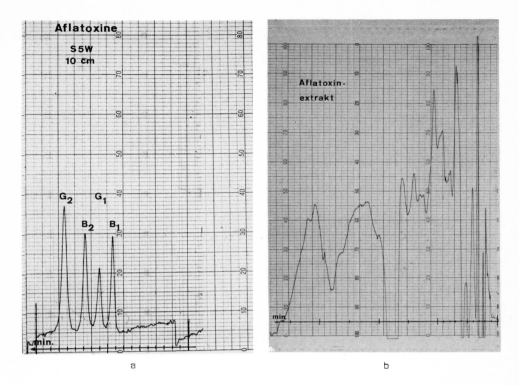

Fig.1 a) Aflatoxin-Standard-Mixture
       b) Peanut-Extract
       Conditions: S5W, 10cm, CH_2Cl_2 satd. with water, 0.6% methanol

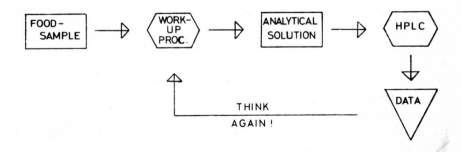

Fig.2   General Structure of an HPLC-Analysis

The most obvious compounds, that offer themselves for LC-analyses, are certainly those of low volatility and high polarity. However, every GLC-specialist knows, that one can easily analyse triglycerides up to 1000 daltons, and extremely polar compounds like carbohydrates and the like could just as easily be converted to derivatives like acetates, silyl ethers etc. and therefore would be suitable for GLC ! A further argument to use HPLC heard very often is, that it can be automated. This is also true for GLC, TLC, enzymatic analysis, titration and colorimetric analysis. Where then, does HPLC really make sense, where is it really the method of choice ? Although the question, whether HPLC could be a good method for a given problem, may sound trivial, once the pure chromatographic conditions are worked out, it remains the one central question, which must be asked and answered at the beginning of every new analytical problem. Not only the problem as such has to be considered, but also it's frequency of occurrence, its urgency as to the time allowed to obtain a result for one sample, the cost of alternative methods and so on.

Urgrinovits (1) presented an evaluation of the different methods to determine sugars in food. He compared the time-consumption and cost-efficiency of GLC, HPLC and enzymatic methods and came to the conclusion, that, depending on the number of samples per day and whether one, two or more sugars had to be determined all three methods had their range of optimal applicability.

Let us assume, that the decision to solve a particular analytical problem with the help of HPLC has been taken. One is then faced with the hard day-to-day problem of how to tackle the analysis. Basically a series of different approaches can be envisaged:

1. Certain chromatographic conditions have been worked out with the given equipment and standard solutions to analyse a compound. Then it is quite clear, that the sample-preparation has to be designed in such a way, as to produce a final solution, that has no interfering components. This approach may lead to an impractical, very complicated and tedious clean-up process.

2. From chemical knowledge a very simple and quick clean-up of the sample might be envisaged. However, it can be foreseen, that this will lead

to a very crude extract with lots of similar compounds in the final
solution. Then, the chromatographic system has to be designed so as
to be specific and selective for the compound in question. This may
mean the choosing of specialized stationary phases, exotic mobile
phases, specific detectors or postcolumn derivatisation.

3. A further problem arises with the analysis of trace components. The
following case may serve as an example: A compound has to be detected
at the ppb-level. The detector is able to give an interpretable signal
for, e.g. 1 ng under the given chromatographic conditions. A reasonable
injection volume is in the range of 1 to 2o µl. It therefore follows
implicitly, that the clean-up of the food sample has to be designed
so as to give a solution, of which 1o µl represent 1 g of food.
This means in other words, that during the clean up, the compound in
question has to be concentrated at least 1oo-fold with respect to
all the other food ingredients. From these considerations it becomes
quite clear, that there can be no general cook-book recipies for
LC-analyses in food. Every single problem will have to be judged on
its own merits, and the evaluation of an analytical procedure for a
given problem has to take into account all possible aspects.

EXAMPLES

Adapt Chromatography

Very often, a cheap and relatively modest column can prove satisfactory
for a certain separation problem. It is, for example, rather easy to
determine food preservatives (2) on a silica column of poor performance
(Fig.3).

The separation of sorbic acid and benzoic acid is not good at all,
however, due to the large differences in their UV-maxima the two compounds can be determined by selecting the proper detecting wavelength.
Nevertheless, this attractive approach is not suitable for the analysis
of these compounds in food. The procedure, which leads to an extract
suitable for chromatography on a silica-column is not only tedious and
time-consuming, but moreover shows catastrophic yields for certain foods !
(Table 1). It therefore follows, that a different chromatographic system
has to be chosen. Like in the majority of LC-analyses from food a

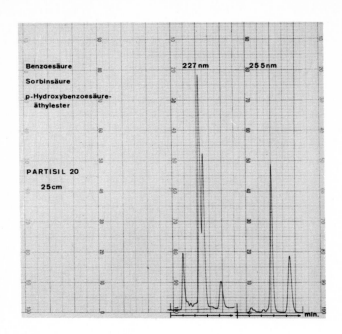

Fig. 3   HPLC of Food-Preservative
          Conditions: Hexane/Tetrahydrofurane

Konservierungsmittel, LC-Analysen

Extraktionsausbeuten nach LMB Kap. 44 B, in %

	Benzoesäure		Sorbinsäure		p-Hydroxybenzoe-säure-aethylester	
	Aether	Chloroform	Aether	Chloroform	Aether	Chloroform
Wein	100		100		100	
Bier	*	100	*	85	64	90
Tafelgetränk mit Orangensaft	47	100	30	76	100	100
Konfitüre	16	100	5	90	70	100
Kuchen	73	100	38	77	85	61
Kaffeerahm	8	88	10	58	19	59
Joghurt	60	78	21	63	95	59
Mayonnaise	100	*	73	58	100	40
Margarine	*	62	5	59	1	20

* : nicht auswertbar, da durch andere Peaks verdeckt.

Table 1   Extraction-Yields of Food-Preservatives

reversed-phase column proves to be the system of choice: sorbic acid in wine (Fig.4) or parabens in cosmetic products (Fig.5) demonstrate the case.

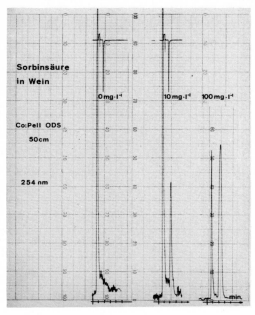

Fig.4   Analysis of Sorbic Acid in Wine
         Conditions: Water

Fig.5   Parabens in skin-cream. Peaks (in order of elution):
         p-Hydroxybenzoic acid-methy, ethyl, propyl-ester. Hypersil-5-ODS
         water: acetonitrile 4:1 with o,1% acetic acid, detection at 254

In these examples, the sample clean-up is virtually non-existent: the sample in question is simply dissolved or suspended and filtered in the mobile phase. Once an analytical system of this remarkable simplicity is found, it is always necessary to check, whether the eluted and assigned peak really represents the compound in question. In other words: its identity has to be proven. Fig.6 shows an example, where this has been done in situ by scanning the UV-spectra of two peaks.

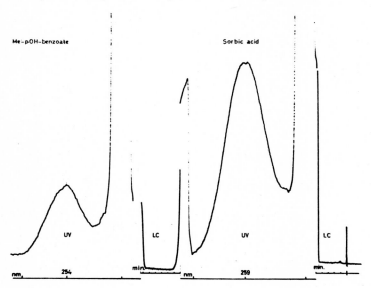

Fig.6   In Situ Spectra of Preservatives.
         Text-mixture of sorbic acid and methyl-p-hydroxybenzoate.
         Detector: Perkin-Elmer LC 55 with scanner

This example shows then, that by adapting the chromatographic system so as to require a very simple sample clean-up procedure, an efficient and fast analysis results. In this way many other applications have been developed, e.g. caffeine in tea and coffee, theobromine in cocoa, saccharine in soft drinks etc.

Adapt Clean-Up

An example, where the sample clean-up has to be adapted to the chromatographic system is the analysis of patulin in apple-juice (2). Patulin is a metabolite of rather high cyto-toxicity of the spoilage-

fungus Penicillium glaucum. It also shows some antibiotic properties
and its presence in apple-juice is limited to ca. 5o ppb in most
European countries. Patulin is easily chromatographed on silica columns;
however its chemical nature on the one hand suggests that it is not
too easily separated from many other neutral components present and the
required detectability in the ppb-range clearly calls for an enrichment
during the sample preparation.

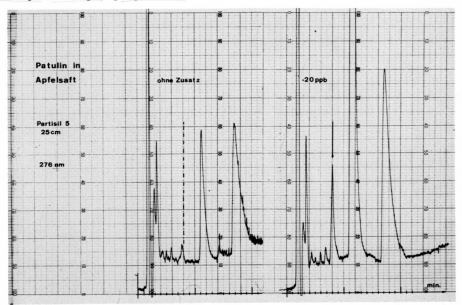

Fig.7   Patulin in apple-juice. Partisil 5. Di-isopropyl-ether:
        Tetrahydrofuram 95:5 with o.1% acetic acid.

Caffeine, Theobromine

Fig.7 shows the chromatogram of an extract of apple-juice. The sample
clean-up in this particular case was carried out by the elaborate
AOAC-method published by Ware (3) and consists in solvent extraction
and pre-chromatography on a preparative silica column. Since then,
numerous modifications of this work-up have been published, which
partly work with solvent-extractions (4,5) or with Merck-Extrelut-
columns (6). However, all these work-up-procedures lead to similar
extracts, which in turn are chromatographed on silica or on reversed-
phase columns. In all cases, the detection is done by UV at about
276 nm. A further example in this field are numerous papers on the
analysis of other mycotoxins like aflatoxin.

Adapt both

At last, two examples are presented, where both the sample clean-up and the chromatographic system had to be taylor-made to solve a certain problem. The first concerns the analyses of histamine and other biogenic amines in alcoholic beverages like wine, beer etc. Biogenic amines may be formed by the action of microorganisms during fermentation. Since they are correlated with various undesirable effects on the blood-pressure of the consumer, and the wine makers for instance try very hard to avoid their occurence, an efficient analysis was called for. The concentrations of the various amines lie in the ppm-range. Theoretically they would be suited for direct LC-analysis, provided they all had a highly absorbing UV-chromophor. Unfortunately this is not the case for many of them. It was therefore decided to convert all the biogenic amines into derivatives with common detecting features (7). It is well known, that phenols and amines are easily reacted with DANSYL-chloride to yield highly fluorescent dansylates. Since the reaction takes place under basic aqueous conditions, it could be foreseen, that an extraction of the reaction-mixture with an organic solvent would be very selective, with respect to the separation from the Dansylates of amino-acids. Finally, a chromatographic system had to be found which allowed the injection of an organic extract, so as to avoid a solvent change after the isolation of the dansylates. This meant of course the choosing of a silica-column and led to the following simple procedure for amines in wine: 5oo μl of wine are reacted with dansyl chloride in acetone at pH lo, the acetone is blown off, the mixture is extracted with ca. 3oo μl of ethylacetate, of which a few microliters are injected into the chromatograph of which the chromatogram in Fig.8 is obtained. The richness of this chromatogram and the knowledge that histamine is not the only amine present led immediately to the application of gradient-chromatography (8) which, in fact, allows to quantitate quite a number of amines in one run (Fig.9). The last example concerns the analysis of synthetic estrogens in calf-urine.

Since we were faced with the problem of rapidly analysing dozens of samples per week with a detection-limit of about lo ppb stilbestrols and ethinyl-estradiol, we decided to adapt one of the several known methods to our lab equipment. We knew how to handle dansylates and therefore we had no problems in getting acceptable chromatograms of standards (Fig.lo).

Applications of HPLC   23

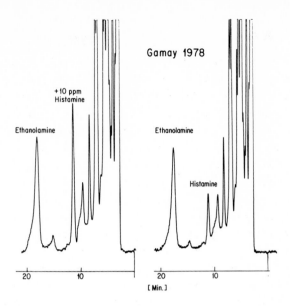

Fig.8   Histamin in Wine. Spherisorb S 5 W. Cyclohexane: Ethylacetate 55:45

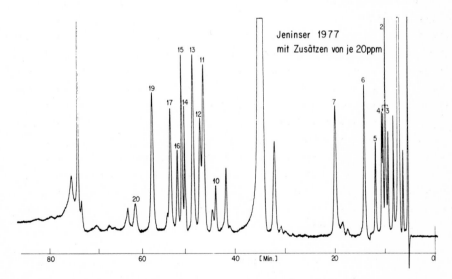

Fig.9   Biogenic Amines in Wine; 2:Isopentylamin;3:Isobutylamin,
4:Butylamin;5:2-Phenylaethylamin;6:Aethylamin;7:Methylamin;
1o:Cystamin 5 ppm;11:Aethylendiamin;12:Cadaverin;13:Putrescin;
14:1-Amino-2-propanol;15:Serotonin 4o ppm;16:Octopamin;
17:Histamin;19:Aethanolamin;2o:Normetanephrin 4o ppm

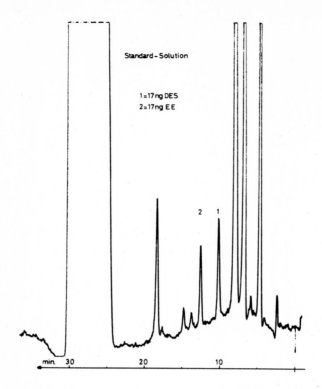

Fig.10   Diethylstilbestrol(DES) and Ethinylestradiol(EE),
Standard mixture, Dansylate. Conditions: S5W 25 cm,
Cyclohexane/Ethylacetate-gradient

As one can see, we are exactly at the point, where one can calculate, that the sample clean-up with this given chromatographic system has to include a concentration-factor for these two estrogens of at least 10 to, if possible, 100. This can, in the particular case, be achieved through a series of chemical solvent extractions designed to isolate **phenols** only. This procedure however led to a completely unsuitable extract (Fig.11)

One is therefore forced, already having adapted the chromatography with respect to sensitivity and selectivity, to further adapt the clean-up. This could in this case be done by pre-chromatography of the enzymatically hydrolised urine on C18-Sep-Pak-Cartridges and led to a solution suitable for screening in the required 10 ppb-range (Fig.12).

Seemingly positive samples are subjected to gc/ms-confirmation.

Applications of HPLC   25

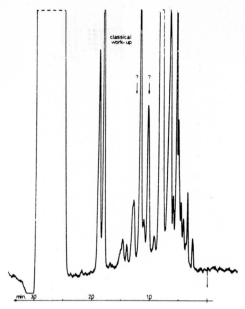

Fig.11   Solvent-Extract of Calf-Urine
          Conditions like Fig.10

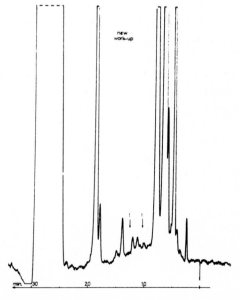

Fig.12   Extract of Calf-Urine, obtained with Sep-Pak-Clean-Up
          Conditions like Fig.10

REFERENCES

1 M.Ugrinovits, Chromatographia 13 (1980), 386

2 R.Battaglia, Mitt.Geb.Lebensm.Hyg. 68(1977), 28

3 G.M.Ware, C.Thorpe, A.E.Pohland, J.AOAC 57(1974), 1111

4 H.Tanner, C.Zanier, Schweiz.Zeitschr.f.Obst und Weinbau 112(1976), 656

5 H.Woidich, W.Pfannhauser, G.Blaicher, Lebensmittelchem. Gerichtl.Chem. 32(1978), 61

6 U.Leuenberger, R.Gauch, E.Baumgartner, J.Chromatog. 161(1978), 3o3

7 R.Battaglia, D.Fröhlich, J.HRC+CC Short Comm.(1978), 1oo

8 D.Fröhlich, R.Battaglia, Mitt.Geb.Lebensm.Hyg. 71(1980), 38

A version of this lecture has also been published in Ernährung 5 (1981) by agreement of the publishers.

Analysis of Vitamins in Food by HPLC

B.HEDLUND

SIK - The Swedish Food Institute, Box 270 22, S-400 23 Göteborg, Sweden

SUMMARY

High performance liquid chromatography (HPLC) has been used for the analysis of different fat- and water-soluble vitamins in human and animal foods. The HPLC methods are compared to standard methods for vitamin analysis.

For vitamin A and E the vitamins are extracted with hexane after alkaline hydrolysis and analyzed by a reversed-phase system with water-methanol as eluent, using UV-detection at 313 and 280 nm, respectively. The HPLC method shows good reproducibility and correlates well to the standard methods.

Ascorbic acid is analyzed by a reversed-phase ion-pair technique, using water-methanol, with the addition of tetra-hexylammonium hydroxide, as eluent. Ascorbic acid is detected at 254 nm. The analysis is rapid, with direct injection of juices and beverages, or a 5 min extraction with 8% meta-phosphonic acid for solid samples.

Thiamine (B_1) and riboflavin (B_2) are extracted by acid hydrolysis and treated with a diastatic enzyme. After the proteins have been precipitated with TCA, the vitamins are separated by a reversed phase ion-pair system, with UV and/or fluorescence-detection.

INTRODUCTION

The use of High Performance Liquid Chromatography (HPLC) has increased rapidly in the area of food analysis during the last five years. HPLC is now a well established analytical technique for the determination of sugars, food preservatives etc. HPLC is also the ideal method for analyzing vitamins in complex samples such as foods, due to the ability to analyze more than one component at the same time, and to the fact that both the separation of the component of interest from interfering substances and the detection/quantitation are carried out in the same system.

Separations of vitamins have been reported by several workers (1). However, the major part of these separations has not been applied to ordinary food products, but to the analysis of heavily fortified foods, pharmaceutical preparations and pure vitamins. Often these methods include the separation of several vitamins by gradient elution, neglecting the important sample preparation step, that is usually different for each vitamin. Also correlation and recovery data between HPLC-analysis and standard methods of vitamin analysis are very seldom, or never presented.

FAT-SOLUBLE VITAMINS

Vitamin A and E

One extraction method used for vitamin A and E is a rapid extraction with an appropriate solvent, hexane, benzene, methanol etc. (2). This method is suitable only for pure vitamin preparations, heavily enriched foods, or foods with a known vitamin A composition, since the fat-soluble vitamins often need to be freed from the entrapping matrix by hydrolysis. For ordinary foods this method gives only a 20-40% recovery. The most versatile extraction method is first an alkaline hydrolysis under reflux, then extraction of the vitamins (3) and, finally, HPLC-analysis.

For separation it is possible to use both straight- and reversed-phase systems (2). Straight-phase systems were shown to be sensitive to "poisoning" of the column, and also to have poor reproducibility. The reversed-phase system used in this study is a C_{18}-bonded phase with methanol/water as eluent (2). Other workers have used THF, acetonitrile etc., however, since it may be hazardous to use these solvents, they should be avoided.

Figure 1 shows a separation of vitamin A and E in a dog-feed formula, with a separation time of about 10-15 minutes depending on the type of packing material. The use of dual detection gives an increase in sensitivity.

The HPLC-method has good reproducibility, $\pm$ 1% for repetitive injections, and correlates well to the official Carr-Price and Emmerie-Engel methods, Table 1. The recovery for vitamin A and E is in the region of 90-98%.

WATER-SOLUBLE VITAMINS

Ascorbic acid

Liquid samples are directly injected if they contain large amounts of particulate matter after filtration. Solid samples are extracted with 8% metaphosphoric acid (2).

Analysis of Vitamins in Food by HPLC

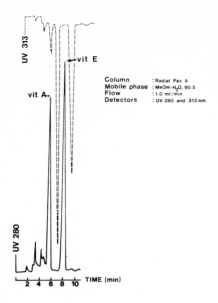

Figure 1. Separation of vitamin A and E in a dog feed formula.

Figure 2 Separation of ascorbic acid in juice and vegetables.

Table 1. Comparison between HPLC and official methods for determination of vitamin A and E.

PRODUCT	VITAMIN A				VITAMIN E			
	HPLC		CARR-PRICE		HPLC		EMMERIE-ENGEL	
	/UG/G[A]	CV,%[B]	/UG/G	CV,%	/UG/G	CV,%	/UG/G	CV,%
INFANT FORMULA	2.9	2.0	2.6	3.4	65.5	1.8	59.6	2.4
GRUEL	3.6	3.1	3.9	4.2	91.5	4.5	93.4	6.1
PORRIDGE	2.4	4.1	2.4	3.7	72.8	5.6	89.7	3.1
MARGARINE	8.6	2.9	7.4	2.1	102.5	2.4	104.0	3.1
DOG FEED	12.5	1.7	12.9	1.9	146.9	2.6	152.2	2.9

A) MEAN OF FOUR ASSAYS
B) COEFFICIENT OF VARIATION

Ascorbic acid that is strongly polar can be separated by ion-exchange chromatography or as an ion pair in a reversed-phase system. The latter system using tetrahexylamine as counter-ion have been shown to be advantageous. Figure 2 shows some separations of juices and vegetables.

What then are the advantages in analyzing ascorbic acid by HPLC? A comparison between HPLC, Tillman titration and fluorometric determination, Table 2, shows that all three methods correlate with one exception: the Tillman titration in the case of red beets and a baby food also containing red beets is not satisfactory, due to difficulties in observing the colour change of the indicator.

Table 2. Comparison between HPLC, Tillman titration and automated fluorometric method for ascorbic acid assay.

FOOD	ASCORBIC ACID					
	HPLC		TILLMAN TITR.		TECHNICON	
	MG/100 G[A]	CV,%[B]	MG/100 G	CV,%	MG/100 G	CV,%
ORANGE JUICE	26.8	2.1	26.8	1.9	26.2	1.8
PEAS	16.2	1.9	18.4	1.4	18.1	2.2
RED BEETS	5.3	2.3	16.4	6.8	5.6	1.2
FRUIT DESSERT	3.6	1.8	4.1	1.5	3.6	2.4
BABY GRUEL	46.5	2.6	48.2	2.4	46.1	1.4
BABY FOOD	13.8	1.8	22.1	4.5	14.8	1.9

A) MEAN OF FOUR ASSAYS
B) COEFFICIENT OF VARIATION

Comparing HPLC and automated fluorometric analysis, we find no differences in sample capacity and that the fluorometric method also has better reproducibility and recovery. One advantage with HPLC is the possibility of directly injecting liquid samples.

Vitamin B_1 and B_2

Thiamine and riboflavin are extracted by acid hydrolysis and treated with a diastatic enzyme to convert the vitamins to their free forms. After extraction, proteins are precipitated by TCA, in order to protect the HPLC column (for several foods this step may be omitted).

Previously reported methods have been utilizing ion-exchange, straight-phase or gradient--elution systems for the separation of B-vitamins (1). These systems are, however, sensitive to poisoning of the column, are irreproducible and require time-consuming column regeneration. Some methods applied to foods have been using ion pairing techniques with heptane-sulfonic acid as counter ion (4). It is also possible to add cationic components to the mobile phase to decrease the retention time of charged substances on a reversed-phase column (5).

Riboflavin is preferably detected by fluorescence, while thiamine, being non-fluorescent, is detected by UV at 254 nm. However, it is also possible to use selective detection of thiamine after post column derivatisation to the fluorescent thiochrome. This is achieved by mixing the column eluate with alkaline potassium ferricyanid.

Figures 3 and 4 show a separation of vitamin B_1 and B_2 in an extruded wheat-flour sample using water/methanol with the addition of TDA (tridecylamine) as eluent and with UV/fluorometric detection (Figure 3) or fluorometric detection (Figure 4) for both vitamins.

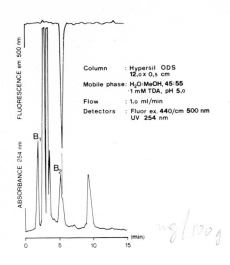

Figure 3. Separation of vitamin B_1 and B_2 in wheat flour.

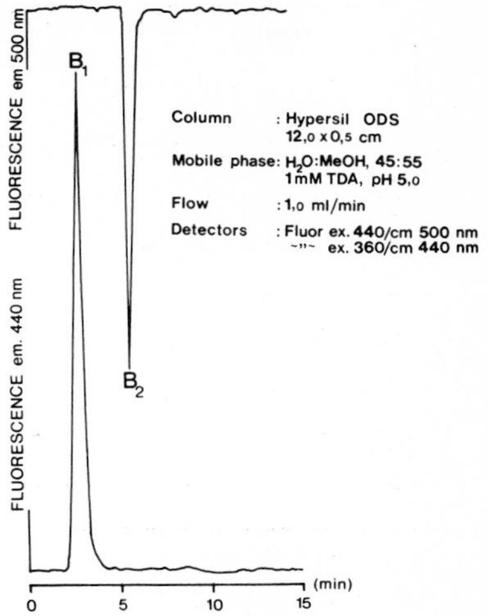

Figure 4. Same as figure 3 but fluorometric detection for B_1.

Table 3 shows a comparison between HPLC, an automated thiochrome method and an automated fluorometric method for the determination of thiamine and riboflavin. The analytical values and coefficient of variation are about the same for HPLC and the chemical methods. Recovery studies after fortification of the food with thiamine pyrophosphate and flavin-adenine-dinucleotide give a recovery of about 95 and 98%, respectively.

Table 3. Comparison between HPLC and automated assay for thiamine and riboflavin determination.

	THIAMINE				RIBOFLAVIN			
	HPLC (FLUOR.)		TECHNICON		HPLC		TECHNICON	
FOOD	UG/G[A]	CV.%[B]	UG/G	CV.%	UG/G	CV.%	UG/G	CV.%
RICE	5.7	2.9	5.6	2.6	1.8	2.6	1.7	1.8
BREAKFAST CEREAL	6.5	1.6	6.5	1.4	10.6	1.9	10.4	0.9
PEAS	1.1	5.4	1.3	4.6	0.7	3.9	0.8	3.1
PORK	7.8	1.6	7.4	1.8	2.3	2.1	2.1	1.9
LIVER PASTE	1.2	5.2	1.6	5.1	14.2	1.1	12.8	0.3

A) MEAN OF FOUR ASSAYS
B) COEFFICIENT OF VARIATION

CONCLUSIONS

What are then the conclusions concerning HPLC for vitamin analysis? First, it is obvious that HPLC needs the same or almost the same sample preparation steps as for conventional analytical techniques, if HPLC is to be applicable to most foods. Since the sample preparation step is usually the most time-consuming, the introduction of HPLC will not cut down the total time of analysis drastically.

The benefits of HPLC for vitamin analysis are:

- of course, the capability of analyzing more than one vitamin at a time, thus saving time and equipment,
- that HPLC is easy to automate, giving the method a high capacity for routine work,
- the possiblity of separating the vitamins from interfering substances,
- that compared to standard methods, HPLC minimizes the laborative handling with chemicals
- and, the most important point, the possiblity to "tailor make" analytical methods by HPLC.

Perhaps it is difficult to make an HPLC-method "official" in the sense that it should work with most foods, but for different products with a known composition, it is possible to make an HPLC method for each product with enormous savings on, for example, sample preparation time as a result.

As an overall conclusion, there are reasons to believe that, with a correct basic approach and with an adequate sample preparation, HPLC will be a powerful method for the determination of vitamins in foods.

REFERENCES

1. C.J. Blake, Food RA. Sci. & Techn. Surveys, $\underline{95}$ (1977)
2. C.B. Hedlund, to be published (1981)
3. R. Strohecher and H.M. Genning, Vitamin Assay-Tested Methods, Verlag Chemie, Weinheim (1966)
4. R.B. Thoma and M.M. Tabekhia, J. of Fd. Sci. $\underline{44}$ (1979), 263
5. C.B. Hedlund, to be published (1980)

Determination of Fat-soluble Vitamins by HPLC

By André ROUGEREAU, Alain GUILLER, Jacques GORE, Odile PERSON.

Institut de Nutrition. 37000 - TOURS - FRANCE -

SUMMARY.

The diversetinning technologies of aliments, may, in different ways, modify the composition of the aliments. The setting of dosage techniques of liposoluble vitamins in H.P.L.C. allows, not only to determine these modifications, but also to observe the Vitaminic Rate during the making.

This present work proposes the extraction and dosage methods of the A, E, D_2 vitamins, for the fresh and tinned vegetals.

INTRODUCTION.

According to the different recent works, (1-2), our concern was to determine an extraction and dosage method of liposoluble vitamins from vegetal products, as well as from animal products.

This technique has been set in order to be able to compare the influence of diverse industrial technologies of the making and tinning of aliments, tinned foods, deep-freezed foods, etc...

METHODS EMPLOYED.

1- Extraction Methods.

a) Vitamin A Extraction.

After having weighed 5 g. of aliment in a phial, we add 20 ml. of NaOH at 50 per 100, and the whole is warmed in a water-bath.
Then, we add 100 ml of ethyl alcohol and 2 ml of an alcoholic solution of hydroquinon at 20 per 100. (p / v). The mixture is put in a water-bath during 30 mn. then, poured in a decanting ampoule with 100 ml of water.

After having added 50 ml. of ethylic ether and stirred, the vitamin extraction is realised in 3 stages, by 50 ml of petroleum ether. (150 ml on the whole). Between each extraction, we strongly stir the ampoule, and we let the 2 phases decant.

The ethereal phase is washed 3 times by 100 ml of water, then filtered, and évaporated dry. The residue is set again by 1 ml. of hexane.

Remark :

Saponification with the help of soda is not necessary for non-fat products such as vegetals.

b) Vitamin D Extraction.

To a 5 g. test-sample of the product, we add 1 g. of pyrogallol, and 90 ml of a mixture of KOH at 50 per 100, and ethanol; (Proportion 60-30).
The whole is put in a water-bath during 30 mn. The extraction is accomplished in a decanting ampoule by 3 times 50 ml of petroleum ether.
The ethereal portion is finally filtered, evaporated dry. The residue is set again by 1 ml. of hexane.

Remark :

Saponification with the help of the mixture potash-ethanol is not necessary for non-fat vegetal products.

c) Vitamin E Extraction.

To a 5 g. test-sample of aliment, we add 100 ml. of a methanolic solution of ascorbic acid; (0,5 g. of ascorbic acid, 4 ml. of condensed water, 20 ml. of ethanol QSP 100 ml. with methanol). The mixture is put in a boiling water-bath during 15 to 20 mn. After having added 15 ml of KOH at 70 per 100, we put it back in the water-bath during 35 to 40 mn. The whole is poured in a decanting ampoule with 50 ml of water.

The extraction is realised in a decanting ampoule by 120 ml. of ethyl ether. The ethereal phase is filtered and dried on Na_2SO_4 then, the extraction is repeated with 120 ml. of ethyl ether. The ethereal phase joined to the previous one is filtered, evaporated dry. The residue is set again by 1 ml. of hexane.

Remark :

Saponification with the help of potash is not necessary for non-fat vegetal products.

2- Standard Solutions.

- Vitamin A

    Preparation of a solution at 200 IU / ml.

- Vitamin D

    Preparation of a solution at 400 IU / ml.

- Vitamin E

    Preparation of a solution at 0,2 mg / ml.

3- Dosage.

We have dosed the vitamins with the help of a Chromatograph in liquid phase, (H.P. 1010), with variable wavelength.
The processing conditions for the 3 Vitamins, A, D, E, are given in Table 1.

4- The Aliments.

The dosages have been realised on carrots, and on oil-sardines.
The results which are given are an average of 10 samples.

Chromatographic conditions.	Vitamin A.	Vitamin D.	Vitamin E.
Flow.	2,50	2,50	2,50
% B.	92	92	92
Column P.	83	83	83
Max P.	400	400	400
Min P.	0	0	0
Oven - T.	40	40	40
S - Temp A.	80	80	80
S - Temp B.	45	45	45
Wavl	330	330	330
Cht Spd.	1,00	1,00	1,00
Zero	10,0	10,0	10,0
Attn.	4	4	4
Area Rej.	50	50	50
Slp sens.	0,20	0,20	0,20

TABLE 1.

CHROMATOGRAPHIC CONDITIONS EMPLOYED FOR THE VITAMINS DOSAGE.

A = Condensed Water.
B = Methanol.
Column = Lichrosorb RP 8 - 10.

RESULTS.

1- Evaluation of the extraction Rate. (Table 2.).

By the mean of different controls and checkings : calorimetric dosage method, adding of vitamins, we came to the following results :

. For the vegetals, which do not need a preliminary saponification in order to get rid of the lipids, the Extraction Rate of Vitamin A is 100 per 100, the one of Vitamin D_2 is 70 per 100, and finally the Rate of Vitamin E is 98 per 100 in average.

. For the animal tissues, which require saponification, we found out the following Extraction Rates : For Vitamin A, 65 per 100, for Vitamin D_2, 63 per 100, and for Vitamin E, 69 per 100.

2- Carrots Vitaminic Rates.

We compared the values of fresh carrots with those of tinned carrots. The results are summed up in Table 3.
We can see that the contents of Vitamin A are of 650 IU for 100 g of dry weight, for the fresh carrots, and of 100 to 400 IU for 100 g of dry weight for the tinned carrots. Concerning Vitamin D_2, we found out for the two types of carrots, fresh and tinned, 10 IU for 100 g of dry weight. Finally, the average Vitamin E contents are of 2,90 mg for 100 g of dry weight, in the case of fresh products, and of 0,60 mg. for 100 g of dry weight in the case of tinned products.

3- Oil-sardines Vitaminic Rates.

Concerning this preparation, we have noticed that most of the liposoluble vitamins are to be found in the oil and not in the tissues. See Table 4.
More, we must underline the fact that we do not find Vitamin E in the examinated samples.

TABLE 2.

EVALUATION OF THE EXTRACTION RATE.

	Vitamin A.	Vitamin D.	Vitamin E.
With Saponification.	65 per 100.	63 per 100	65 per 100
Without Saponification.	100 per 100.	70 per 100.	98 per 100.

	NUMBER OF DOSAGES.	Vitamin A IU p.100 (retinol)	Vitamin D_2 IU p.100 if presence.	Vitamin E (tocopherol). mg p.100.
N° 1	5	400	10	0,30
N° 2	5	400	10	0,60
N° 3	5	400	10	0,60
N° 4	5	100	10	0,70
N° 5	5	100	10	0,50
N° 6	5	100	10	0,80
N° 7	5	200	10	0,005
N° 8	5	100	10	0,60
N° 9	5	100	10	0,65
N° 10	5	100	10	0,65
FRESH CARROTS.	5	650	10	2,90

TABLE 3.

CONTENTS OF LIPOSOLUBLE VITAMINS

IN TINNED CARROTS.

	NUMBER OF DETERMINATION.	VITAMIN A. IU p.100 (retinol)	VITAMIN D_3 IU p. 100	VITAMIN E. (tocopherol) mg p. 100.
1	5	40	170	Absence.
2	5	160	500	Absence.
3	5	Absence	60	Absence.
4	5	130	220	Absence.
5	5	60	250	Absence.
6	5	90	Absence	Absence.
7	5	60	Absence	Absence.
8	5	270	250	Absence.
9	5	90	100	Absence.
10	5	Absence	Absence	Absence.
OIL	10	320	1200	Absence.

TABLE 4.

CONTENTS IN VITAMINS A, E, D.

IN TINNED SARDINES.

CONCLUSIONS.

The extraction devices realised for this work show that they are satisfactory for the vegetals, for they do not require a previous saponification.

However, with regard to animal products, saponification produces a certain damage to the liposoluble vitamins.

On the other hand, the Chromatographic Technique proves its reliability and above all, allows a very satisfactory limit detection, that is to say of 5 IU for the Vitamin A per ml. of extract, of 20 IU per ml of extract for Vitamin D_2 or D_3, and of 0,05 mg per ml of extract for the Vitamin E.

BIBLIOGRAPHY.

1- P. SODERHJELM and B. ANDERSSON, J. Sci. Fd. agr. 29 (1978), 697-702.

2- R. SCHUSTER, Doc. H.P. (1978), 232.

Determination of Methyl and Ethyl Caffeate in Vegetables by HPLC with Electrochemical Detection

G.SONTAG, F.I.SCHÄFERS and K.HERRMANN

Institute of Analytical Chemistry, University of Vienna, Austria
Institute of Food Chemistry, Technical University Hannover, W.Germany

SUMMARY

A method for quantitative determination of methyl caffeate and ethyl caffeate in vegetables was developed. First the voltammetric behaviour of these compounds was investigated. It was found that these substances are oxidized at the same potential. Therefore the esters were separated by HPLC and detected with an amperometric detector. The detection limit of methyl caffeate is 4 ng and of ethyl caffeate 5 ng. Selected ion monitoring mass spectrometry was applied for the identification of these compounds, too.

INTRODUCTION

Some time ago Cannon (1) and Bohlmann (2) found methyl- and ethylesters of hydroxycinnamic acids in leaves and roots of various plants. The investigations of Vösgen (3) led us to suppose that such esters are present in vegetables and fruit. He extracted and derivatized the esters, separated the trimethylsilyl derivatives by gaschromatography and detected these compounds with a flame ionization detector. The lack of selectivity of this detector led to chromatograms which made the evaluation difficult.
The aim of our work published recently (4) was:

a) to work out a method, which enables us to determine these esters without derivatizing them and
b) to use a detector with a high selectivity
For this purpose HPLC with an amperometric detector seemed to be suitable.

ELECTROCHEMICAL BEHAVIOUR

For the use of an amperometric detector it is necessary to know the electrochemical behaviour of the esters. Therefore the electrochemical behaviour of the methyl- and ethyl esters of hydroxycinnamic esters was investigated with direct current and cyclic voltammetry at a glassy carbon electrode at different pH values.
It was found that these esters are detected well in acid medium and that the esters of different hydroxycinnamic acids are oxidized at different potentials. On the contrary there is no difference between the peak potentials of methyl- and ethyl ester of the same acid. Table 1 shows the peak potentials of these substances at pH 3.

Table 1

Compound	Peak potential (volt)
Methyl caffeate	+ 0.495
Ethyl caffeate	+ 0.495
Methyl sinapate	+ 0.530
Ethyl sinapate	+ 0.530
Methyl ferulate	+ 0.680
Ethyl ferulate	+ 0.680
Methyl p-coumarate	+ 0.880
Ethyl p-coumarate	+ 0.880

This behaviour can be explained as follows: the oxidation of phenolic compounds depends first on the number of hydroxy substituents. Ortho diphenolic compounds are more easily oxidized than mono-phenolic compounds. This can be seen in table 1 by comparing the peak potentials of caffeic esters with those of p-coumaric esters. Furthermore the oxidation of these compounds depends on the fact whether one or more hydroxy groups are substituted by methoxy groups. Methoxy groups cause

an increase of the peak potential.
To elucidate the electrochemical behaviour we investigated the pathways
for the electrochemical oxidation with cyclic voltammetry. The
following figure shows the oxidation of the esters of caffeic acid
and ferulic acid.

Figure 1

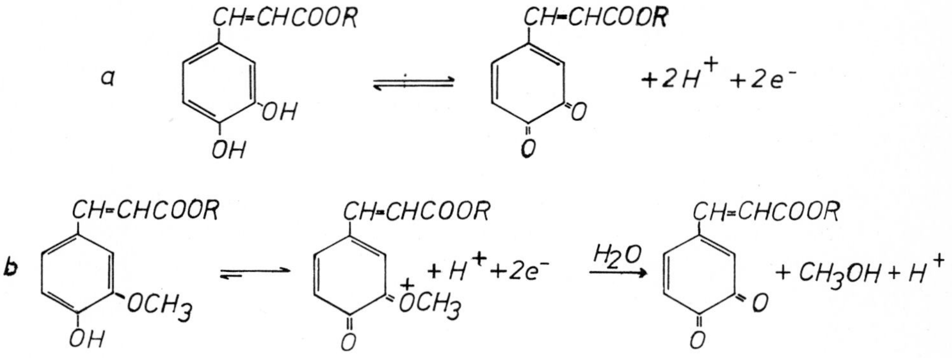

a) Catechol ester ⇌ ortho-quinone ester + $2H^+ + 2e^-$

b) Guaiacol-type ester ⇌ cationic intermediate + $H^+ + 2e^-$ $\xrightarrow{H_2O}$ quinone ester + $CH_3OH + H^+$

Methyl caffeate or ethyl caffeate (a) are easily oxidized in a two
electron process to the corresponding quinone. This reaction is
reversible. The esters of ferulic acid (b) form an ionic intermediate.
This reaction is irreversible. Then the intermediate is hydrolyzed to
the corresponding quinone. The ester group is not oxidized itself,
therefore methyl and ethyl esters of the same acid are oxidized at the
same potential. This means, that a seperation is necessary if methyl
caffeate and ethyl caffeate should be determined.

LIQUID CHROMATOGRAPHY WITH AMPEROMETRIC DETECTION

The esters were separated by reversed phase chromatography employing
isocratic elution from Nucleosil-C_8 with methanol, acetic acid and
bidistilled water (3oo:2o:68o, v:v:v). A pH 3 was established with
1 M sodiumhydroxide. The flowrate was 1.8 ml/min., the injection
volume 2o μl and the column eluent was monitored with an electro-

chemical cell (Metrohm EA 1o96). Under these conditions only ethyl caffeate and methyl p-coumarate were not separated. But choosing the appropriate working potential for the electrochemical cell we could differentiate between these compounds without separating them.
In order to find out the optimal working potential for the electrochemical cell, we have investigated the hydrodynamic curves of the esters. These experiments are carried out by repeated injection of the compounds at different potentials. The peak heights are plotted as a function of potential for each compound resulting in the usual voltammetric wave. The working potential is chosen in the plateau of the hydrodynamic curve. The following figure shows the working potentials of the esters under investigation.

Figure 2

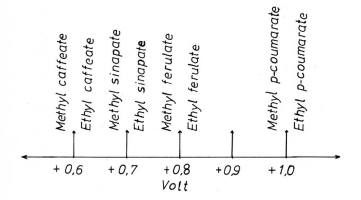

In this way we can regulate the selectivity of the detector. At + o.6 V only methyl caffeate and ethyl caffeate with highest sensitivity are indicated. The current response of methyl sinapate and ethyl sinapate do not reach the limiting plateau. Methyl and ethyl esters of ferulic acid and p-coumaric acid leave the elctrochemical cell undetected. At + 1.o V all esters with highest sensitivity are detected.

APPLICATION

In order to demonstrate the applicability of the method vegetable and fruit samples were examined for methyl and ethyl esters of caffeic acid, sinapic acid, ferulic acid and p-coumaric acid. Because of the

complex nature of these samples a clean up procedure was necessary.
Clean up procedure: vegetable and fruit samples are reduced to small pieces with a knife, then methanol is poured over them and the mixture is homogenized, centrifugated and filtered. Methanol is evaporated and the remaining aqueous solution is poured on a polyamide column. Anorganic material, dyes, sugars and several acids are eluted with water, phenolics are retained. Then the esters are eluted with methanol. Methanol is evaporated again and the remaining aqueous solution is extracted with diethylether. After evaporation of the ether the residue is solved in methanol. This is the parent solution.
Liquid chromatography: 2o µl of the parent solution are injected and the chromatograms are evaluated qualitatively and quantitatively. For the identification of the esters the retention times (k'-values) are compared with those of a standard solution (methyl caffeate: k' = 3.5; ethyl caffeate: k' = 6.2). Additionally, to control these results, the trimethylsilyl derivatives of the esters are separated by gaschromatography and identified by selected ion monitoring mass spectrometry.
For quantitative evaluation methyl sinapate is added as an internal standard and the peak heights are measured.

RESULTS AND CONCLUSIONS

The results are summarized in table 2. The detection limit of methyl caffeate is 4 ng and that of ethyl caffeate 5 ng. The recovery fluctuates between 8o and 85%. In addition to these samples we have examined black currants, plums, blackberries and onions for methyl caffeate and ethyl caffeate but these substances were not present. The investigation of the potato peels was very interesting. We have analyzed three samples. Sample 1 and 2 were stored over a longer period, sample 3 was harvested freshly. It is seen that freshly harvested potatoes show a higher content of methyl caffeate than potatoes stored longer.

Table 2

Sample	Compound	Concentration (ppm)
celery	methyl caffeate	0.13
	ethyl caffeate	1.16
carrots	methyl caffeate	0.07
	ehtyl caffeate	0.09
potato peels, 1	methyl caffeate	6.32
potato peels, 2	methyl caffeate	7.18
potato peels, 3	methyl caffeate	18.2

Finally some conclusions about the function of these compounds. It seems that caffeic esters act antioxidatively like propyl gallate which is often added to food. Furthermore our results let us suppose that methyl caffeate inhibits the germination of the potatoes as long as it occurs in the potato peels.

REFERENCES

1　J.R.Cannon, P.W.Chow, M.W.Fuller, B.H.Hamilton, B.W.Metcalf and A.J.Power, J. Chem. 26 (1973), 2257.
2　F.Bohlmann and Ch.Zdero, Phytochemistry 15 (1976), 131o; 17 (1978), 487; 18 (1979), 95.
3　W.Vösgen, unpublished
4　G.Sontag, F.I.Schäfers and K.Herrmann, Z. Lebensm. Unters. Forsch. 17o (198o), 417.

High Performance Liquid Chromatographic Analysis of Short-Chain Carboxylic Acids

C.GONNET and M.MARICHY

Laboratoire de Chimie Analytique 3, (CNRS, ERA 474, M.PORTHAULT) Université Lyon I, 43 Boulevard du 11 Novembre 1918, 69622 VILLEURBANNE CEDEX.

SUMMARY

Short chain carboxylic acids are analyzed by HPLC with and without pre-column derivatization. Direct analysis (without derivatization) of acids can be performed using pure aqueous mobile phases and chemically bonded stationary phases. The best selectivity is obtained with ODS modified silica (high loading Carbon content). The influence of pH and ionic strength on acid retention is investigated and mobile phases adjusted at pH = 1.0 - 1.5 ($HClO_4$ - $NaClO_4$ 0.2 M) are shown to be the most convenient.

The second part of the investigation describes the HPLC separation of acids as their phenacyl esters.

These procedures were used in the analysis of organic acids in red wine. Results indicate that the direct analysis of wine extract does not give sufficient selectivity and it is better to use a derivatization technique.

INTRODUCTION

Most of short chain carboxylic acids, especially those possessing four carbon atoms are of biological importance and it is, therefore, of interest to have a technique suitable for the analysis of this class of compounds. Since most acids absorb UV radiation in the range 190-220 nm, detection of small quantities can be difficult. Carboxylic acids are often converted to their derivatives (UV absorbing or fluorescent derivatives) before chromatographic analysis, to improve detectability and selectivity.

The derivatization of carboxylic acids as phenacyl esters is well known (1-6) and has been shown to be quantitative especially in the presence of crown ethers as catalysts ; more recently, 4-bromomethyl-7-methoxy-coumarin (Br-Mmc) was found to form strongly fluorescent esters with fatty acids (7,8) as well as with dicarboxylic acids (9,10).

The purpose of this paper is to describe and compare direct analysis and precolumn derivatization techniques in the case of organic acids in wine.

EXPERIMENTAL

Reagents and Chemicals

All chemicals and solvents used in this investigation were reagent grade. Water for HPLC was deionized and distilled in glass. All chromatographic solvents were degassed in an ultrasonic bath. p-bromophenacylbromide PBPB (INTERCHIM,Montluçon,France) was recrystallised from ethanol. Dicyclohexyl 18-Crown-6 was purchased from INTERCHIM and used without purification.

Apparatus

The liquid chromatograph used in this study consisted of the following equipment : an ALTEX model 380 pump for solvent delivery (TOUZART & MATIGNON,Paris), a damping system, a RHEODYNE injection valve 70.10. The UV detector was a PYE UNICAM LC3 spectrophotometer. Several commercial reversed-phase packings were used : R SIL C 18 HL-5 , R SIL C 18 LL-5 (high loading C 18 % and low loading C 9 %) were obtained from INTERCHIM (Montluçon, France), LICHROSORB RP 8, NUCLEOSIL C 18 and C 8 5 µm from TOUZART & MATIGNON (Paris). All of them were packed into columns of 20 cm x 4.6 mm and 15 cm x 4.6 mm tubings according to a technique previously described (11).

Derivatization

The method of GRUSHKA (2,3) was used in derivatization reaction.

Preparation of a red wine extract

50 ml of red wine are treated on charcoal then eluted on a cation (H^+ form) exchanger and fixed on an anion (carbonate form) exchanger. Organic acids fraction is eluted with a 10 % ammonium carbonate solution then evaporated to dryness (45°C under vacuum). The dry extract is dissolved in 20 ml mobile phase before injection.

RESULTS AND DISCUSSION

Direct analysis

HPLC of underivatized acids was performed using a C 8 or C 18 reversed-phase packing and an acidified ($NaClO_4$ - $HClO_4$) mobile phase. The acids were monitored at 220 nm. Figure 1 shows that the best resolution is obtained on a NUCLEOSIL C 8 column or on a R SIL C 18 HL-5 column (fig. 4).

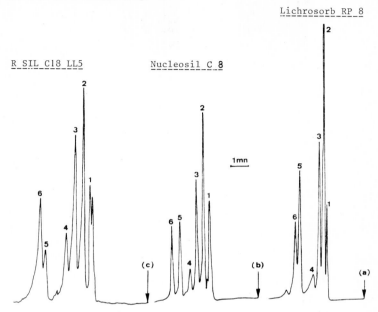

Figure 1 - Columns : (a) Lichrosorb RP 8 ; (b) Nucléosil C 8 , (c) R SIL C 18 LL5. Mobile phase : $NaClO_4$ 0.2 M - $HClO_4$; pH = 1.0 ; Flow-rate = 1 mℓ/mn. Solutes : 1-galacturonic acid, 2-tartric acid, 3-Lmalic acid, 4-L(+) lactic acid, 5-citric acid, 6-succinic acid.

The influence of pH and ionic strength is illustrated in Fig. 2 and 3. It can be seen that for pH values higher than 1.5-2.0 , acid retention decrease ; the optimal pH is 1.0-1.5 in all cases. Mobile phase ionic strength has a very slight influence on acid retention : for almost all the acids, the retention does not change as sodium perchlorate concentration increase.

Figure 4 is a representative chromatogram of a wine extract analysed using the optimal conditions. Tartric and lactic acid can be identified but the resolution is not good enough.

Derivatization

The scheme of preparation of PBPB derivatives is the following :

R-COOH $\xrightarrow[\text{Br-}C_6H_4\text{-CO-CH}_2\text{-Br}]{\text{KOH}}$ R-COO-CH$_2$-CO—⟨⟩—Br

Dicyclohexyl 18 - Crown-6   p-bromophenacylester

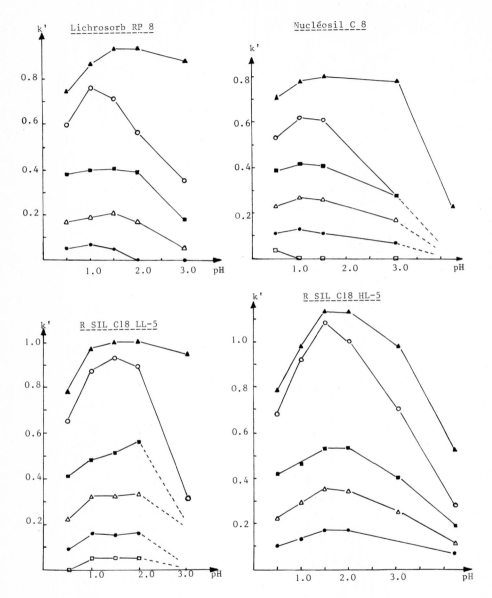

Figure 2 - Mobile phase : $NaClO_4$ 0.2M - $HClO_4$; Flow-rate : 1ml/mn.
Solutes : □ galacturonic acid , ● tartric acid , △ L-malic acid
■ L(+) lactic acid , ○ citric acid , ▲ succinic acid.

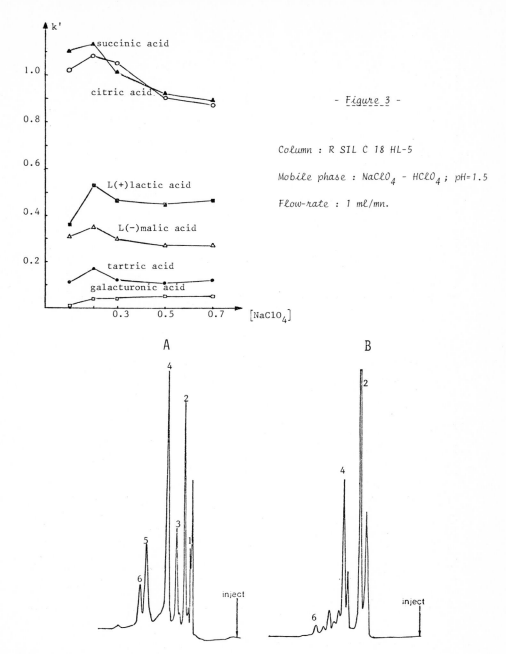

Figure 4 - Column : R SIL C 18 HL-5
Mobile phase : $NaClO_4$ 0.2 M - $HClO_4$; pH = 2 ; Flow-rate = 0,6 ml/mn.
Solutes : See fig. 1.
(A) Standard mixture ; (B) Wine extract.

The wine dry extract, dissolved in methanol is neutralized by a potassium hydroxyde solution. The solvent is removed, an excess of alkylating reagent and crown-ether are added. The solution is heated to 80°C for 1h30 mn. The derivatives were monitored at 254 nm.

This reaction was carried out on a wine dry extract dissolved in methanol. Chromatograms of standard mixture of acids and dry extract are shown in Fig.5.

It seems that the direct HPLC separation of short chain carboxylic acids does not give sufficient resolution particularly in the case of hydroxyl acids so that a pre-column derivatization is necessary to improve selectivity and detectability.

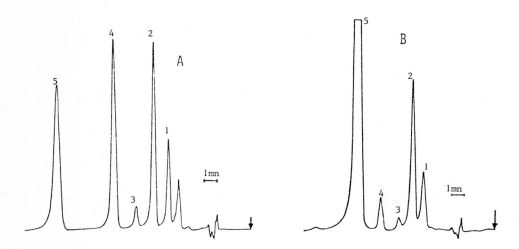

Figure 5 - Column : Spherisorb ODS ; Mobile phase : MeOH-H$_2$O(LiCl 0.1M)60/40 ; Flow-rate 0,6ml/mn. ; Solutes (A) - derivatives of : 1 - formic acid ; 2 - acetic acid ; 3 - propionic acid· 4 - reagent excess ; 5 - butyric acid. - (B) - derivatization of a wine extract : 1 - tartric acid ; 2 - lactic acid ; 3 - acetic acid ; 4 - unidentified - 5 - reagent.

REFERENCES

1 - R.F. BORCH, Anal. Chem., 47 (1975), 2437.

2 - E. GRUSHKA, H.D. DURST and E.J. KIKTA, J. Chromatogr. 112,(1975), 673.

3 - H.D. DURST, M. MILANO, E.J. KIKTA, S.A. CONNELLY and E.GRUSHKA, Anal. Chem. 47, (1975), 1797.

4 - H.C. JORDI, J. Liquid Chromatogr. 1, (1978), 215.

5 - R.A. MILLER, N.E. BUSSELL, C. RICKETTS, J. Liquid Chromatogr. 1, (1978), 291.

6 - N.E. BUSSEL, R.A. MILLER, J. Liquid Chromatogr. 2, (1979), 697.

7 - W. DÜNGES, Chromatographia, 9, (1976), 624.

8 - S. LAM and E. GRUSHKA, J. Chromatogr. 158, (1978), 207.

9 - E. GRUSHKA, S. LAM, J. CHASSIN, Anal. Chem. 50, (1978), 1398.

10 - C. GONNET, M. MARICHY and N. PHILIPPE, Analusis, 7, (1979), 370.

11 - B. COQ, C. GONNET and J.L. ROCCA, J. Chromatogr. 106 , (1975), 249.

Separation and Estimation of Galacturonic Acid by HPLC

E. FORNI, R. GIANGIACOMO, A. POLESELLO

Istituto Sperimentale per la Valorizzazione Tecnologica dei Prodotti Agricoli, Milano, Italy

SUMMARY

In view of developing a HPLC method for the quantitative evaluation of the galacturonic acid content in the pectic substances, the chromatographic behaviour and the conditions for the separation and the estimation of the galacturonic acid alone and in presence of other uronic acids and carbohydrates were investigated.
Separation tests were carried out using both strong and weak anion-exchange resins columns, eluting by different pH and ionic strengths buffer solutions under different solvent flow rates. The results show that is possible to achieve a very good separation of the galacturonic acid from other uronic acids and carbohydrates in a very short time by using a normal amino-bonded phase column functioning as a weak anion-exchanger in the pH range 3-5, eluting by potassium dihydrogen phosphate buffer solution and detecting the separated bands by UV at 200 nm.
The range of sensitivity and reliability for the quantitative estimation of the galacturonic acid together with the possibility of developing a method for the determination of the pectic substances in fresh and processed fruit and vegetables are discussed.

INTRODUCTION

Among the methods for the evaluation of the total content of pectic substances in fruit and vegetables, those based on the colorimetric or spectrophotometric determination of the galacturonic acid originated by the chemical or enzymic hydrolysis of the polyuronide chains of the pectin are widely used (1). However they often are time consuming and not sufficiently accurate or satisfactory, so that they are still subjected to modifications for improving their specificity and reliability (2,3).

On the other hand galacturonic acid has resulted easily separated both from uronic and other organic acids by ion-exchange column chromatography (4). However the classical liquid chromatography is not suitable for routine analysis, due to its well known limitations as summarized by Snyder and Kirkland (5).

With the advent of HPLC these limitations have been mostly overcome.

As a matter of fact Palmer and List (6) achieved a very rapid separation of the galacturonic acid from a mixture of thirteen organic acids by HPLC on strongly basic anion exchange resin eluting by 1N Sodium formate and detecting fractions refractometrically. By this method the galacturonic acid eluted after about six minutes, sharply separated from the other components of the mixture, with a very good accuracy and reproducibility, and it could be quantitatively estimated since the standard curves of the individual acids resulted linear over at least a 50-fold range in concentration.

The present paper deals with the researches on the possible use of HPLC for the rapid and accurate estimation of the galacturonic acid of the pectic substances.

It was studied:

a) the conditions for separation of the galacturonic acid from the most frequent uronic acids, sugars and other components present in a pectin enzymic hydrolysate;

b) the optimum conditions for the separation of the galacturonic acid in view of its quantitative estimation;

c) the use of an internal standard to check the quantitative estimation and the evaluation of reliability and the range of sensitivity of the method;

d) the development of a method for the determination of pectic substances in fruit and vegetables.

MATERIALS AND METHODS

The liquid chromatograph used in this research was a Trirotar II Jasco, equipped with a variable loop injector and an Uvidec II spectrophotometric variable wavelength detector. The chromatogram was recorded and processed on a Shimadzu Chromatopac C-R1A Data processor. A column Policolumn 1/4' i.d. x 25 cm l. packed with Lichrosorb-NH_2 10 μ, an amino bonded phase column functioning as a weakly basic anion-exchanger in the pH range 3-5 was chosen after a comparison test with other strong and weak anion-exchange resins columns.

As a mobile phase, Potassium Dihydrogen Phosphate solution was preferred to Acetate buffer solution because, at the same ionic strength, it does not absorb at 200 nm, the U.V. wavelength range suitable to detect the compounds lacking any chromofore other than the carboxyl groups. This mobile phase was also used by Wehr (7) for the quantitative analysis of organic acids in wine.

RESULTS AND DISCUSSION

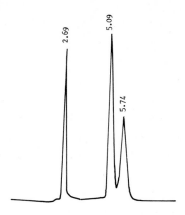

Fig. 1

a) Fig. 1 shows the good separation of the galacturonic acid (T_R = 5.74) from the glucuronic acid (T_R = 5.09) and mannuronic acid (T_R = 2.69), eluting with 0.05M KH_2PO_4 at a flow rate of 1ml/min. (T_R = retention time)

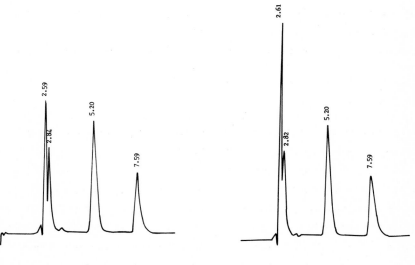

Fig. 2 Fig. 3

Fig. 2 shows the separation of the galacturonic acid ($T_R = 5.20$) from the glucose ($T_R = 2.59$) and xylose ($T_R = 2.84$) and Fig. 3 the one from the glucose ($T_R = 2.61$) and galactose ($T_R = 2.82$). It appears that in the chosen conditions, sugars do not interfere with the determination of the galacturonic acid, as these non-ionic compounds are directly eluted from the column.

In the same chromatograms it is possible to observe that the succinic acid ($T_R = 7.59$) added as an internal standard is eluted about 2.5 minutes after the galacturonic acid at 1ml/min flow rate and does not interfere with any other compounds (A) present in the solution of the enzymic hydrolysates of the pectin as seen in Fig. 4.

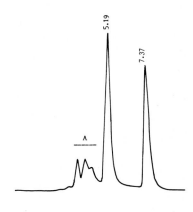

Fig. 4

The succinic acid was chosen after a screening with other organic acids not present in the enzymic hydrolysates of pectin, because its retention time is close to the one of the galacturonic acid, so as to cut the time for the analysis.

b) Then the best molar concentration of the KH_2PO_4 aquaeous solution was studied to achieve the most efficient isocratic separation of the galacturonic acid from the other substances present in the pectin enzymic hydrolysates in the least possible time, in view of its quantitative determination.

Molarities from 0.5M to 0.05M were tested. The most suitable one appeared 0.075M, because the galacturonic acid peak was sharply separated from the other substances in a reasonably short time (5.25 min at 1ml/min flow rate).

c) The quantitative estimation was carried out according to the internal standard method, so as to check the influence of some possible operative changes on the galacturonic acid peak area. The concentration was calculated using the scale factor according to the formula :

$$\frac{A_G}{A_{IS}} \times F_c \times \frac{W_{IS}}{W_S} \times 100 = \text{content \%}$$

where : A_G = Area of the galacturonic acid peak
A_{IS} = Area of the internal standard peak
F_c = Scale factor
W_{IS} = Weight of the internal standard
W_S = Weight of the sample

as processed by C-R1A Data processor.

The calibration curve resulted linear from 5 ppm to 40 ppm of monohydrate galacturonic acid.

The standard deviation for 24 ppm of galacturonic acid was $\pm$ 0.6 corresponding to an error of $\pm$ 2.5%.

d) To develop an analytical method for pectins, the procedure of McCready and McComb for the extraction of total pectic substances (8) was followed, consisting of the alkaline de-esterification of pectins followed by the enzymic depolymerization of polygalacturonic acid chains by pectinase, with some modifications, so as to get an extract ready to inject into the Liquid Chromatograph.

The method tested on standard polygalacturonic acid and some commercial pectins appears very reliable and rapid enough for routine purposes. According to the preliminary comparison tests in progress, it seems possible to anticipate that the HPLC method is more reliable and quicker than colorimetric and spectrophotometric ones, as the HPLC separation and determination of galacturonic acid of the extract does not take more then ten minutes.

The above summarized method has been developed in view of its applications on a research about the natural pectic substances composition in fruit and vegetables. Due its specificity and simplicity it may become a rapid and reliable tool to characterize the individual pectic compounds, during their fractioning process.

REFERENCES

1 R.M. McCready, in: "Methods in food analysis" ed. M.A. Joslyn, 2nd ed. Academic Press New York, 1970, p. 565
2 N. Blumenkrantz, G. Asboe-Hansen, Anal. Biochemistry, 54 (1973), 484
3 R.W. Scott, Anal. Chem. 51 (1979), 936

4   J.K. Khym, "Analytical ion-exchange procedures in chemistry and biology" Prentice Hall Inc., Englewood Clift, N.J., 1974, p. 160
5   L.R. Snyder, J.J. Kirkland, "Introduction to modern liquid chromatography" J.Wiley & Sons, New York, 1974, p. 4
6   K. Palmer, D.M. List, J.Agri.Food Chem. 21 (1973), 903
7   C.T. Wehr, "Liquid chromatography at work" Varian Bull. n° 85
8   R.M. McCready, E.A. McComb, Anal.Chem. 24 (1952), 1986

The Determination of Food Additives by Ion-Pair Liquid Chromatography

G. VAN DE HAAR, J.P. CORNET

Food Inspection Service, P.O. Box 465, Groningen, Holland and Food Inspection Service, Nieuwe Gracht 3, Haarlem, Holland

ABSTRACT

First a short introduction is given about the Dutch Food Law and the organization of HPLC-investigations of the 16 Food Inspection Service Laboratories. Methods are given for sample preparations and determinations of additives in different products. Most of the separations are performed on a reversed-phase system with a mixture of acetonitrile and a buffered cetyltrimethylammoniumbromide solution as mobile phase. The additives are divided in 5 groups: a. colour preservatives (e.g. nicotinic acid); b. flavor intersifiers (e.g. glutamic acid); c. food colours (e.g. Sunset Yellow FCF); d. preservatives (e.g. benzoic acid); e. miscellaneous (creatinine, quinine and NTA). For some of the methods the results of circular tests are given.

INTRODUCTION

Additives and their limits for use are specified for many foodproducts under the Dutch Food Act. This Act not only comprises food and beverages, but also cosmetics, toys, packings and other commodities. The demands of the Act are checked by sixteen Food Inspection Service Laboratories (F.I.S.L.). They co-operate in many ways. One is the HPLC-investigation-group, in which at least one food-chemist of each of the 16 laboratories participates. A new method developed by one of the laboratories is usually checked by another laboratory before it is submitted to a circular test of 5 or 6 laboratories. If the first circular test is successful than a full-scale circular test is carried out by 18 laboratories (16 F.I.S.L. + National Health Institute - R.I.V. Bilthoven + Central Institute for Food Research - C.I.V.O. -

T.N.O. Zeist). New methods etc. are published mainly in the "Ware(n)-Chemicus", the Dutch Journal of Food Science (English abstract for each paper) (1). Not only HPLC, but all techniques in relation to the work of the Food Inspection Service is published in this journal.

CHROMATOGRAPHIC SYSTEMS

The first chromatographic system consisted of the following components:
Model 6000A solvent delivery system (Waters), Wisp automatic sample injector (Waters), M440 absorbance detector (Waters), LC-UV 3 (Philips), Fluorimeter (Vitratron) + 18 μl cell, Data Module (Waters).
The second chromatographic system consisted of the following components:
L.C. Serie 3 solvent delivery system (Perkin Elmer), L.C. 420 autosampler (Perkin Elmer), L.C. 65T variable wavelength detector/oven module (Perkin Elmer), Fluorimeter 204A (Perkin Elmer) + 18 μl cell, Sigma 10 Data System (Perkin Elmer).

Columns: μ Bondapak C18 (Waters)   300 x 4.0 mm I.D.
Lichrosorb 10RP8 (Chrompack)   250 x 4.6 mm I.D.
Lichrosorb 10RP18 (Chrompack)   250 x 4.6 mm I.D.
Lichrosorb 10 NH_2 (Chrompack)   250 x 4.6 mm I.D.

Main eluents:
Eluent A: acetonitrile (Lichrosolv - Merck)
Eluent B: Dissolve 2.00 g cetyltrimethylammoniumbromide in 1 l of water in a volumetric flask of 2 l. Add 100 ml buffer pH 7.0 (Merck 9887). Fill to the mark with water.

RESULTS

a. colour preservatives.

In the Netherlands nicotinic acid or niacin is allowed as an additive to meat products up to 0.015%. It prevents the loss of the red colour of the meat. The conditions for the separation of nicotinamide, ascorbic acid and nicotinic acid are as follows: Column: Lichrosorb 10 RP8. Mobile phase: 35% eluent A - 65% eluent B. Flow-rate: 2.0 ml/min. Temperature: $35°$ C. Detection: UV 262 nm.
Retention times : nicotinamide   1.8 min.
ascorbic acid   3.2 min.
nicotinic acid 4.0 min.
The chromatographic system for the determination of nicotinic acid in minced meat (2) is as mentioned above.
Sample preparation: Nicotinic acid is extracted from a homogenized sample (15 g) with 50 ml ethanol (96%) and heated for 30 min. on a boiling waterbath. The extract is filtered and the filtrate is transferred quantitatively into a volumetric flask of 100 ml, left overnight and filled to the mark with ethanol (96%). A part is filtered through a microfilter for HPLC-analysis. The relative standard deviation for n=6 determinations in one minced meat sample (level 0.015%)

was 2.6%. The recoveries ranged from 96 to 102% for levels of 0.015; 0.032 and 0.075%.

It was suggested that freezing of meat or adding of nicotinic acid as a solution might affect the recovery, but tests showed that neither 2, 5 or 9 days freezing at -24° C, nor adding of nicotinic acid as solid or as a 1 or 2% solution, had significant effect on the recovery (recoveries 94 - 107%; relative standard deviation 5%). A circular test of 6 laboratories showed relative standard deviations from 2 to 5%. We also apply this method for the determination of nicotinic acid and ascorbic acid in mixed herbs, used by butchers for spicing meat products. For these mixes there are only limits through their use in preparation of food (products).

In co-operation with the F.I.S.L. Enschede we developed a method for the separation of ascorbic acid and iso-ascorbic acid in these products. In the Netherlands the use of iso-ascorbic acid is forbidden. The conditions for the separation are: Column: Lichrosorb 10 NH_2. Mobile phase: 150 ml buffersolution (dissolve 0.89 g Na_2HPO_4.2aq in 500 ml of water in a volumetric flask of 1 l. Add a citric acid solution (5%) until pH 5.5 and fill to the mark with water) in a volumetric flask of 1 l.

Fill to the mark with acetonitrile. Flow-rate: 2.0 ml/min. Detector: UV 254 nm. Retention times: iso-ascorbic acid   9.2 min.
                   ascorbic acid   12.4 min.

A simular method was published by Hoffman (3).

b. flavor intensifiers.

An important flavor intensifier is glutamic acid, also known as ve-tsin and widely used in oriental kitchens. It may cause the Chinese restaurant syndrome. Glutamic acid is allowed in several products, in soups up to 1%. Glutamic acid is determined by a precolumn derivatization with dansylchloride and fluorescence detection (4). The following conditions are used for this determination: Column: Lichrosorb 10RP8. Mobile phase: 45% eluent A - 55% eluent B.

Flow-rate: 2.0 ml/min. Detector: Fluorimeter, excitation 360 nm, emission 520 nm. Retention time: glutamic acid (dansylated) - 8.5 min. Sample preparation: Glutamic acid is extracted from a homogenized sample (containing 10 - 100 mg glutamic acid) with 200 ml water and 0.5 g charcoal. One ml of the filtered extract is treated with 1 ml 0.1 M $NaHCO_3$ and 1 ml dansylchloride (3 mg/ml acetonitrile). The reaction is left to proceed for two hours in the dark at 40° C. The method is linear for the measured concentration range 5 - 50 mg glutamic acid/100 ml water. The relative standard deviation for n=6 determinations in one sample mixed herbs (level 8%) was 1.7%. The recoveries ranged from 94 to 102% for levels of 2 and 4%. A circular test of 5 laboratories showed recoveries from 93 to 108% and a relative standard deviation of 5.4%. Two other flavor intensifiers are inosinic acid and guanylic acid. They are allowed in several products like soups and bakery products up 0.05%.

After extraction with hot water these compounds are chromatographied under the same conditions as in the nicotinic acid determination.

Retention times: guanylic acid   11.6 min.
                 inosinic acid   12.4 min.

c. food colours.

A different group of food additives are food colours. Only approved colours (regulated in the colouring degree of the Dutch Food Act) are allowed. The separation of Sunset Yellow FCF (E 110) and orange GGN (nog allowed) illustrates the power of ion-pair LC. These are azocompounds, which only differ in the substitution place of the sulphonate group on the benzenering. The conditions for the separation of these compounds: Column: Lichrosorb 10RP8. Mobile phase: 41% eluent A − 59% eluent B. Flow-rate: 2.0 ml/min. Detector: UV 254.

Retention times: sunset yellow fcf   7.8 min.
                 orange ggn   9.0 min.

The determination of amaranth and azorubine in (red) fruit-wine can be done under slightly different circumstances: mobile phase: eluent A 50% − eluent B 50%. Amaranth is not allowed in some countries because of suspected carcinogenic properties. In Holland the use is limited to 30 mg/kg by an agreement between government and industry. A circular test of 18 laboratories for the analysis of anaranth in a beverage gave the following results:

average recovery	101%
repeatability	5%
reproducebility	12%
relative standard deviation	8%

d. preservatives.

Sorbic acid and benzoic acid (and derivatives of benzoic acid) are allowed in different quantities in several products (beverages, jams, wine, mustard etc.). For confirmation of the presence of sorbic acid and benzoic acid an ion-pair system can be used. The chromatographic conditions are: Column: Lichrosorb 10RP18. Mobile phase: 55% methanol − 45% H_2O (0.19% CTAB and 10% phosphate buffer ph 7 Merck 9887). Flow-rate: 2.0 ml/min. Detector: UV 230 NM.

Retention times: sorbic acid   19.0 min.
                 benzoic acid   21.5 min.

Usually an ionic suppression system is used for routine analysis. (5) Conditions: Column: Lichrosorb 10RP18. Mobile phase: 20% acetonitrile − 80% H_2O (pH 4.5 acetate/acetic acid buffer). Flow-rate: 2.0 ml/min. Detector: UV 230 nm.

Retention times: benzoic acid   5.8 min.
                 sorbic acid   7.2 min.

For beverages and wine the only sample preparation is filtration.

A circular test of 18 laboratories for the analysis of sorbic acid and benzoic acid in a beverage gave the following results:

	benzoic acid	sorbic acid
average recovery	99.2%	99.3%
repeatability	8 %	6 %
reproducebility	14 %	13 %
relative standard deviation	4	4.5%

e. miscellaneous.

An example of the use of octylsulphate as a counter-ion in ion-pair L.C. is the determination of creatinine. (6).

Chromatographic conditions: Column: Lichrosorb 10RP18. Mobile phase: 95% octylsulphate (0.02%) solution (pH=5) - 5% methanol.

Flow-rate: 2.0 ml/min. Detector: UV 235 nm.

Retention time: creatinine 4.9 min.

Another example is the determination of quinine in tonic. Instead of a sulphate we here used a sulphonate.

Filtration was the only sample preparation (te remove to carbondioxide).

Chromatographic conditions: Column: Lichrosorb 10RP18. Mobile phase: dissolve 0.5 g sodium 1-octanesulphonate in 100 ml of water in a volumetric flask of 500 ml. Add 50 ml phtalatebuffer (pH 3, 0.05 M), 225 ml acetonitrile. Fill to the mark with water. Flow-rate: 1.5 ml/min.

Detector: Fluorimeter excitation 350 nm, emission 435 nm.

Retention time: quinine 4.1 min.

An example of the analysis of a non-food additive by ion-pair liquid chromatography is the determination of nitrilo triacetic acid (NTA) in phosphate free washing powder. Although NTA itself is not easily detected, many metal chelates of NTA absorb visible and UV light. NTA is determined by formation of a copper-NTA complex.

Chromatographic conditions: Column: Lichrosorb 10RP8. Mobile phase: 40% eluent A - 60% eluent B. Flow-rate: 2.0 ml/min.

Temperature: 30° C. Detector: UV 254 nm.

Retention time: Cu (NTA) 3.6 min.

REFERENCES

1 Editor C/O, Nijenoord 6, Utrecht, Holland
2 H.P. Bolhuis and G. van de Haar, De Ware(n)-Chemicus 8 (1978), 188.
3 B. Hoffmann, Lebensmittelchemie u. gerichtl. Chemie 34 (1980), 114
4 B.A. Douwes and G. van de Haar, De Ware(n)-Chemicus 9 (1979), 107
5 J.P. Cornet and M.J. Duin, De Ware(n)-Chemicus 7 (1977), 130
6 R. van Dijk and G. van de Haar, De Ware(n)-Chemicus 10 (1980), 181

High Performance Liquid Chromatography of Food Dyes

J. BRICOUT

Institut de Recherches Appliquées aux Boissons, 120,avenue Foch, 94015 Créteil
France

SUMMARY

A study was conducted on the chromatographic behaviour of the synthetic food dyes, using ion pairing and reverse phase liquid chromatography with variable U.V-visible detector. The influence of the chain length of the tetraalkylammonium ion and of the composition of the mobile phase was investigated. Similarly liquid-liquid extraction of the food dyes as ion pair was devised. Using these techniques, it becomes possible to determine the nature and the concentration of any synthetic dye in a food product, to check the impurities of some dyes and to study the interaction between food components and food dyes during technological process. Particularly the degradation of E 124 and E 126 was monitored during heat treatment of model solutions and characterized by the apparition of new U.V absorbing compounds.

INTRODUCTION

Among the different food additives, synthetic food dyes were particularly scrutinized for their safety and the general tendancy in the world is to reduce the number of permitted food dyes. In the European Economic Community, a Scientific Committee has reviewed the safety in use of all compounds proposed for inclusion in a Community List of coloring matters authorized for use in foodstuffs. A special attention was given to the sulfonated azodyes. Some of these synthetic dyes were toxicologically acceptable and an "Acceptable Daily Intake" could be established. As an A.D.I represents an estimate of the daily intake of a substance, this means that one must be able to evaluate the concentration of this substance in the foods. The necessity of an analytical procedure for qualitative and quantitative evaluation of synthetic food dyes becomes obvious. Most of the known methods involve extraction by woolyarn and paper or thin layer chromatography. High performance liquid chromatography has been used for dye analysis mostly for the detection of impurities. Anion exchange columns and gradient elution were generally used for this purpose (1, 2) but reversed phase liquid chromatography was also suggested for the analysis of tartrazine as ion pair (3). The separation of synthetic food dyes on reversed phase column was developed by Chudy and al. (4) using Cetrimide as counterion, by Noda and Nishiki (5) using ammonium carbonate and by Martin and al. (6) without any counterion. The main objectives of this study were to determine the most critical parameters for the separation of synthetic food dyes on reversed phase (octadecyl bonded to silica) column and to devise a process for the extraction of these dyes from different beverages. These two procedures were evaluated for monitoring the chemical degradation of some dyes during heat treatment.

EXPERIMENTAL

Apparatus

A Varian 8500 liquid chromatograph with a stop flow injection system and a Varichrom variable wavelength U.V visible detector was used. The column was 25 cm x 4.6 mm ID packed with 10 μ R.P 18 (Merck).

Reagents

Dyes were obtained from different manufacturers and were used as 1 % water solution. All solvents were analytical reagents grade and tetra-alkylammonium bromides were Merck products. Mobile phases were freshly prepared each day and filtered through Millipore filter.

Model solutions

They were prepared with 100 g/l glucose, 10 g/l citric acid (pH 2), 0.1 g/l of E 124 or E 126 and 0 or 0.5 g/l ascorbic acid. After filling and capping 250 ml bottles with these solutions different heat treatments were applied : 15 minutes at 80, 100 and 120°C. Controls were stored at 0°C.

RESULTS

I. Separation of food dyes

Three parameters were studied : nature and concentration of the counterion, and composition of the mobile phase. The capacity factors of the different dyes increase with the concentration of the tetra-alkyl ammonium counterion until an approximate plateau which occurs at about 20 mM. This concentration was chosen for all the other experiments. The nature of the tetra-alkyl ammonium counterion has a strong influence on the capacity factor of the dye. We have considered 3 counterions: tetrabutylammonium bromide, N dodecyl N-ethyl NN dimethyl ammonium bromide (Laudacit) and N cetyl NNN trimethylammonium bromide (Cetrimide) and we have observed an enhancement of the capacity factor of the synthetic dyes with the carbon number of the alkyl chain (table I). This effect is more pronounced for dyes which contain three or four sulfonic groups (E 123, E 126).

TABLE I - Dependence of the capacity factor of synthetic dyes on the carbon number of the greatest chain of tetra-alkylammonium bromide
(column RP 18 - Mobile phase : acetonitrile 60 vol. water 40 vol. containing 20 mM counterion)

E.E.C number	Capacity factor		
	C_4 alkyl chain	C_{12} alkyl chain	C_{16} alkyl chain
E 102	0.5	3.4	60
E 103	0.35	0.7	2.5
E 110	0.5	1.6	9.4
E 123	0.6	4.4	92.5
E 126	0.9	21	

The composition of the mobile phase has a strong influence on the capacity factor which decreases when the percentage of organic solvent increases. The elution strength increases in the order : methanol, acetonitrile, acetone and the selectivity of these 3 organic solvents is slightly different. Finally the separation of the synthetic food dyes can be achieved without solvent gradient with an RP 18 column and two possible mobile phases (figure 1 and 2).

This method of separation can be used for the determination of the impurities of some dyes. For example, E 104 was resolved in four components which can be detected at either 254 or 430 nm.

The quantitative evaluation of a synthetic dye in a beverage can be easily done by direct injection of the liquid ; the use of a variable wavelength visible detector allows a very good characterization and quantitative evaluation of the investigated dye. But this favorable situation does not occur very often and it appears necessary to extract selectively the synthetic food dyes before their analysis by HPLC.

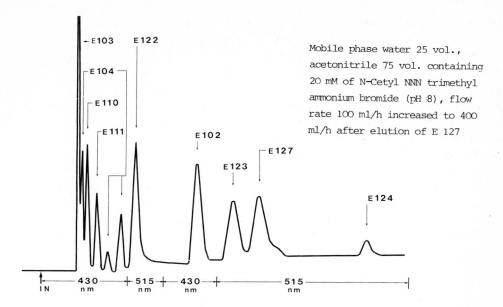

Figure 1. Separation of yellow and red synthetic dyes on RP 18 column with a variable wavelength detector (430 or 515 nm)

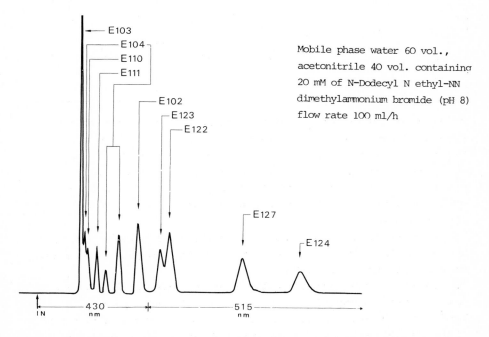

Figure 2. Separation of yellow and red synthetic dyes on RP 18 column with variable wavelength detector (430 or 515 nm)

II. Selective extraction of synthetic food dyes

The possibilities of extracting the dye as an ion pair with a tetra-alkylammonium ion was investigated. Different counterions were tried : tetraethylammonium, tetrabutylammonium, N dodecyl N ethyl NN dimethylammonium (Laudacit) and N cetyl NNN trimethylammonium (Cetrimide). Only the last one allows a good extraction by chloroform from a model solution of each of the synthetic dyes. The procedure was the following : 0.25 g of Cetrimide was added to 50 ml of model solution containing 10 % of saccharose, 1 % of citric acid and 0.01 % of synthetic dye and the pH was adjusted to 10 with ammonia. This solution was extracted three times with 50 ml of chloroform-methanol (80 vol-20 vol). The organic phase was dried over natrium sulfate concentrated under vacuum and taken up in a few milliliters of methanol. The yield of extraction was evaluated for each dye by spectrophotometry and exceeds 95 % in each case. The extracted dyes can be separated by high pressure liquid chromatography using the RP 18 column and a mobile phase : water 30 vol., acetonitrile 70 vol. containing 20 mM of Laudacit. Each dye gives only one peak, the height of which is proportional to the concentration of the dye in the beverage. The accuracy of this quantitative evaluation is about 10 to 15 %.

III. Stability of E 124 and E 126 during heat treatments

The model solutions containing these two dyes were extracted by chloroform methanol in the presence of Cetrimide. The extracts were analyzed by HPLC at 520 nm and 254 nm. At 520 nm only the pure dye was observed but at 254 nm new substances were detected in some cases. These products do not appear when the model solutions do not contain ascorbic acid; but when the model solutions contain ascorbic acid and are heated at temperature of 100° or 120°C, these U.V absorbing compounds are present. Their concentration increased considerably from 100°C to 120°C and consequently the final concentration of the dye declined. We can estimate that these new U.V absorbing compounds are degradation products of E 124 or E 126. From the chromatogram of the extracts of the model solutions containing ascorbic acid and heated 15 minutes at 120°C, we can estimate that 60 to 80 % of E 124 and E 126 are degraded to new non identified compounds. Figure 3 shows, as an example, the chromatograms of the extracts of model solutions containing E 124 and ascorbic acid without heat treatment (A) or heated at 120°C 15 minutes (B). The composition of the mobile phase was 70 % acetonitrile, 30 % water containing 20 mM of N dodecyl N ethyl NN dimethyl ammonium bromide (pH 8). The flow rate was 100 ml/h and the detection was at 254 nm.

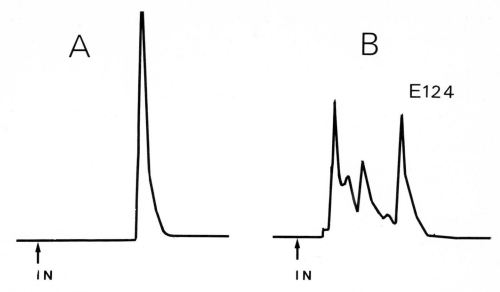

Figure 3. Chromatograms on RP 18 column of extracts of model solutions containing E 124 (for conditions see text)

Conclusion

The principle of liquid liquid extraction of synthetic dyes as ion pair and high performance reversed phase liquid chromatography with ion pairing agent appears as an efficient tool to study the nature and concentration of synthetic sulfonated azo-dyes in beverages and to monitor the behavior of these dyes during technological processes.

REFERENCES

1 E.A Cox, J.Assoc.Off.Anal.Chem. 63, (1980), 61
2 JE Bailey and E.A Cox, J.Assoc.Off.Anal.Chem., 58, (1975), 609
3 D.P Wittmer, N.E Nuessle and W.G Haney, Anal.Chem., 47, (1975), 1422
4 J. Chudy, N.T Crosby and I. Patel, J.of Chromatog., 154, (1978), 306
5 A. Noda, S. Nishiki, Shokuhin eiseigaku zasshi, 19, (1977), 321
6 G.E Martin, M. Tenenbaum, F. Alfonso and R.H Myer, J.Assoc.Off.Anal.Chem., 61, (1978), 908

Acknowledgments

This work was supported by Research Grand 77 70 409 from Délégation Générale à la Recherche Scientifique et Technique

Analysis of Phenols in Smoke, Smoke Preparations and Smoked Meat Products

L. TOTH[1], R. WITTKOWSKI[2] and W. BALTES[2]

[1] Bundesanstalt für Fleischforschung, Kulmbach, Bundesrepublik Deutschland
[2] Institut für Lebensmittelchemie der Technischen Universität, Berlin (West)

SUMMARY

The results show that in the field of food research only the combined use of preparative methods with modern analytical procedures leads to a sufficient examination of complex, aromatic flavor mixtures. Our phenol extract turned out to be a mixture of about 240 substances, instead of the original assumed 60. Their identification will take quite a few years.

INTRODUCTION

Smoking, along with drying, is the oldest procedure to preserve food. Smoke ist produced by the smouldering of wood or the thermic pyrolysis of wood. The smoke components are deposited on the surface of the treated food, which are usually fish, meat and meat products. This simple process produces important sensoric and technical changes. In addition to the well known flavoring and coloring properties of smoke, the product is also preserved by smoking. Smoke contains bactericide, fungicide and also antioxidative ingredients by which it counteracts the microbial as well as chemical spoilage. More important technological effects are the hardening of natural skin and the formation of a secondary skin.

The procedure of smoking is a technology which has been empirically developed over thousands of years (1). An intense scientific investigation into the smoking process started 20 years ago. The present results show more the particular complexity of the process rather than providing an explanation. The reason is seen in the developing of modern analytical techniques. It proves that the present results have been wrong or at least insufficient. The number of assumed combinations in smoke has increased from a few hundred to 10000 . But until now only 280 different components have been identified (2). Thus there remains a large field of activity for the use of modern analytical techniques.
Wood, which is the basic material for smoke production, is of simple composition. It consists of approximately 50 per cent cellulose and 25 per cent of hemicellulose as well as 25 per cent lignin, including small amounts of resins, oils etc. During the thermal pyrolysis of wood, these macromolecules are split, the cleavage products are further pyrolysed and oxidized. Active primary products react together and form many new smoke components. Of the identified 280 components hydrocarbons and phenols predominate, at least in number. Most of the identified components are polycyclic aromatic hydrocarbons and phenols.
In the last ten years they have become the target group of intense research work because numerous constituents of these groups, especially benzo(a)pyrene, have carcinogenic attributes. The phenolic compounds are responsible for the smoke flavor and also for it's preservable effects. This makes the interest on their examination understandable.
A few phenols are suscepted to be carcinogenic or at least to be co-carcinogenic (3). Our task consists of isolating the phenolic components of smoke, preparative to analysis and to making them accessible for toxicological studies.

RESULTS AND DISCUSSION

1. Formation and Isolation
Phenols are predominantly formed by pyrolysis of lignin. Lignin is a polymerized from the following zimtalcohol derivatives: cumar-; coniferyl- and sinapinalcohol, of which sinapinalcohol is predominantly found in lignin of hardwoods.
The present conception is that lignin is degradated during its pyrolysis and ferula- and sinapinacid are produced (4,5). The propenyl-groups which are positioned on the p-position are further decomposed and the main components of the phenols are produced: These are guajacol and syringol. We have been able to prove that soft-wood produces predominantly guajacol and its homologous substitutes which are positioned in the p-position; syringol and its derivate appear predominantly in the smoke

of hard woods (6). This decomposition represents only the main pathway of phenol-formation. The evidence of numerous different phenols - there are about 60 according to the literature - point to some more pyrolytic mechanism.

In order to isolate phenolic components completely, it was necessary to prepare a method for the condensation of smoke (7). After cooling the smoke at about $20^{\circ}C$, the smoke constituents were condensed by directing them through a water bath. A complete condensation of the phenols and numerous other smoke constituents was achieved when we absorbed the particle phase of smoke with the help of several filters made of glass wool. This smoke condensate was separated as a watery and a tarry fraction. About 70 % of the phenols are found in the water phase (7).

In the literature are descriptions of several methods for isolation of phenols from diluted solutions. The most recommended method has been the extraction of phenols from alkalized watery solutions with organic solvents. We were able to demonstrate that such methods lead to the loss of all phenolic compounds with two free hydroxy groups: for example, brenzcatechin, resorcin and their homologs.

In comparing several methods we found that the phenols, excepting the phenolcarbonacids, can be isolated from a watery solution adjusted to pH 6,5 - 7,0 by $NaHCO_3$ by extraction with ethylacetate. About 80 % of the phenols have been recovered. The rough-extract consists of about 70 % of phenols. A further clean up may be done by a water distillation at $170^{\circ}C$ and with the help of column chromatography. By this procedure (extraction by ethylacetate and distillation) very different losses of singular phenols occur. The average recovery rate is about 60 %. From smoked food, the phenols have been isolated directly using water distillation at $170^{\circ}C$.

Column chromatography silanised silica gel has proved to be an advantagous tool. The clean up by column chromatography is necessary for the separation of polycyclic hydrocarbons which disturb toxicological investigations. The silanised silica gel was prepared from Kieselgel 60 (Merck, Darmstadt) by adding a solution of trimethylchlorsilane in chloroform and forcing the silylation reaction at $190^{\circ}C$ after evaporating the chloroform. The preparation of water cooled column is done by pouring the material into a solution of 40 % ethanol, transfering it to the column and washing it with a solution of 40 % ethanol. From this column - 2 cm thick and 8 cm high - all phenols as well as phenol-aldehydes and -ketones are eluted with 200 ml solution of 40 % ethanol. The column can be used again after rinsing the polycyclic aromatic hydrocarbons with an absolute alcohol and an equilibrating with a solution of 40 % ethanol.

2. Gaschromatography of phenols

For the separation and identification of the phenols found in smoke, the gaschromatography of their silylether on glass capillaries coated with OV 17 has proved successful (7,8). The recommendation of the literature to use carbowax columns for the separation of underivated phenols is disadvantagous because of the loss of the dihydroxyphenols. Thus it is not astonishing that brenzcatechin, resorcin and numerous other phenols with two or three OH-groups have been very seldom found in smoke.

The investigation of the trimethylsilylether of phenolic compounds from a rough extract and the gaschromatography of the cleaned phenol-fraction on an OV 17 glass capillary shows that our cleaning procedure does not lead to the elemination of special phenols (7). Moreover, it shows that besides the 20 predominant components about 80 more phenolic compounds exist in low concentrations. Their identification with the help of GC/MS on line system has been difficult most of the time, if not completely impossible, because numerous peaks turned out to be a signal of two or more components. This heterodyning shows us again that the most productive separation method, the capillary gaschromatography, is not acceptable for the analysis of many mixtures of natural compounds.

3. Fractionation

In order to reduce the number of substances in a gaschromatogram, we separated first the monohydroxyphenols from di- and trihydroxyphenols with the help of a Na-borat solution (9). By the use of different evaluation techniques, we were able to identify by GC/MS-coupling 62 phenols, among them numerous phenols which are still unknown. Further more 57 mass-spectra of unknown compounds have also been registered.

In order to enable a complete identification of this complex mixture, an advanced fractionation method by means of the reversed phase-liquid chromatography has been worked out. The phenol extract was fractionated in a preparative reverse phase (RP-8) column. A gradient elution using a methanol water-mixture in a ratio 40 : 60 up to 100 : 0 enabled us to separate 10 fractions. The ten fractions were separated by gaschromatography in ten different runs (see Figure).

From these gaschromatograms one can see that several signals, which in the gaschromatogram of the unfractionated extract appeared to be only one homogenous substance, can be separated into different peaks. Through the fractionation, a strong enrichment was achieved at the same time, so that under these analytical conditions small amounts of components appear as very intense signals. Following this procedure the mass-spectrometric identification of phenols is much easier and in some cases only possible under such conditions. A particular advantage is that the appearance of substances in special fractions allow their identification by structural analysis because of high selectivity of the fractionating procedure. For example: phenol (peak 1), guajacol (peak 7) and syringol (peak 16) appear in the 4th and 5th fraction. The presence of further hydroxy-, as well aldehyd- and acetogroups in molecules increase the solubility of phenols in water. Therefore brenzcatechin (peak 8) and 3-methoxybrenzcatechin (peak 18) are found in the 2nd fraction, phenolaldehydes and ketons in the 3rd fraction, whether or not these compounds represent the same rf-values as the above mentioned ones. On the other hand, methyl-, ethyl- and propylgroups need a higher elution volume. We found the kresols (peaks 2,3,4), 4-methylguajacol (peak 12) and 4-methylsyringol (peak 17) in fraction 7. Dimethyl- and ethylphenols were determined in fractions 8 and 9, isoeugenol (peak 19/3 and 21) in fraction 10.

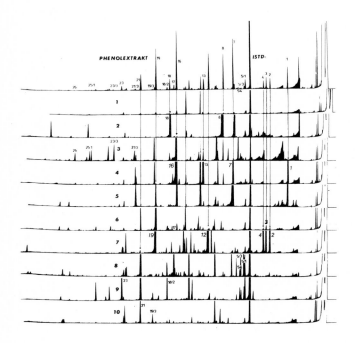

Figure

Gaschromatographic separation of 10 phenol extract fractions from a OV 17 glass capillary (25 m, Temp. cycle: 80 °C, 8 min. isotherm, 2 °C to 220 °C)

REFERENCES

1 D.J. Tilgner, Fleischwirtschaft 57 (1977), 45
2 K.Möhler, "Das Räuchern", Verlag der Rheinhessischen Druckwerkstätte, Alzey (Germany), 1978
3 L. Tóth, Fleischwirtschaft 60 (1980), 461
4 W. Fiddler, W.E. Parker and A.E. Wasserman, J.Agric. Food Chem. 15 (1967), 757
5 T. Kossa, Dissertation der TU-Berlin, Nr. 55/FB 13 (1976)
6 L. Tóth, Fleischwirtschaft 60 (1980), 1472
7 L. Tóth, Fleischwirtschaft 60 (1980), 728
8 M.R. Kornreich and Ph. Issenberg, J. Agric. Food Chem. 20 (1972), 1109
9 R. Wittkowski, L. Tóth and W. Baltes, Z. Lebensm.Unters.Forsch. (in preparation)

Investigation on Contamination of Food by Polycyclic Aromatic Hydrocarbons

K. TIEFENBACHER, W. PFANNHAUSER, H. WOIDICH

Forschungsinstitut der Ernährungswirtschaft, Blaasstraße 29
A-1190 Vienna/Austria

SUMMARY

Some aspects of the contamination of food by PAH are discussed. These aspects are their origin, their distribution into environment and how they contaminate food. The different kinds of food contribute to PAH uptake with human nutrition are mentioned.
The determination of a pattern of eight PAH is proposed to be sufficient for most problems of food control.
Selected analytical results are presented and discussed.

INTRODUCTION

The determination of benzo(a)pyrene (BaP) and of other PAH in food has become more and more important during the last years which is shown by the increasing number of publications and the different national and international regulations and recommendations about the limit concentration of BaP and PAH in food.
This is firstly due to the fact that some of these compounds have been proven to be cancerogenic in animal tests. In order to prevent potential cancer risks for humans, some limit concentrations have been approved for certain products.

Secondly the PAH could be indicators for other undesirable substances. For example, the regulations of BaP in smoked food imply also restrictions for other suspected components of smoke (1).

Many authors suppose some relationship between consumption of smoked food and higher incidence of gastric cancer (2), but epidemiological studies do not allow such conclusions with certainty (3).
Metabolism of PAH leads to the ultimate reactive species. The different steps of this reaction have been clarified in the last years (4), but the connection with the later occurrence of cancer is not yet clear today.

ORIGIN AND DISTRIBUTION OF PAH

Most of PAH result from anthropogenic combustion sources e.g. residental heating, industrial processes (especially coking and thermal power plants), waste combustion, automobile exhausts etc. They also result from cigarette smoking or from food technology processes like smoking, grilling and in some cases direct drying with combustion gas.
The occurence of PAH in coal and petroleum is not of direct interest in food contamination. Biosynthesis as a source of the worldwide contamination with a mixture of some 100 compounds of this type is not very reasonable. The analysis of sediment cores moreover shows a dramatic rise of PAH concentration during the last 100 years in good proportion to industrialisation and increasing use of fossil fuels. In older sediments only few compounds were found which originate for instance from diagenesis of triterpenoids. The single occurrence of perylene in such sediments has been shown, possibly it comes from diatomees (5).
So combustion is the main source for PAH. If one accounts for the release of BaP and related substances he finds a striking dependence on the kind of fuel. Coal contributes substantially to overall emissions as like in domestic heating as in its industrial use (6). Petroleum products or even gas give far lower emissions.

The substantial extent of domestic heating can be seen in the maximum concentration of PAH in the air in the winter. Another part of the emissions comes from refuse burning, above all open burning of refuse, wood or straw.

Only 1-2 percent of emissions are supposed to come from motor vehicles (6). To a high extent the emitted PAH are bound to the particle phase of the emissions, they are distributed in the air over wide distances and are sedimented to soil, water and plants.

CONTAMINATION PATHWAYS FOR FOOD

Foodstuffs of plant origin are primarily contaminated by air pollution, which is sedimented. The uptake of PAH by plants via roots seems to be not significant.
In table 1 the main groups of foodstuffs are listed, where a contamination can occur. The first groups are arranged according to their contribution to overall PAH-uptake with human nutrition, which depends from (a) consumed quantities and (b) average content of PAH. It is to be remarked that the level of environmental pollution at the location of growth is essential for foods of plant origin (7).

Table 1

Potentially contaminated foodstuffs and possible origins of PAH
     E... environmental pollution
     T... food technology or food preparation

FOOD	MAIN PATHWAYS OF CONTAMINATION
Vegetables	E - deposition (air pollution)
Cereals	E - deposition (air pollution); (T - direct
flour, bread, etc.	(baking) drying)
Fruit	E - deposition (air pollution)
Smoked food	T - outdated smoking technology
meat, sausages, fish, cheese, etc.	
Vegetable fats	E - deposition (air pollution); (T - direct
oils, fats, margarine, etc.	drying of raw material)
Drinking water	E - wash out (air pollution), runoff
Grilled food	T - grilling (broiling)
Spices	E - deposition (air pollution)
	T - direct drying
Coffee	T - roasting
Tea	E - deposition (air pollution)
	T - direct drying
Sea food	E - water pollution (waste water, oil
fish, mussels, oysters, etc.	spills, sedimentation of air pollution)

In table 1 we can see, that foods of plant origin give an essential contribution to PAH-uptake, which is higher than that of food technology or food preparation.

In highly populated or industrialized districts the contamination with PAH can rise by a factor 10 or higher than the values or rural regions. Of course in Austria the main regions for grain, fruit or vegetables production are contaminated very low.

Washing environmentally contaminated products can remove only 10 to 20 percent of PAH (7).

At next the technological causes for PAH contamination are to be discussed.

Direct drying of grain with combustion gas has been - at most in Austria - replaced by better techniques.

Drying with combustion gas also can contribute to contamination of oils and fats, of certain spices and teas. By raffination of oils most of the PAH can be eliminated by a special treatment.

Smoking of foods in connection with PAH-contamination is one of the fields best studied. The uptake of PAH by smoked food is only about 5 percent of overall load, but there are some persons with higher consumption of such products. A substantial decrease of PAH content is possible by optimization of smoking technology.

Now I want to discuss some special cases which are not important for overall consumption, but are interesting for different reasons:

Grilling: Especially in charcoal broiled food PAH can arise if the glowing charcoal is not beside the food to be grilled but under it, so that fat drips off and undergoes pyrolysis.

Frying and roasting in normal kitchenlike preparation poses no problems in respect to PAH-formation.

Sometimes certain dried spices show higher concentrations. The environmental pollution of the location of growth and sometimes direct drying with combustion gas in the supplier countries are to be reflected by the PAH-levels.

With tea the situation is similar. The PAH-transfer into the infusion is very low. According to our investigations about 1-2,5 % of PAH in tea leaves are found in the infusion.

REFERENCE COMPOUNDS FOR DETERMINATION OF PAH IN FOOD

The term "polycyclic aromatic hydrocarbons" denominates a complex mixture of many substances. Benzo(a)pyrene, often denoted as "reference compound", in many cases is not a true representative to judge the extent of contamination with PAH. The relations in concentration of PAH are depending substantially on their origin and are different for instance in automobile exhausts and in curing smoke. Thorough information about such an mixture of many substances and an identification of a great number of compounds gives analysis by capillar -GC/MS. Nevertheless often only BaP is evaluated because it is easy to determine and renders not so much analytical work. Besides the results for BaP give an first overview to contamination.

In practical food control the determination of a limited number of main substances is sufficient in most cases and represents a good compromise in respect to analytical expenditure.

For our determinations we use in situ fluorescence measurement after separation by TLC on acetylated cellulose (8).

Advantages of this method are specific identification of substances, high sensitivity and low costs. The high sensitivity of fluorescence determination enables miniaturizeing of the clean up.

Details of the analytical method are presented elsewhere (9).

Usually eight PAH are determined quantitatively :

fluoranthene
pyrene
benzo(a)anthracene
chrysene

benzo(a)pyrene
benzo(b)fluoranthene
perylene
benzo(g,h,i)perylene

RESULTS AND DISCUSSION

In table 2 some results of our determinations of PAH in food are presented. For the sake of clarity only the values for BaP are listed. Table 2 comprises only determinations of PAH for monitoring studies. Samples we get and which are suspected to contain a high PAH-contamination are not listed here.

With vegetables and fruit most of the samples are contaminated very low despite of single higher values. Only a greater number of results - we are going on to finish studies about PAH in 250 samples of vegetables, grain and fruit - will describe the actual situation in Austria better.

For a long time smoked food has been of major interest so that we have many results. Most of commercial samples of smoked meat and sausage are low in PAH content.

The situation with home smoked products is quite different. Home smoking in many regions of Austria is common and poses certain problems.

Table 2

Determination of PAH in food: Results for benzo(a)pyrene

	number of samples	median	range
		microgram/kg (ppb)	
Vegetables	22	0,10	0,01 - 9,46
Fruit	17	0,01	0,01 - 0,20
Smoked food:			
Meat products	76	0,27	0,11 - 3,63
"home"smoked	11	1,60	0,95 - 10,46
Sausages	21	0,19	0,01 - 1,96
Fish, canned	38	0,64	0,05 - 16,40
other fish	29	0,10	0,01 - 5,13
Cheese	17	0,17	0,01 - 1,76
Oysters, mussels	13	2,27	1,29 - 12,83
Oils, fats	21	0,60	0,01 - 3,72
Spices	6	5,69	1,67 - 19,11
Grilled meat	7	0,49	0,33 - 10,15
Paraffins, waxes	4	0,43	0,25 - 3,14
		nanogram/kg (ppt)	
Mineral(table)water	55	0,05	0,01 - 0,95
other drinking water	17	0,40	0,05 - 2,20

Table 3

Distribution of PAH in smoked products canned in oil

a) SMOKED SPRATS ("Kieler Sprotten"): 75,1 % by weight "fish"
                                                    24,9 % "oil"

		FL	PY	B(a)A	CH	B(b)F	B(a)P	PER
PAH in "fish"	ppb	8,9	5,5	1,0	0,7	0,5	0,4	0,1
	%	31	35	31	28	33	32	34
PAH in "oil"	ppb	59,5	31,2	6,7	5,4	3,0	2,5	0,6
	%	69	65	69	72	67	68	66
average concentration	ppb	21,5	11,9	2,4	1,9	1,1	0,9	0,2

b) SMOKED OYSTERS: 89,9 % by weight "meat", 10,2 % "oil"

		FL	PY	B(a)A	CH	B(b)F	B(a)P	PER
PAH in "meat"	ppb	240,3	77,0	18,3	24,9	10,5	6,2	1,5
	%	73	78	74	78	74	67	71
PAH in "oil"	ppb	799,0	189,7	57,1	62,4	33,0	27,2	5,3
	%	27	22	26	22	26	33	29
average concentration	ppb	297,3	88,5	22,3	28,7	12,8	8,3	1,9

In smoked fish, especially with canned products, sometimes higher concentrations of PAH are to be found.
In our experience <u>oysters</u> and <u>mussels</u> regularly show higher concentrations of PAH, which to a substantial extent is not due to a wrong smoking technology but to water pollution.

In table 3 a detailed study on canned smoked sprats respectively oysters is shown. The content of the cans was separated in "oil" phase and "meat" by dripping. It can be seen that PAH are enriched in the oil. In the case of the sprats only about one third of the PAH are consumed with the "fish", but with oysters (with low "oil" content) about two thirds.

In the category: vegetable fats and oils(table 2) the raffinated products, which represent most of the consumed quantities are low in contamination. Not raffinated special products often show higher values of PAH.

For spices we have only few analyses at present. Here the PAH are enriched relatively because of dehydration, but the consumed quantities of spices are very low.

The results for grilled meat and for paraffins (used for food packaging materials) are in the normal range.

At last in table 2 bottled mineral(table)water and other samples of drinking water(well, spring, company's water) are compared. The first mentioned waters, often coming from higher depth are less contaminated by about a factor 10 on average.

REFERENCES

1. Deutsche Forschungsgemeinschaft, Mitteilung VII der Fremdstoff-Kommission (1972)
2. W. Fritz and K. Soos, Nahrung 21 (1977), 951
3. L.J. Dunham and J.C. Bailar, J. Nat. Cancer Inst. 41 (1968), 155
4. D.M. Jerina, D.R. Thakker and H. Yagi, Pure Appl. Chem. 50 (1978), 1033
5. S.G. Wakeham, Ch. Schaffner and W. Giger, Geochim. Cosmochim. Acta 44 (1980), 403, 415
6. National Academy of Sciences, Particulate Polycyclic Organic Matter, Washington DC (1972)
7. G. Grimmer and A. Hildebrand, Dtsch. Lebensm. Rdsch. 61 (1965), 237
   W. Fritz and R. Engst, Z. Ges. Hyg. 17 (1971), 271
8. H. Woidich, W. Pfannhauser, G. Blaicher and K. Tiefenbacher, Chromatographia 10 (1977), 140
9. K. Tiefenbacher, H. Woidich and W. Pfannhauser, Ernährung/Nutrition 4 (1980), 346

A version of this lecture has also been published in Ernährung 5 (1981) by agreement of the publishers.

Degradation and Metabolism of Relevant Pesticide Residues

R. ENGST

Central Institute of Nutrition, Potsdam-Rehbrücke,
German Democratic Republic

SUMMARY

With the degradation and metabolism of pesticide residues on crops and in foods and the decrease of original active compounds connected therewith, the importance of food hygienic-toxicological problems does not disappear. The essential difficulties often begin only then, because secondary products are formed or transformation takes place, which in abiotic or biotic systems can cause detoxifications or, on the other hand, toxifications. These conditions are demonstrated by organochlorine compounds like DDT and lindane, but also by chlorophenols. Possibilities of dioxin formation are taken into consideration. The fungicidal ethylenebisdithiocarbamates and likewise the manifold possibilities of toxifying reactions of the organophosphate active compounds are especially discussed. Photo-induced reactions and nitrosations are mentioned. In connection with the significance and relevance of pesticide residues, basic principles concerning the establishment of acceptable residue amounts are interpreted.

+
 Original version:
 R. Engst: Abbau und Metabolismus relevanter Pestizidrückstände,
          Ernährung/Nutrition 5 (1981) 2, S. 61 - 67
 see also:
 R. Engst: Chemical Toxification of Pesticides in the Environment,
          Advances in Pesticide Science, Part 3, p. 590 - 597.
          Pergamon Press Oxford and New York 1979

INTRODUCTION

The world population is now over the 4 milliard mark, 15 - 20 % of them being acutely and constantly troubled with hunger. About one-third of mankind suffers from undernourishment and malnutrition. This fact and, currently, the annual addition of about 80 million to the world population underline insistently the urgency of increasing food production. Manifold efforts are being made to reach this aim. In agriculture especial weight is given to the application of pesticides and growth regulators in plant production. According to information from the Food and Agriculture Organization (FAO) we should take into consideration that about 35 % of the world harvest is lost owing to pests and plant diseases. The losses in the various countries in Europe and North America come to 20 - 28 %, in Africa and Asia they even average more than 40 % with top losses of far more than 50 % (1).

The application of pesticides and other agrochemicals is unavoidable but not without problems. Nowadays, mankind has to cope with a general toxicological situation, the present complexity and significance of which cannot be overlooked entirely. According to current knowledge, the residues of agrochemicals, especially of the ubiquitous pesticides, which are taken up into the food chain, become a decisive factor in it. Pesticides are used because of their toxicity to the pests of plants and animals. For that very reason it cannot be expected a priori that they are harmless or indifferent to higher living creatures, especially mammals. Great efforts are made to achieve a broad selectivity of the toxic effect against the pests; in this way the selection of the active principles is greatly influenced. However, a risk remains that must always be taken into account when substances which are foreign matters within the physiological processes are used in obtaining, producing or storing food.

In former conceptions the evaluation of pesticides consumed as residues by man and animals in the course of food intake was based on toxicological and other evaluations of the original, unchanged and pure active principles. With the disappearance of a pesticide principle in its original form from a contaminated medium the problems of toxicological residues are not solved in any way. However, it must be taken into account that the active principle has been translocated within bioaccumulative processes into another system or converted into more or less dangerous chemicals. The reaction products formed in the environment or being accumulated can be an additional source of danger within the ecosystem.

In the context of pesticide transformation, a differentiation can be made between <u>metabolism in biotic environment</u> and <u>transformation in abiotic environment</u>. Certain reactions were found, indeed, to take place exclusively in one system or the other. Abiotic genesis of photodieldrin and dioxin formation are just two examples. Nuclear hydroxylation of aromatic systems in the presence of cytochrome P 450 and subsequent conjugation, as reported for the metabolism of Lindane, Carbaryl, and Benzimidazole, are clearly linked to biotic environment. Nevertheless, there are many examples in which transformation is neither exclusively biotic nor of abiotic nature only.

SPECIFIC DEGRADATION- AND METABOLISM-PROCESSES

Reactions involving the chlorophenols are the most remarkable toxifying reactions of abiotic occurrence in pesticide chemistry. Chlorophenols are base products for synthesis, such as the formation of phenoxyacetic acid herbicide derivatives. They are even commercially manufactured, for example, as fungicidal pentachlorophenol. They have a built-in capacity for intermolecular and intramolecular elimination. For example, dehydration, dechlorination, and dehydrohalogenation may take place in parallel, when polychlorophenols are heated to something in excess of 300 ^{o}C, say, by incineration of wood which had been exposed to pentachlorophenol treatment. Such processes will result in the generation of hydroxylated, polychlorinated biphenyls (PCB) and predioxins as well as of polymeric polychlorophenyl ethers (Fig. 1).

As well known, particular attention has to be given to 2,3,7,8-tetrachlorodibenzo-p-dioxin for several reasons, including its high LD_{50}, between 0.022 mg and 0.066 mg per kilogram body weight (rat), yet, even more for its embryotoxicity and teratogenicity (2). The LD_{50} of 2,4,5-trichlorophenol, a possible precursor of this compound, is 820 mg/kg body weight (rat).

Chlorination of phenols is one of the reactions with parallel abiotic course. They are likely to take place by elimination of hydrogen chloride from two molecules of chlorophenols, with that hydrogen chloride then affecting dimeric intermediate products.

Higher chlorated phenols formed in this way have a higher acute toxicity than lower chlorated ones.

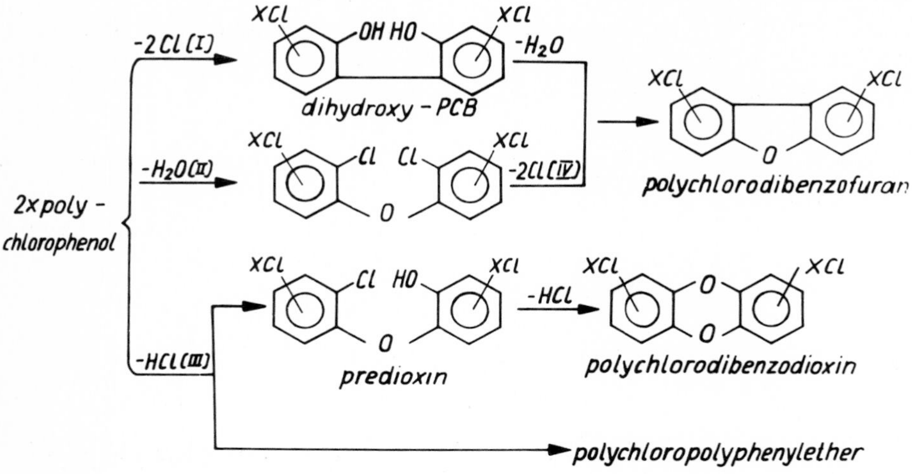

Fig. 1. Intermolecular reaction of polychlorophenols

Reference should be made, in this context, to special self-chlorination of pentachlorophenol, which then results in the formation of hexachlorobenzene (HCB).

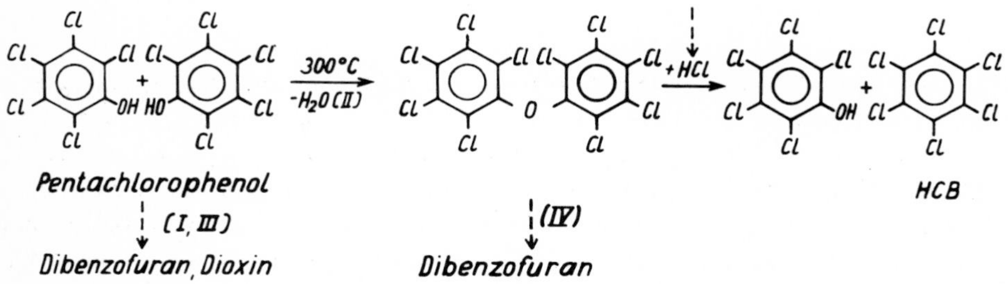

Fig. 2. Abiotic synthesis of hexachlorobenzene

Another mode of hexachlorobenzene formation has been elucidated more recently. It is based on microbial transformation of Lindane. Those abiotic and biotic modes of environment-generated HCB formation are confronted with degradation. Both Lindane and HCB can be transformed into chlorophenols which, in turn, are precursors to HCB formation again (3, 4). In that context a somewhat remarkable phenomenon is recordable. A detoxifying step in metabolism will first be followed by a toxifying step. It, however, plays only a minor role in the general background of detoxification (Fig. 3). Polychlorophenols actually derive their own relevance to the process from their capacity to form dibenzofuranes and dibenzodioxins. That relevance is further underlined by higher occurrence of HCB, not only with Lindane use, but even more in the presence of polychlorophenols.

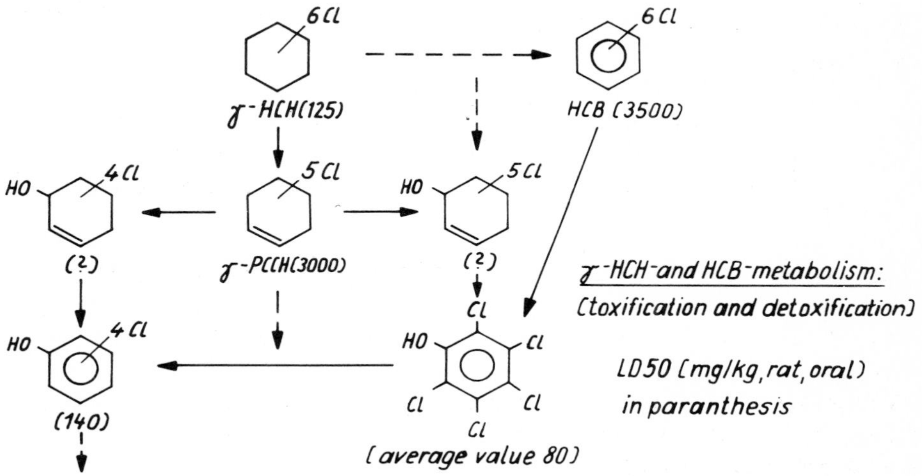

Fig. 3. Gamma-HCH and HCB metabolism
(toxification and detoxification)

In the past the most frequently used pesticide was p,p'-DDT. The estimated amount of DDT that has been produced since 1944 is much more than 3 million tons, with an average annual production of about 1 000 000 tons. The measures in the developed countries to limit the use of DDT have not yet led so significant changes in its total annual application (5, 6). In this connection it must be mentioned that DDT, today as before, maintains its value as an economical, effective and available pesticide for the special situation of the developing countries. In considering its persistence and accumulation in the food chain and the human organism, information on the degradation processes of DDT as affected for the most part by enzymatic, that means biotic influences, seems to be important. Greater amounts of DDE (dichlorodiphenyldichlorethylene), additional to DDT, are more or less regularly observed in the environment and particularly in warm-blooded animals. In the human fatty tissue DDT and DDE are to be found nearly constantly with a ratio of 3 : 7; thus DDE prevails. This ratio, however, does not generally apply to body organs and other tissues (7). Traces of DDT (dichlorodiphenyltrichlorethane) are evident in mammals. In the environment it is often observed in greater amounts. We observed in the visceral fat of 6 perished whitetailed eagles up to 400 mg DDD/kg besides 20 mg DDT and 1200 mg DDE. Considerable amounts of DDT and its metabolites were fairly regularly found not only in the different tissues and organs of these birds of prey but also in the eggs of wild birds. Small amounts of DDA (dichlorodiphenyl acetic acid) occur in human and animal excretions. The same applies to the water-soluble DDOH (dichlorodiphenylethanol). Investigations with ^{14}C-labelled DDT showed that in organs of mammals (rats) with a high metabolic activity DDD, DDA and DDT plus small amounts of DDOH and DDE are to be found.

According to information from literature, a further substance, namely the water-soluble DCB (dichlorobenzophenone), has been found in resistant body lice. From this the conclusion may be drawn that the resistance of the body louse is partly due to its acquired capacity of quickly transforming the lipophil DDT into DCB, being water soluble and capable of being excreted (8, 9). In warm-blooded animals the degradation of DDT occurs only tentatively. This is supported by the well-known fact that between the diminishing of the DDT load and the stabilisation of the new load level 1 - 2 years may pass. The explanation of a complete degradation cycle concerning the behaviour of DDT in the organism of mammals is therefore complicated. We were, however, able to trace the microbial degradation of DDT and compose a scheme for the metabolism of DDT as early as 1967 (Fig. 4).

We were able to prove the enzymatic character of the degradation and to indicate the differentiated effects of enzyme fractions. The degradation cycle confirms that the metabolic changes cause a hydrophilisation of the original lipophilic principles, which enables elimination from the organism and therewith detoxification.

Ethylenethiourea (ETU), which occurs in conjunction with dithiocarbamate degradation, may be formed by reactions of elimination, if the bisdithiocarbamate is heated with no exposure to any oxidative effects at all, for example, when food with dithiocarbamate is boiled. Bisdithiocarbamate residues, therefore, can be a potential source of ETU (10).

Elimination reactions of alkylenebisdithiocarbamates are characterized by two phenomena:

1. Separation of sulphocarbon from the carbamide acid group and formation of an amine;
2. Splitting off of hydrogen sulphide from the carbamide acid group and formation of isothiocyanates.

Secondary reactions will then give rise to the formation of ETU and poly-ETM (ethylene thiurammonosulphide). The substance classified in the scheme as ethylenebisisothiocyanate sulphide (EBIS) is identical with a product of degradation that used to be called monomeric soluble ETM. The author's own findings are likely to prove that ETM actually coexists with EBIS, doubtless in polymeric insoluble form. EBIS and ETM both are of pronounced fungicidity, and their acute toxicity to warm-blooded organisms is higher than that of dithiocarbamate originally used. We have found the peroral LD_{50} of EBIS for rats to be 1,570 mg/kg live weight. This actually differs from the values of 4,500 mg/kg found for Maneb, or more than 5,200 mg/kg found for Zineb. The acute LD_{50} of ETU for rat has been found to be 900 mg/kg. Its thyreotoxicity as well as suspected mutagenicity and carcinogenicity are additional factors of potential hazard (fig. 5).

Degradation of dithiocarbamates is accelerated by oxidation, as well. EBIS as well as ETU may be formed via thiuram sulphides. The catalytic oxidative effect of the metal fixed in the molecule is a decisive influence. Oxidation caused by the catalytic effect applies to manganese rather than to zinc. The zinc-dithiocarbamate complex (Zineb), consequently, is more persistent than the manganese complex (Maneb). That discrepancy in stability between Maneb and Zineb will be borne out with particular eminence, if both active ingredients get into contact with plant sap (11).

ETU is of relatively high stability to hydrolysis. Yet, it is readily susceptible to photo-oxidation, with the latter being strongly favored by sensitizers (flavonoids, acetone). ETU is of short life when exposed to the environment of account of its high oxidation capacity. In such cases, it is ethylene urea (EU) which is mainly formed. ETU residues in plants, harvested crops, and animal products will be of relatively low level in response to adequate use, except for possible consequences of thermal effects. In addition, ETU is microbially oxidized to CO_2 in soil. Intraplant sequels following ETU degradation were not only EU but 2-imidazoline, as well (12). It is formed by oxidative degradation of ETU, but intermediate occurrence of ethylenethiourea-S-oxides has to be expected. Hence, ETU, when exposed to various environmental factors, will form a larger number of degradation and sequential products by oxidation.

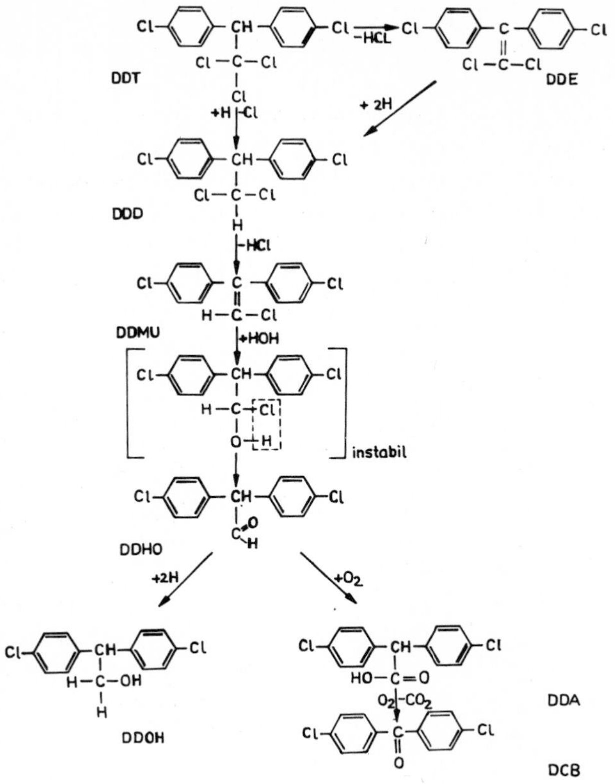

Fig. 4. Reaction scheme of the enzymatic degradation of DDT

Residues of non-decomposed dithiocarbamates on crops with freeland treatment usually form surface coats. Residue dynamics, therefore, will be greatly influenced by light or radiation energy, precipitation, temperature, and humidity. Acids excreted by the plants concerned, do have an additional effect in terms of degradation. It should be born in mind that even at the time of application the agents used may contain different portions of degradation products, above all ETU and EBIS or ETM, due to certain factors relating to manufacture and storage. However, most of these products are secondarily formed by unchanged application of the proper agents. That is normally the main amount.

Consequently, there is no reason for concern about possible contamination of the environment by non-decomposed ETU. The conclusion may be shown that ETU formation has to be taken into serious consideration in the unavoidable thermal technological processes. It will be accompanied by an unambiguously toxifying effect which then will determine degradation of fungicidal dithiocarbamates.

A positive response to this situation has been recorded by the competent WHO-Committee. At its Joint Meeting in 1974, it has recommended to lower the temporary ADI for dithiocarbamates from 0.025 to 0.005 mg/kg body weight. Tolerances were also recommended for ETU residue assessments according to which ETU levels in harvested crops for foodstuffs should not exceed the order of 0.01 mg/kg.

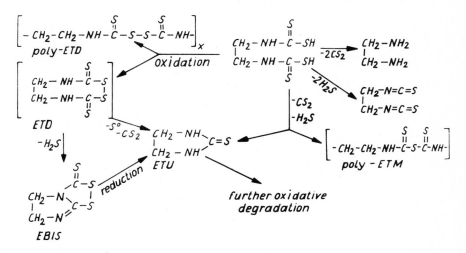

Fig. 5. Degradation and metabolism of ethylenebisdithiocarbamates (simplified)

Photo-oxidation of dien-pesticides will clearly cause severe toxification via epoxidation. Biotic reactions may occassionally result in final products of the very same kind, and elimination, discussed earlier in this paper, may produce toxification, in this context as well (13).

Oxidation is the prerequisite for toxicity of many organophosphates. There are phosphoric acid esters and thiophosphoric acid esters which inhibit cholinesterase. In addition, thionophosphoric acid esters become strong esterase inhibitors only after oxidation (Fig. 6, 1).

This type of reaction is likely to take place in abiotic systems but even more in biotic systems. Several authors have suggested that

thionophosphoric acid esters cannot inhibit cholinesterase. They
felt that the low amount of inhibition was attributable merely to
contamination by the P-O analogue (oxone) (14). Several authors
produced evidence to the effect that photochemically induced oxone
formation was possible under the direct impact of atmospheric oxygen
(15). Yet, oxone formation was observed also in anaerobic solutions,
which led to the assumption that water might be involved (16). It
is strongly felt that photochemical reaction cannot be caused effectively by the presence of oxygen sent. Both are essential to that
type of reaction studied by Crosby. Traces of oxone often are found
as residues in foodstuffs. They do not accumulate in the environment
on account of their continued photochemical transformation into non-
toxic products, for instance on the way of hydrolysis.

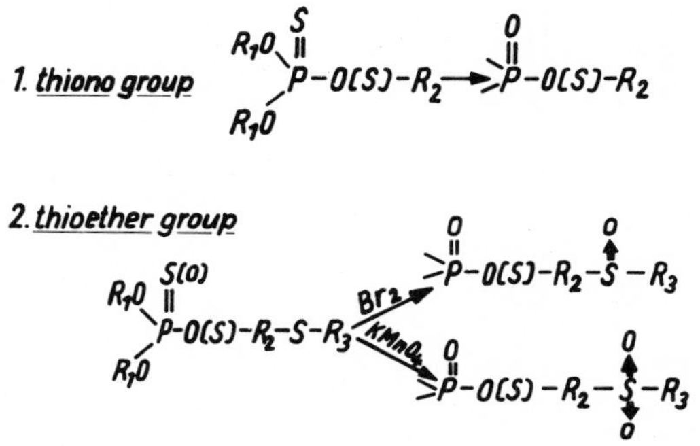

1. *thiono group*

$$R_1O\diagdown\overset{S}{\underset{\|}{P}}-O(S)-R_2 \longrightarrow R_1O\diagdown\overset{O}{\underset{\|}{P}}-O(S)-R_2$$

2. *thioether group*

$$R_1O\diagdown\overset{S(O)}{\underset{\|}{P}}-O(S)-R_2-S-R_3 \xrightarrow[KMnO_4]{Br_2} \begin{array}{l} >P(=O)-O(S)-R_2-\overset{O}{\underset{\uparrow}{S}}-R_3 \\ >P(=O)-O(S)-R_2-\overset{O}{\underset{\underset{O}{\downarrow}}{S}}-R_3 \end{array}$$

Fig. 6. P-esters, toxification by oxidation

The thresholds specified for active ingredients have been clearly
separated from those given for pertinent oxones in the tolerance lists
of several countries, including the GDR; so far as those are significant and, consequently, relevant to form residues of oxidable organophosphorous compounds.

The thio-ether group is often more vulnerable to oxidation than
the thio-phosphoryl group, for instance, in the case of parathion. The
example of demeton-O has shown that the thio-ether group, with a
toxifying effect which is purely chemical, can be readily oxidized to
sulphoxide and further to sulphone (Fig. 6, 2) (17). Externally formed ionic oxidants (sensitizers) are believed to be responsible also
for a wide range of isomerizations. In other words, oxone of the
original or parent substance is not the only form of parathion with
strong inhibition of cholinesterase. There are isomeric substances
of parathion with increased cholinesterase inhibition when compared

to the parent product, but reduced acute toxicity to houseflies. They were identified as thiolesters of parathions (18).

Other transformations of organophosphorous compounds were found to be based on their own alkylation capacity. For example, storage of demeton-S-methyl may lead to abiotic formation of the sulphonium compound with thousandfold toxicity. In that reaction one molecule of P-ester does function as donor and another as acceptor (19).

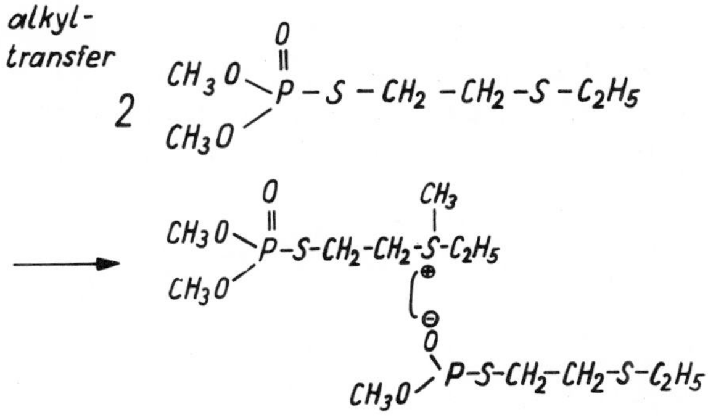

$$2 \begin{array}{c} CH_3O \\ CH_3O \end{array} \!\!\!\!\!\!\!\!\!\!\!\!\!\!\!\!\!\!>\!\!\!\overset{O}{\underset{}{P}}-S-CH_2-CH_2-S-C_2H_5 \quad \text{alkyl-transfer}$$

$$\longrightarrow \begin{array}{c} CH_3O \\ CH_3O \end{array} \!\!\!\!\!>\!\!\!\overset{O}{\underset{}{P}}-S-CH_2-CH_2-\overset{CH_3}{\underset{\oplus}{S}}-C_2H_5$$

$$\begin{array}{c} \\ CH_3O \end{array} \!\!\!\!\!>\!\!\!\overset{\overset{\ominus}{O}}{\underset{}{P}}-S-CH_2-CH_2-S-C_2H_5$$

Fig. 7. Toxification by abiotic alkyl-transfer

Such transalkylations take place in biotic systems, as well, with glutathion or nucleic acids being possible acceptors. This particular process should be kept under further observation, since the occasionally recorded cancerogenicity of thiophosphates might be attributable to alkylation of nucleic acids and formation of irreversible alkyl-acceptor products. It may again be seen that many abiotic reactions take place on the basis of unchanged reactionprinciples when occurring in a living environment (20).

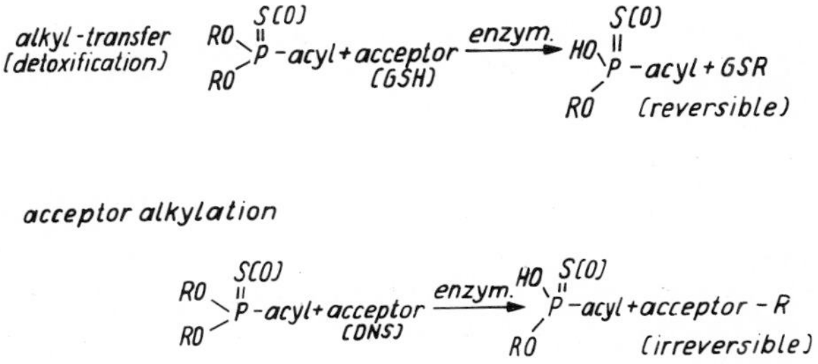

$$\text{alkyl-transfer} \atop \text{(detoxification)} \quad \begin{array}{c} RO \\ RO \end{array} \!\!\!\!>\!\!\!\overset{S(O)}{\underset{}{P}}-acyl+acceptor \atop (GSH) \xrightarrow{enzym.} \begin{array}{c} HO \\ RO \end{array} \!\!\!\!>\!\!\!\overset{S(O)}{\underset{}{P}}-acyl+GSR \atop \text{(reversible)}$$

acceptor alkylation

$$\begin{array}{c} RO \\ RO \end{array} \!\!\!\!>\!\!\!\overset{S(O)}{\underset{}{P}}-acyl+acceptor \atop (DNS) \xrightarrow{enzym.} \begin{array}{c} HO \\ RO \end{array} \!\!\!\!>\!\!\!\overset{S(O)}{\underset{}{P}}-acyl+acceptor-R \atop \text{(irreversible)}$$

Fig. 8. Toxification by biotic alkyl-transfer

Hydrolysis, another basic reaction, usually has a detoxifying effect on phosphoric acid esters, since it causes loss of cholinesterase inhibition to the substances involved. Particular reference should be made, however, to the rather toxic nitrophenol generated temporarily when active principles of the parathion group undergo hydrolysis. However, their further biotic transformation into aminophenol and the possibility of glucuronidation are likely to rule out serious misgivings in toxicology.

The N-nitroso compounds belong to the most objectionable substances consumed by man in the course of food intake. Most of them have proved to be extremely cancerogenic in animal experiments. Therefore, greater attention has been given in recent years to nitrogenous pesticides and their nitrosability. Abiotic nitrosation has been established for a number of active principles, such as dialkyldithiocarbamates, carbamine acid esters (Carbaryl, Propoxur), triazines, and benzthiazurone derivatives. Some of these processes have been confirmed in vivo. The reaction was found to take place with (I) or without (II) degradation of the pesticide molecule (21).

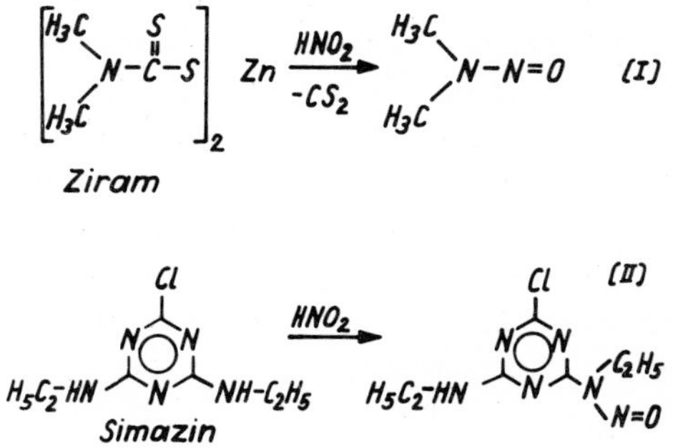

Fig. 9. Nitrosation of pesticides

The reaction calls for favorable conditions (pH) but also for excess nitrite, which is usually lacking. This might be the reason why the potential danger resulting from the use of the pesticide is less dramatic, since under field conditions the nitrosable residues are merely traces, and only small quantities of these will be actually nitrosated. The nitroso pesticide problem, however, can be of considerable dimensions, if the agent has been contaminated in the process of manufacture by higher amounts of reactants or by nitroso compounds which have already been formed in the process of production.

CONCLUSION

To sum up it can be stated that pesticide residues are susceptible to chemical change also in a abiotic environment. Elimination, oxidation, isomerisation, and condensation play an important role and may be induced or advanced by external effects, such as light, oxygen, and temperature. Biotically induced modifications, including those caused by enzyme systems or animal or plant organisms, often cannot be differentiated from the above effects. Both categories of effects may lead to modifications which usually result in detoxification in the course of the physiological process. However, they might cause toxification, as well, under abiotic conditions. Residue assessment should be modified to account for possible residues which might be more suspicious than the active principles are. The assessment of tolerances for terminal pesticide residues, consequently, is influenced with consideration of the relevance of the residues concerned, motivation of pesticide application, and good agricultural practice.

The relevance or severity of terminal residues depends on material factors as the occurrence of recordable quantities, type of residue and its chemo-physical properties, such as persistence and reactivity. Ecological factors which might be of detrimental impact on the ecosystem concerned. Toxicological factors, such as acute and chronic toxicity; suggestions as to potential delayed damage and behaviour in organisms of warm-blooded animals. The motivation of pesticide application as based on the concept of optimum pest control, preservation of crops, and favourable cost-benefit ratio. Good agricultural practice means emphasis of adequate application to achieve optimum results at minimum input and risk.

In conclusion, the question should be raised, if and to what extent the degradation or transformation product of a pesticide has to be taken into consideration as part of residues and to what extent it may be tolerated at all. Any isolated treatment of biotically or abiotically generated substances obviously is impossible.

Any decision, finally, should proceed from the benefit of a given pesticide to human health and the wellbeing of society. Even if the formation of transformation products has to be reckoned with, that potential benefit has to be kept widely in excess of the remaining risk. Hence, residues of degradation and transformation products should be judged not only by toxicity data, although they certainly are of priority relevance. Due consideration rather has to be given to the totality of all criteria mentioned in this paper which are of complex influence on permissible residue tolerances.

REFERENCES

1. H.H. Cramer, Zur wirtschaftlichen Bedeutung des Pflanzenschutzes. In: Chemie der Pflanzenkunde und Schädlingsbekämpfungsmittel, Bd. 1; hrsg. von R. Wegler, Springer-Verlag, Berlin, Heidelberg and New York (1970)
2. Conference on Dibenzodioxins and Dibenzofurans, Nat. Inst. Environm. Health Serv. April 2-3, 1973. Environm. Health Perspectives 5 (1973)
3. W. Sandermann, Naturwissenschaften 61 (1974), 207
4. R. Engst, R.M. Macholz and M. Kujawa, Residue Reviews 68 (1977), 59
5. G. Zweig, Qualitas Plantarum - Plant Foods Human Nutrition 23 (1973), 77
6. L. Fishbein, J. Chromatogr. 98 (1974), 177
7. R. Engst, R. Knoll and B. Nickel, Pharmazie 24 (1969), 673
8. R. Engst and M. Kujawa, Nahrung 11 (1967), 751
9. R. Engst and M. Kujawa, Nahrung 12 (1968), 783
10. IUPAC, Pure and Appl. Chem. 49 (1977), 675
11. R. Engst and W. Schnaak, Residue Reviews 52 (1974), 45
12. J.W. Vonk, Chemical Decomposition of Bisdithiocarbamate Fungicides and their Metabolism by Plants and Microorganisms. Thesis, Univ. Utrecht (1975)
13. G.T. Brooks, Chlorinated Insecticides, Vol. 2, CRC Press, Cleveland (1974)
14. H. Ackermann, Arch. Toxikol. 24 (1969), 325
15. D.G. Crosby, Ber. IUPAC-Commission on Terminal Residues, Madrid (1975)
16. J.R. Grunwell and R.H. Erickson, J. agric. Food Chem. 21 (1973), 929
17. H. Niessen, H. Tietz and H. Frehse, Pflanzenschutz-Nachr. Bayer 15 (1962), 129
18. G. Hilgetag and H. Teichmann, Angew. Chem. 77 (1965), 1001
19. D.F. Health and M. Vandekar, Biochem. J. 67 (1957), 187
20. R.E. Menzer, Residues Reviews 48 (1973), 79
21. IUPAC, Pesticide Nitrosamines, A Special Report. Commission on Terminal Pesticide Residues, Warsaw (1977)

Possibilities of the Amino Acid Analyser in the Fruit Juice Analysis

W. OOGHE

Labo Bromatologie, De Pintelaan 135, B-9000 Gent (Belgium)

SUMMARY

During the last years much scientific work has been done to establish the falsifications of fruit juices.

Any fruit, and so any fruit juice, has a typical amino acid pattern. It depends on the genetic properties, the climate, the place and growth conditions, the technological processes, etc.

A too extreme dilution with water or the addition of other juices can be detected on the basis of discrepancies from those patterns. The addition of protein hydrolysates, cheap amino acids or other products, with the aim of enhancing the formol index, can be established directly from the amino acid pattern as a result of the presence of unusual peaks in abnormal quantities. This survey shows the necessity of the differentiated amino acid analysis.

A description of the instrument used, a Technicon NC-2P, and the experimental conditions is presented.

Experimentally we have determined the free amino acid composition of more than hundred and fifty fruit juices, self pressed as well as commercially available ones. Based on a statistical treatment of the obtained absolute amino acid values, minimum values for 5 different juices are proposed. Further a statistical χ^2-test is elaborated, based on the per cent values of 7 amino acids. This test permits to check the authenticity of an unknown juice by comparison with the proposed standard composition.

Finally we have tried to assay the quality of a fruit juice as a function of the total amino acid quantity of each juice.

1. INTRODUCTION

During the last years much scientific work has been done to establish the falsifications of fruit juices.

Any fruit, and so any fruit juice, has a typical amino acid pattern. It depends on the genetic properties of the material, the climate, the place and growth conditions, the degree of ripeness, the technological processes, etc (1,2,3).

A too extreme dilution with water, the addition of masking agents or other juices can be detected on the basis of discrepancies from those patterns.

This study is limited to the most common juices of the belgian market, namely lemon juice, grape-fruit juice, orange juice, apple juice, white and red grape juice.

The main type of falsification has always been and still is a too profuse addition of water. This can be ascertained on the basis of certain chemical parameters such as ash, potassium, phosphate, betaine, sugar free extract content and the formol index, minimal values of which have been fixed. In the past the formol index has been considered as a very important parameter, as it was a measure of the total amino acid content. Because a too low formol index is to be considered as an indication of an excessive dilution, it is artificially manipulated by some unfair producers. In this context we mention:

- the addition of protein hydrolysates: such an addition however can be demonstrated by the presence of uncommon amino acids like isoleucine, leucine, tyrosine and phenylalanine, which usually occur only in trace amounts in fruit juices (3,4,5);

- the addition of cheap amino acids like glycine, aspartic and glutamic acid or the addition of compounds which have a positive influence on the formol index value, as ethanolamine and ammonium (6,7);

- the dilution with other cheap fruit juices or juices of inferior quality such as pulp wash extracts: in most instances this kind of falsification is difficult to find out (8).

This survey shows that a normal formol index is not a guarantee at all, but indicates the necessity of the differentiated amino acid analysis.

Not only minimal amino acid values, as required in some countries for orange juice, but equally maximal values can be fixed. Other possibilities are offered by the amino acid composition, expressed in terms of relative abundances or by the ratios of certain amino acids.

2. THE AMINO ACID ANALYSER

For applications in the field of food analysis it mostly suffices to dispose of an amino acid analyser that can separate the so-called hydrolysate amino acids or the amino acids obtained after protein hydrolysis.

For the determination of the free amino acids a compromise should be found between analysis time and separation efficiency which has made that the serine and asparagine peaks are not resolved in our system.

The sample preparation for juices, without added sugar, can usually be limited to the dilution in a buffer solution at pH 2.0 and to an ultracentrifugation before injecting 200 µl of sample on the column.

3. EXPERIMENTAL PART

Experimentally we have determined the free amino acid composition of more than hundred and fifty fruit juices, home-made as well as commercially available ones.

The purpose of our investigations was double: first we wanted to develop a method which would unambiguously detect falsifications and secondly we have tried to establish standards for the main fruit juices available on the belgian market.

By means of freshly prepared juices from fresh fruit we have investigated which free amino acids are characteristic for the different types of fruit. It seems that seven amino acids are always present, but in varying quantities. These are aspartic acid, Σ(serine + asparagine), glutamic acid, proline, alanine, γ-amino butyric acid and arginine.

Let us return now to our investigation that schematically can be presented as follows:

A. Addition of hydrolysates, cheap amino acids and ninhydrin positive compounds.

In most cases these falsifications are easy to distinguish because they are generally accompanied by too low concentrations of the characteristic free amino acids.

Experimentally neither the addition of hydrolysates, neither the presence of an excess of cheap amino acids has been established. Otherwise abnormal high quantities of ethanolamine or ammonium salts have been showed in six samples. Because for these samples much too low values for the different amino acids are established, they can be considered as falsified.

B. Addition of cheap juices of another species or products of inferior quality.

The authenticity of a commerical fruit juice can be controlled based on the relative abundances of the seven per cent values:

- either by comparing to the per cent values of home-made juices,

- either by comparing by means of a statistical χ^2 test to a standard containing the commerical juices of the same species.

Because the comparison with home-made juices has not given a decisive answer for each type of juice, the seven mean per cent amino acid values and standard deviations have been calculated for all analysed commercial juices. Only the seven higher mentioned amino acids were considered by setting their sum equal to hundred. Their mean per cent values and their standard deviations are summarized in Table I for each kind of juice considered.

Further for each fruit juice the χ^2 value has been calculated by means of the equation:

Table I : Relative abundances of the seven amino acids considered (sum = 100).

Sample Amino acid	lemon	grape-fruit	orange	apple	grape (white)	grape (red)
ASP acid	21.69 ± 3.78	22.77 ± 3.36	9.94 ± 1.04	16.06 ± 3.42	3.59 ± 1.05	2.82 ± 0.61
Σ (SER + ASP)	31.66 ± 5.54	24.72 ± 5.66	17.97 ± 2.28	73.41 ± 3.73	5.09 ± 1.65	5.05 ± 1.21
GLU acid	8.22 ± 0.66	4.99 ± 0.98	3.29 ± 0.34	5.46 ± 1.17	5.27 ± 1.24	4.97 ± 1.18
PRO	18.66 ± 4.27	17.20 ± 3.71	32.19 ± 2.90	—	27.29 ± 8.37	33.62 ± 5.58
ALA	9.23 ± 1.96	6.97 ± 2.49	4.07 ± 0.52	3.56 ± 1.10	10.73 ± 1.52	12.15 ± 2.17
γ-AMB	7.48 ± 2.91	12.52 ± 2.77	12.01 ± 1.11	1.86 ± 0.77	10.76 ± 1.41	10.24 ± 1.89
ARG	3.06 ± 2.71	10.84 ± 1.52	20.54 ± 1.29	—	37.28 ± 4.97	31.16 ± 2.83

$$\chi^2 = \sum_{i=1}^{7} (\frac{x_i - \bar{x}_i}{\sigma_i})^2$$

where: x_i : the per cent value of an amino acid;

$\bar{x}_i$: the mean per cent value of the same amino acid in the considered standard;

σ_i : the standard deviation on the mean value.

There is a relation between the so obtained χ^2 value and the per cent probability with which a fruit juice belongs to a particular group. The greater the χ^2 value the smaller becomes the probability; so, for 6 degrees of freedom, the probability becomes 0,05 % for a χ^2 value of 18.9 and less than 0.01 % for a χ^2 value greater than 28.0. However we state that last value as an absolute maximum, a commercial juice already seems suspect for a value greater than 18.9.

The maximum value of 28.0 is only exceeded for 2 orange juices. All other juices with a certain reservation for one grape-fruit concentrate, one apple juice and one grape juice, could be considered as authentic.

C. Addition of water or too foregoing dilution of a concentrate.

This type of falsification can only be evidenced by absolute minimum values, which are to be established for the amino acids. By means of a statistical approach we have tried to derive standards from the obtained absolute amino acid values of the commercial juices analysed.

Based on Student's t distribution, for all juices the mean value $\bar{x}$ and the dispersion s of each amino acid is calculated by a computer.

Our procedure is the following:

after elimination of the suspect values we have calculated standards for each amino acid based on the equation:

$$\text{minimal value} = \bar{x} - t_c \cdot s \cdot \sqrt{\frac{N+1}{N-1}}$$

where: $\bar{x}$: the corrected mean value;

t_c : the confidence coefficient used, depending on the confidence level and the sample size;

s : the dispersion;

N : the number of the sample.

Because N was to small for lemon juice, no standards have been derived for that juice.

First to obtain minimal amino acid values, we have eliminated all values lower than the limit obtained introducing a t_c value corresponding to the 68.27 % confidence levels. After calculation of the corrected mean value $\bar{x}_2$ and dispersion s_2 we have established the minimal amino acid values introducing a t_c value corresponding with the 80 % confidence intervals.

Secondly quality norms have been established too, by introducing, as well as for the elimination as for the calculation, t_c values corresponding to the 60 % confidence levels.

Both values, as well the minimal value x_M as the quality standards x_O, are presented in Table II for each amino acid. Because glycine is a very cheap amino acid with a great impact on the formol index, we have established maximal values instead of a minimum for that amino acid.

However, in order not to be too absolute, we have built in some security procedure and so we have defined that a juice cannot be accepted as a fruit juice when:

- either one of the amino acids differs more than 20 % of the standard;

- either two standards are not met, of which one at least differs more than 10 %;

- either three standards are not met, of which one at least differs more than 5 %;

- either four or more standards are not met.

Experimentally this approach has seemed to be realistic for "honest" juices.

Based on the minimum norms a relatively great number of the analysed juices is to be considered as falsified by water addition. So one orange juice, four grape-fruit juices, three apple juices, four red and one white grape juice have been rejected.

4. FINAL CONSIDERATIONS AND CONCLUSIONS

A comparable investigation can be extended to fruit lemonades and fruit nectars. After a clean-up and concentration step on a Dowex 50X8 cation exchange column, it seemed to be possible to investigate whether fruit lemonades contain 10 % of the corresponding fruit juice at minimum, as legally required in Belgium.

It was even possible to control the data mentioned on fruit nectars.

When the quality of a juice can be defined a as function of the natural fruit content, it can be expressed in relation to the total amino acid content. As already proposed for orange juice by some authors, we can state that the higher the amino content, calculated as the sum of the amino acids present in an acceptable aminogram, the better the quality of the fruit juice.

Based on the observed dispersion of the total amino acid values, we can state that the quality of the analysed fruit juices, and mainly of grape juice, was very different.

With this lecture I hope to have demonstrated that the amino acid analyser opens many perspectives for the detection of falsifications in the fruit juice industry.

Table II : Proposed amino acid standards (mmol/l).

Juice / Amino acid	grape-fruit (N=15) min. value x_M	grape-fruit (N=15) qual. value x_Q	orange (N=18) min. value x_M	orange (N=18) qual. value x_Q	apple (N=28) min. value x_M	apple (N=28) qual. value x_Q	grape(white) (N=11) min. value x_M	grape(white) (N=11) qual. value x_Q	grape (red) (N=16) min. value x_M	grape (red) (N=16) qual. value x_Q
ASP acid	3.1	3.7	1.95	2.0	0.6	0.65	0.1	0.15	0.15	0.2
THR	—	—	—	—	—	—	tr	0.25	0.15	0.25
Σ(SER+ASP)	3.65	4.15	3.25	3.6	2.45	3.1	0.15	0.25	0.3	0.4
GLU acid	0.7	0.8	0.6	0.65	0.15	0.2	0.15	0.25	0.3	0.4
PRO	2.1	2.65	5.4	6.5	—	—	1.20	1.45	1.4	2.25
GLY	<0.3	<0.3	<0.3	<0.3	<0.1	<0.1	<0.15	<0.15	<0.20	<0.20
ALA	0.75	0.9	0.65	0.75	0.1	0.15	0.55	0.65	0.7	0.95
VAL	tr	0.1	tr	tr	—	—	tr	0.1	0.15	0.2
MET	—	—	—	—	—	—	—	—	tr	tr
ILE	—	—	—	—	—	—	tr	tr	tr	0.1
LEU	—	—	—	—	—	—	tr	0.1	tr	0.1
γ-AMB	1.9	2.1	2.0	2.3	tr	tr	0.55	0.6	0.7	0.85
HIS	—	—	—	—	—	—	0.1	0.15	0.1	0.15
LYS	tr	0.1	tr	tr	—	—	tr	tr	tr	tr
ARG	1.6	1.85	3.65	4.05	—	—	2.0	2.25	1.9	2.5
tot. AA cont.	15.2	17.7	18.0	19.8	3.4	4.3	6.15	6.8	6.4	8.7

REFERENCES

1 H.J. Bielig, A. Askar, Chem. Mikrobiol. Technol. Lebensm., $\underline{1}$ (1972),183

2 E. Fernandez-Flores et al., J.A.O.A.C., $\underline{53}$, (1970), 1203

3 P.D. Niedmann, Deutsche Lebensmittel-Rundschau, $\underline{72}$ (1976), 119

4 H. Nootenboom, Voedingsmiddelentechnol., $\underline{10}$ (1977), 14

5 C.E. Vandercook, R.L. Price, J. Food Sci., $\underline{37}$ (1972), 384

6 J. Weits et al., Z. Lebensm. Untersuch. u. -Forschung, $\underline{145}$ (1971), 335

7 S. Wallrauch, Flüss. Obst, $\underline{44}$ (1978), 386

8 S. Wallrauch, Flüss. Obst, $\underline{45}$ (1979), 39

Determination of Rare Sugars in Fruits After Separation by Gel Chromatography

H. Scherz

Deutsche Forschungsanstalt für Lebensmittelchemie, Lichtenbergstr. 4
D-8046 Garching

SUMMARY

A gel chromatographic procedure using Biogel P-2 had been developed for the enrichment of oligosaccharides, which occur as small amounts in fruits beside the main sugars (glucose, fructose, sucrose). Such compounds, obtained from the fruits, which are important in the field of food technology as well as those ones, obtained from several fruit processing products were characterized hereafter by thin layer chromatography.

INTRODUCTION

The presence of small amounts of oligosaccharides beside the main sugars glucose, fructose and sucrose in a few fruits had been described in literature (1,2,3,4). For getting a general survey about the occurence of such compounds in fruits, which are used as raw materials in food technology, a simple gel chromatographic procedure was developed, based on studies of Dellweg (5) and Kainuma (6). This procedure allows an effective enrichment of these compounds from the dominant fruit sugars.

EXPERIMENTAL

Homogenisation of the fresh fruit materials with methanol (weight-ratio 1:5 to 1:7); centrifugation; addition of 10 % aqueous solution of Pb-acetate to the clean supernatant for precipitating the fruit acids; centrifugation; removing the excess of Pb-ions in the clear solution by H_2S; filtration; neutralisation of the clear solution with an anion exchanger (Dowex 1x8; CO_3-form).
Two fold fractionation of the concentrated solution on Biogel P-2 columns (eluent: distilled water; temperature: 65°C; L: 1000 mm; Ø: 50 mm and 20 mm); collection and concentration of the fractions, containing the oligosaccharides; finally separation and characterisation by thin layer chromatography on precoated silicagel plates (solvent: ethylacetate - methanol - water 68:23:9 v/v; continuous flow technique (7); visualisation: aniline-diphenylamin-phosphoric acid (8).

RESULTS

It was found, that according to the composition of oligosaccharides, two groups of fruits are existing.
The fruits of the first one contain oligosaccharides, consisting mainly of glucose and fructose. The thin layer chromatographic separation patterns were in general similar to that one, obtained from the sugars of the reaction between sucrose and invertase. To this group belong apricots, plums, pears, oranges, peaches, bananas, apples and strawberries.
The fruits of the second group (cherries, grapes and tomatoes) contain oligosaccharides consisting mainly of aldoses.
Additionally some experiments had been carried out with fruit processing products such as marmelades and fruits juices. In the most cases, a lot of different oligosaccharides had been found. Although those of the raw fruits were detectable, the most of them derived from added sugar materials during processing.

REFERENCES

1   A.S.F. Ash and T.M. Reynolds, Nature 174 (1954), 602
2   A.S.F. Ash and T.M. Reynolds, Austr.J.Chem. 8 (1955) 276, 445
3   R.W.Henderson, R.K. Morton and R. Rawlinson, Biochem.J. 72 (1959), 340

4 G. Häseler and K. Misselhorn, Z.Lebensm.Unters.Forsch. 129 (1966), 222
5 M.W. Dellweg, M. John and G. Trenel, J.Chromatogr. (Amsterdam) 57 (1971), 89
6 K. Kainuma, A. Nogami and G. Mercier, J.Chromatogr. (Amsterdam) 121 (1976), 361
7 E. Bancher, H. Scherz and K. Kaindl, Mikrochim.Acta 1964 652
8 J.C. Buchan and P. Savage, Analyst 77 (1952), 401

Correlation of Gel Chromatographic and Sensory Profiles used for the Evaluation of Food Products

J. POKORNÝ, N. N. LÝ, J. KARNET, J. PAVLIŠ, A. MARCÍN AND J. DAVÍDEK

Department of Food Chemistry, Prague Institute of Chemical Technology, CS - 166 28 Prague, Czechoslovakia

SUMMARY

The positions and areas of various gel chromatographic peaks of extracts from food materials are significantly correlated with their sensory values and with the intensities of various odour and flavour notes in sensory profiles of some food products. The three following applications were studied and found useful:
1° The gel chromatography of orange oil or extracts from orange-flavoured beverage on cross-linked polystyrene gels using tetrahydrofuran as a solvent. The areas of GPC peaks corresponding to oxygen-containing mono- and sesquiterpenoid compounds correlated with the sensory value.
2° The gel chromatography of appetizers on Sephadex LH-20^R using 50 % ethanol as a solvent. The multiple regression analysis of all peaks enables the prediction of sensory quality, and gives information on the formula used for the preparation of appetizer.
3° The gel chromatography of aqueous extracts from defatted dry soups on Sephadex G-15 using a buffered sodium chloride solution as a solvent. The peaks correspond either to flavour precursors or to off-flavour products formed during the storage, in some cases, they are due to flavour-neutral products, such as polymers, which are correlated with the quality of aroma.

INTRODUCTION

Organoleptic properties are correlated either with the amount of various flavour substances present in fresh food or with the amount of off-flavour compounds formed during the storage of food, and in some cases, with the amount of some flavour-neutral substances if they are correlated in their turn with some flavour-active compounds. The correlation is only indirect in the last case but still may be highly significant in spite of the lack of sensory activity of the compounds involved.

The flavour-active fraction is often characterized by its spectral or chromatographic profile which is related to the sensory quality even without the necessity to identify the chromatographic peaks. Gas chromatographic profiles are widely used for the prediction of sensory quality and flavour profiles. Infrared, ultraviolet and NMR or mass spectra profiles are often used for the evaluation of essential oils for perfumers. In the last few years liquid partition chromatography profiles have been used as a measure of sensory quality with success.

In the present paper, we try to show the applicability of GPC /gel chromatographic/ profiles for this purpose. We shall show three cases of applications of the method as an example.

EXPERIMENTAL

In case of non-alcoholic beverages flavoured by orange oil, the beverage was extracted with diethyl ether or dichloromethane, the extract was dried and concentrated. In case of appetizers, the sample was directly injected without any preparation. In case of dry soups, the sample was extracted with chloroform, and the defatted sample was extracted with a phosphate buffer, pH = 7.9 /0.238 g dipotassium hydrogen phosphate, 5.314 g of potassium dihydrogen phosphate, and 11.691 g of sodium chloride in 1 litre/ /1/.

The GPC profiles of orange oil were determined using five 1.2 m x 8 mm columns in series, packed with cross-linked polystyrene gel

/S-232-Gel/, and tetrahydrofuran as a solvent /2/. For the preparation of GPC profiles of an appetizer, two 600 mm x 50 mm columns in series packed with Sephadex LH-20^R /Pharmacia, Uppsala, Sweden/ were used, and 50 % aqueous ethanol as a solvent. In case of dry soup extracts, columns K 26/4 /Pharmacia, Uppsala/ packed with Sephadex G-15 /Pharmacia, Uppsala/ were used. The above phosphate buffer was used as a solvent.

RESULTS AND DISCUSSION

A. Evaluation of Orange oil in Non-alcoholic Beverages

The gel chromatogramme of a fresh natural orange oil contains a large peak of D-limonene /about 90 % of the total extract/ followed by few minor peaks of oxygenated monoterpenoic and of sesquiterpenoic components. In the gel chromatogramme of a sample of rancid orange oil of poor sensory quality, containing higher amount of hydroperoxides and ketonic oxidation products, the peak following the peak of D-limonene was much larger than in a fresh oil /the relative retention time = 1.45, the retention time of D-limonene = 1.00/ because limonene hydroperoxides and other oxidation products are eluted in this peak /peak B/.

The correlation between areas of peak B and the sensory quality /as determined by hedonic scoring with use of a 5-point scale/ is statistically significant /the correlation coefficient k = 86.5 %, N = 20, P = 99.9, the tabellar value = 67.9/. Higher point scoring corresponds here to better sensory quality of a beverage prepared with addition of the respective orange oil. Therefore, the higher is the area of peak B, the lower is the sensory quality of the beverage.

As the area of peak B in fresh deterpenized orange oil is higher, a separate correlation curve is necessary to establish in case of the application of deterpenized oils.

When diethyl dicarbonate is used as a preserving agent, various products are formed which interfere with the analysis.

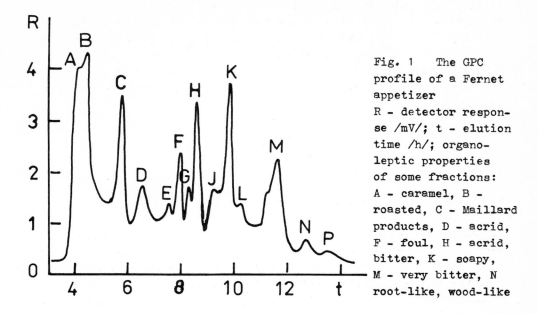

Fig. 1 The GPC profile of a Fernet appetizer
R - detector response /mV/; t - elution time /h/; organoleptic properties of some fractions: A - caramel, B - roasted, C - Maillard products, D - acrid, F - foul, H - acrid, bitter, K - soapy, M - very bitter, N root-like, wood-like

B. Evaluation of Appetizers

In case of appetizers the organoleptic properties of fractions were determined by separating a 400 ml sample and by determining the sensory quality of the fractions; a profile consisting of a set of 12 descriptors /3/ and the hedonic scoring were used; 30 experienced judges, males and females, aged 23 - 30 years were chosen for the evaluations.

An example of a Fernet appetizer separated by GPC is shown in Fig. 1. The intensities of some partial flavours determined by the profile method were significantly correlated with some GPC peaks. In 89 % cases the sensory quality corresponded to the expected value based on the correlation with the GPC profile.

Both the GPC profiles and the flavour profiles of extracts of various drugs used for preparation of appetizers were determined. Significant linear correlations were found suitable both for the prediction of flavour of the mixture of extracts and for the estimation of formula of appetizers which have been chromatographed.

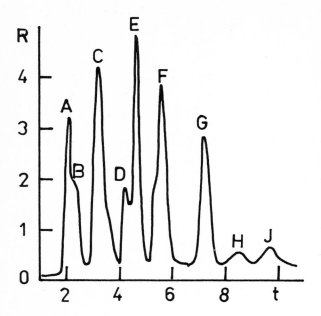

Fig. 2 The GPC profile of extract from beef broth
R - detector response /mV/; t - elution time /h/; A to J - main peaks

C. Evaluation of Dry Soups

Two types of reactions proceed in dry soups on storage, namely non-enzymic browning reactions of reducing sugars and various aroma constituents with free amino acids, and autoxidation reactions of lipids followed by reactions with amino acids or proteins. Flavour compounds are produced, modified, or bound by these reactions.

An example of a GPC profile of beef broth extract is shown in Fig. 2. The high-molecular mass fraction near the front consisted mainly of proteins. Their molecular mass increased by various free-radical reactions /4/, mainly by irradiation /5/ or by interactions with oxidized lipids /6/ or by Maillard reactions during storage. They do not directly affect organoleptic properties but their changes are correlated with the degree of deterioration processes. The increase of their concentration was correlated with the increasing intensity of the off-flavour. Correlation coefficients between the peak area and the sensory quality /hedonic scoring, 7-point scale, higher ratings corresponding to better quality/ were -96.9 % and -90.4 % for the peaks

B and C, respectively /N = 9; P = 99.0 %; the tabellar value = 87.5 % /, therefore, the increase of peak area corresponded to the decrease of sensory quality.

The low-molecular mass peaks eluted in the latter stage of chromatography, were identified as lower peptides, amino acids or sugars. They disappeared during the storage as a result of Maillard reactions or by interaction with oxidized lipids /7/. Their disappearance was correlated with the appearance of off-flavours. Correlation coefficients between the peak areas and the sensory quality were 97.2 %, 95.7 %, and 87.8 % for the peaks D, F, and G, respectively.

There was also a significant correlation found between the areas of some GPC peaks and the fullness of odour and flavour of the broth /scoring with use of a 5-point scale, lower scores corresponding to fuller and richer flavour/, e. g. 81.8 % and 87.4 % in case of peaks B and H, respectively /N = 9; P = 95.0 %; the tabellar value = 75.4 %/. Similar relations were obtained for other beef and chicken soups.

REFERENCES

1 K. D. Schwenke, B. Ender and B. Raab, Nahrung 17 /1973/, 579
2 S. Pokorný, J. Čoupek, N. T. Luan and J. Pokorný, J. Chromatog. /Amsterdam/ 84 /1973/, 319
3 J. Pokorný, J. Karnet and J. Davídek, Nahrung, in press
4 V. G. Malshet and A. L. Tappel, Lipids 8 /1973/, 194
5 W. T. Roubal and A. L. Tappel, Arch. Biochem. Biophys. 113 /1966/, 150
6 B. A. El-Zeany, J. Pokorný, E. Šmidrkalová and J. Davídek, Nahrung 19 /1975/, 327
7 J. Pokorný, Riv. Ital. Sostanze Grasse 54,/1977/, 389

Separation of Histamine Reverse Phase by HPTLC and Soap-TLC and its Determination in Tuna Fish

V. COAS

Laboratory of Commodity Science, University of Florence, Florence, Italy.

L. LEPRI

Institute of Analytical Chemistry, University of Florence, Florence, Italy.

SUMMARY

The use of RP-2 and RP-18 plates impregnated with dodecylbenzene-sulphonic acid (HDBS) and of ammonium tungstophosphate layers in the separation of 21 nitrogenous compounds was investigated. The best elution conditions were: a) 1 M acetic acid + 0.25 M hydrochloric acid in 50% methanol on RP-2 impregnated with 3% HDBS; b) 0.5 M acetic acid + 0.5 M sodium acetate in 50% methanol and 0.5 M ammonia + 0.5 M ammonium chloride in 60% methanol on RP-18 impregnated with 4% HDBS; c) 2 M ammonium nitrate on ammonium tungstophosphate layers.
As a result, we pointed out the presence of hystidine, glycine and lysine in the tuna fish extracts and the absence of cadaverine and putrescine which account for a food-borne hazard and whose determination is of peculiar importance. Histamine can be univocally determined at levels far below that which would cause symptoms of toxicity.

INTRODUCTION

Chromatographic screening methods for histamine in tuna fishes were proposed from several Authors[1-4]. Characteristics common to all the papers are the relationship among detection limit of histamine on the chromatogram, preservation state of food and, therefore, food

poisoning. The researchs are based on a non-selective extraction of histamine from fishes; in fact, in the methanolic or trichloroacetic acid extract, a serie of aminoacids, peptides and biogenic amines may be present together with histamine.
Since the right statement of the problem lies in the previous separation of all the possible nitrogenous compounds in the extracts, we deemed it useful to study the problem from this point of view. Furthermore, the simultaneous determination of compounds such as cadaverine, putrescine and so on, holds a peculiar importance since such products can give more exact informations on the product quality.
To our knowledge, only Lieber and Taylor[3] examined the problem on this basis.
We used reversed-phase high performance thin-layer chromatography (HPTLC) on plates of RP-2 and RP-18 alone or impregnated with an anionic surfactant, such as dodecylbenzene-sulphonic acid (HDBS), and chromatography on layers of ammonium tungstophosphate (AWP).

EXPERIMENTAL

Standard solutions of amino acids, dipeptides and biogenic amines (SERVA, Heidelberg, G.F.R. and SIGMA, St. Louis, U.S.A.) were prepared by dissolving the compounds in a 3:1 (v/v) mixture of methanol and 0.2 M trichloroacetic acid. 0.2 - 0.5 μl of the standard solutions (concentration 0.5 - 1 μg/ml) were deposited on the layers.
The compounds were detected by spraying the plates with a solution of 1% ninhydrin in a 5:1 (v/v) mixture of pyridine and glacial acetic acid and then heating the layers at 100°C for 5 min.
Preparation of fish samples was performed according to the method described by Foo[4]: ten grams of compound were extracted thrice with 10 ml of 5% trichloroacetic acid solution; the combined extracts were concentrated under reduced pressure and dissolved in 5 ml of methanol-water (4:1). Aliquots of extracts (0.5 - 2 μl) were deposited on the layers.
The layers of AWP were prepared according to a previous work[5] while RP-2 and RP-18 plates (Merck, Darmstadt, G.F.R.) were impregnated by dipping them for 5 min. in a 96% ethanol solution containing HDBS in the required percentage and, successively, air-dried.
Unless otherwise stated, the migration distance was 6 cm on impregnated plates and **12** cm on AWP layers. All the measurements were carried out at 25°C with the Desaga thermostatic chamber. The migration time is about 75 min. on RP-2 plates impregnated with 3% HDBS and decreases surprisingly to 45 min. on RP-18 impregnated with 4% HDBS. The elution time on AWP layers is about 60 minutes.

RESULTS AND DISCUSSION

In Table 1 the first column lists the R_f values of the compounds on RP-2 plates. Under these elution conditions these plates are not useful, since the compounds are very little retained. Better results can be achieved by impregnating the plates with a solution of HDBS, as the data of column 2 show, since the substances are stronger retained under the conditions used. The successive column indicates that the retention of the different compounds decreases as the eluent acidity increases, pointing out in this way that an ion-exchange process is operating. Among the separations which can be foreseen on the basis of the R_f values, we effected that among Hm, His, Opm, Gly, Tyr, Lys, Cad,

TABLE 1 - R_f values of standard compounds on plates of RP-2 and RP-18 alone or impregnated with an ethanolic solution of HDBS. Eluents: (1) 1M CH_3COOH in 50% CH_3OH; (2) 1M $CH_3COOH+0.5M$ HCl in 50% CH_3OH; (3) 0.5M $CH_3COONa+0.5M$ CH_3COOH in 50% CH_3OH; (4) 0.5M $NH_4Cl+0.5M$ NH_4OH in 60% CH_3OH.

Substance		RP-2 (1)	RP-2+3%HDBS (1)	RP-2+3%HDBS (2)	RP-18+4%HDBS (2)	RP-18+4%HDBS (3)^	RP-18+4%HDBS (4)	Amount (μg)
Histamine	(Hm)	0.80	0.05	0.46	0.35	0.26	0.59	0.2
Agmatine	(Agm)	0.80	0.05	0.43	0.31	0.30	0.26	0.2
Cadaverine	(Cad)	0.80	0.06	0.50	0.36	0.33	0.23	0.2
Ala-His		0.80	0.09	0.56	0.43	0.37	0.72	0.2
Glycine	(Gly)	0.94	0.37	0.81	0.72	0.63	0.79	0.1
Histidine	(His)	0.86	0.09	0.54	0.42	0.44	0.85	0.2
Lysine	(Lys)	0.88	0.09	0.59	0.46	0.55	0.57	0.2
Octopamine	(Opm)	0.90	0.23	0.78	0.71	0.63	0.75	0.2
β-Phenylethylamine	(β-Ph)	0.87	0.14	0.44	0.32	0.31¨	0.20	0.2
Putrescine	(Put)	0.80	0.08	0.50	0.38	0.30¨	0.22	0.1
Serotonine	(Srt)	0.88	0.28	0.70	0.66	0.63	0.65	0.1
Spermidine	(Sd)	0.63¨	0.02	0.31	0.22	0.05	0.06	0.1
Spermine	(Sm)	0.40¨	0.02	0.25	0.16	0.01	0.02	0.1
Tryptamine	(Tryp)	0.85	0.16	0.46	0.33	0.30	0.23	0.1
Tryptophan	(Trp)	0.90	0.19	0.55	0.46	0.62	0.65	0.1
Tyramine	(Tyr)	0.90	0.27	0.70	0.62	0.63	0.59	0.1
Thiamine	(Thm)	0.88	0.25	0.53	0.45	0.41	0.72	0.5
His-Leu		0.80	0.03	0.30	0.22	0.41	0.84	0.2
His-Ser		0.81	0.10	0.55	0.47	0.46	0.94	0.2
Tyrosine	(Tys)	0.78	0.28	0.75	0.71	0.63	0.86	0.4
Glutamic acid	(Glu)	0.96	0.41	0.84	0.75	0.63	0.87	0.3

^ = R_f value of the first solvent front 0.63; ¨ = Tailing

Sd and Sm on plates of RP-2+3% HDBS eluting with 1 M CH_3COOH+0.5M HCl in 50% CH_3OH.

The affinity of the stationary phase towards the different compounds remarkably increases changing from RP-2 to RP-18 plates. On these last layers the spots are much narrower and therefore more suitable for analytical applications.

TABLE 2 - R_f values of standard compounds on layers of AWP+$CaSO_4 \cdot 1/2H_2O$ (4:2 ratio). Eluents: (1)1M HNO_3; (2)1M NH_4NO_3; (3)2M NH_4NO_3.

Substance	Eluent (1)	(2)	(3)	Amount (μg)
Hm	0.02	0.18	0.30	0.5
Agm	0.02	0.16	0.22	0.5
Cad	0.02	0.19	0.32	0.6
Car	0.03	0.09	0.15	0.4
Gly	0.58	0.76	0.79	0.1
His	0.05	0.32	0.40	0.5
Lys	0.02	0.32	0.44	0.5
Opm	0.24	0.52	0.62	0.5
β-Ph	0.04	0.33	0.43	0.2
Put	0.00	0.28	0.46	0.5
Srt	0.03	0.18	0.27	0.2
Sd	0.00	0.03	0.12	0.2
Sm	0.00	0.02	0.03	0.5
Tryp	0.07	0.03	0.15	0.1
Trm	0.07	0.34	0.36	0.3
Tyr	0.08	0.40	0.47	1.0
Thm	0.02	0.41	0.47	0.5
His-Leu	0.02	0.16	0.19	0.5
His-Ser	0.00	0.17	0.24	0.5
Tys	0.44	0.63	0.66	0.5
Glu	0.51	0.73	0.76	0.5

Along with the increase of the pH eluent (see columns 4,5 and 6 which refer to RP-18+4%HDBS plates),the affinity sequence of the different compounds remarkably changes and the difference between the R_f values of histidine and histamine sharply increases. Such improvements are partly canceled eluting with acetate buffer(col.5)since some amines give rise to elongated spots and a double front,the first of which has not a high R_f value(see Tab.1),is found.

Notwithstanding the good separations which may be achieved on RP-18+4% HDBS plates,histamine can not be univocally determined under the conditions used.

In order to solve this problem, we used layers of AWP which were recently set up by our research group(5).

As the results of Tab.2 show,on these layers the retention order of the compounds is different from that observed on impregnated RP-2 and RP-18 plates.

The eluting power of an ammonium nitrate solution seems more marked than that of nitric acid solution at the same concentration (see col.s 1 and 2). This occurrence can be ascribed to the stronger adsorption of ammonium nitrate than nitric acid by the layer and was already pointed out in the case of aromatic amines (5).

The best conditions for the saparation of the test compounds are obtained eluting with 2M ammonium nitrate (Col.3 of Tab.2).

The use of aqueous-organic solvents does not involve any improvement from an analytical standpoint.

The presence of histamine can be univocally stated on the basis of its chromatographic characteristics on impregnated plates and on AWP layers.

In fact, serotonine, which is the only compound with an R_f value similar to that of histamine on AWP plates, is sharply separated from histamine on the layers impregnated with HDBS.

We studied twentytwo samples of canned tuna fishes and mackerels, taking ten grams of each for the extraction with trichloroacetic acid (see experimental). Less than 5 mg of histamine per 100 g of fish can be detected on RP-2 and RP-18 layer impregnated with HDBS and less than 12 mg of histamine per 100 g of fish on AWP layers. These limits are well suited to the study of the quality of fishes, since in fresh fishes media levels of 5 mg of histamine per 100 g are found and concentrations of 100 mg per 100 g are considered to be the critical level for histamine poisoning(3).

After the chromatography of the 22 samples, it has been found that only two tuna fishes had a detectable histamine content. The chromatograms which refer to two mackerels (samples C_1 and C_2) and to the tuna fishes (samples C_3 and C_4) and to standard mixtures of glycine, histamine, histidine, lysine, spermine, spermidine and glutamic acid are reported in Fig.1. It should be noted that not recently prepared histidine solutions give rise to two spots on this layer; in the case of the samples two spots are always obtained.

From this figure it comes out that the four extracts contain many nitrogenous compounds which give rise to a positive reaction with ninhydrin and therefore the univocal separation of histamine from all the other compounds was necessary also on the basis of the high percentages of the different compounds. From the reported chromatogram it comes out that glycine, spermine and spermidine are present in the two tuna fishes, that histidine is one of the major and constant components of this kind of fishes together with lysine and that histamine is present in a much lower ratio with respect to histidine and glycine or glutammic acid. The areas relative to histamine in the samples C_3 and C_4 (1 μl) are of the same order than that relative to the histamine standard. Since 1 μg of histamine was deposited in the standard mixture, the concentration of histamine in the two samples can be estimated to be about 5 mg in 5 ml of extract. Such value involves an histamine content in the fishes of 50 mg in 100 g.

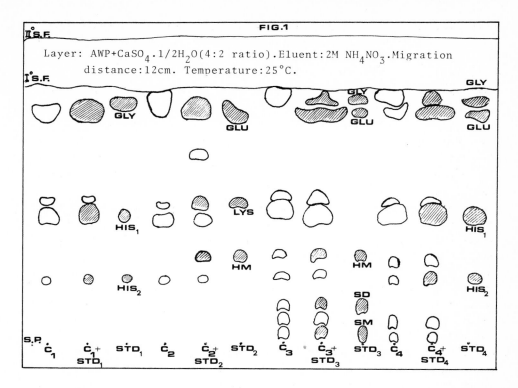

Fig. 1

The presence of histamine, lysine, spermine, spermidine and glycine in the samples, and the absence of toxic compounds, such as cadaverine and putrescine, is confirmed by the chromatogram on RP-18+4% HDBS plate on eluting with 0.5 M NH_4Cl+0.5 M NH_4OH in 60% methanol.

REFERENCES

1) D.E.SCHUTZ,G.W.CHANG,L.F.BYELDANES - J.A.O.A.C. 59 (1976) 1224;
2) J.S.LIN,J.D.BARANOWSKI,H.S.OLCOTT - J.Chromatogr. 130 (1977) 426;
3) E.R.LIEBER,S.I.TAYLOR - J.Chromatogr. 153 (1978) 143;
4) L.Y.FOO - J.A.O.A.C. 60 (1977) 183;
5) L.LEPRI,P.G.DESIDERI,D.HEIMLER - J.Chromatogr. 207 (1981) 29

Gas Chromatography in Food Analysis

L. BONIFORTI 1); S. LORUSSO 2)

1) Istituto Superiore di Sanità, Rome, Italy
2) Istituto di Merceologia, Università, Rome, Italy

SUMMARY

This review brings together recommended GC methods for the analysis of foods in quality assessment and for the determination of food additives. GC is, therefore, a qualitative and quantitative analytical technique, capable of revealing whether foodstuffs have been properly grown and processed, and of detecting what current legislation defines as "fraudulent products".

INTRODUCTION

In the food field, GC is primarily used in the analytical checking of final food products. Surveillance is, in fact, aimed at ascertaining the presence and exact quantities of components thereby verifying that products comply with legal requirements so as to guarantee their genuineness.
GC analytical methods are, however, also aimed at detecting extraneous matter which may basically be broken down into: - additives, i.e., substances deliberately added to a food product, such as antioxydants, dyes, etc. and, contaminants, i.e., substances that are accidentally included (basic, or "environmental" contamination), because of crop treatment (pesticides etc.) or due to foodstuff processing and preserving.
GC analysis of food products is, therefore, capable of revealing whether foodstuffs have been properly grown and processed, and of detecting what current legislation defines as "fraudulent products".

An instance of fraud is, in fact, connected to a series of factors, not least of which is that of the extraneous matter's being detectable and quantifiable. (G.C. is, therefore, a qualitative and quantitative analytical technique with a series of advantages: It is simple, accurate and highly sensitive) Our report mentions some of the most effective GC methods for food analysis reported in the literature. Some useful indications are also given for the GC operator.

APPLICATIONS

Though historically important but more of a milestone for its technological significance, it should remembered that GC was first ever used in the food-field to be more specific: in connection with fatty substances. The original work by Martin and James /1/ concerned the separation of fatty acids and gave rise to an analytical process which within a few years, revolutionised the analysis and knowledge of fatty substances. Before the advent of GC, separation and determination of fatty acid content was a complex and difficult problem. It was only by distillation first and then later by column chromatography that a rough separation of saturated fatty acids was possible. It was even more difficult and less satisfactory to separate saturated acids from unsaturated ones.

Establishing the features of various oils and fats was done exclusively by chemico-physical measurements such as Index of Refraction, Iodine Absorption Number, Acidity Value, Saponification Number, and these were completely inadequate in the case of mixed products. With the arrival of GC, saturated and unsaturated acids, from butyric to erucic acid, could be separated and contents determined as could branched and cis- or trans- isomers.

GC also enabled routine analysis of fatty substances to be conducted in a very short time. These features led to the application of GC to the study of fatty food substances, enabling a series of frauds to be discovered that would otherwise have gone unmasked on the basis of the chemico-physical features revealed by "official" analysis methods. It may be of interest to briefly review the early Martin and James papers /1, 2/. These two scientists separated fatty acids, from formic to lauric acids, in columns containing silicone oil and stearic acid at 137 °C. They also separated alkyl-amines in columns containing liquid paraffin at 100 °C. They did this using apparatus that titrated the emerging peaks with either a standard acid or base. The amount of reagent used was recorded: if the resultant curve is differentiated, we obtain the now familiar gas chromatographic picture of peaks.

This early instrument could only be used for the chromatography of titratable substances. However, for many other aspects, their techniques and columns were as modern as any used today and have never been improved upon.

If we consider GC development, we see that its range of applications in the field of foodstuffs has been increasing ever since the first works by Martin and James. A look at the literature and papers that deal with GC in foodstuffs testifies to such a growth. Table 1 lists pertinent works published over the past 10 years and reviewed by GC. Literature Abstracts and Index, Preston Publ., Inc. A survey of the GC methods and conditions used in the analysis of various foodstuff types follows. At the end of each foodstuff section, the class of compounds examined and corresponding GC stationary phases are given.

FATS AND OILS

Animal Fats: GC analysis enables different animal fats, such as pork, beef, lamb or chicken fats, to be distinguished on the basis of their fatty acid composition /3/. This may be done either by quantitative determination of selected fatty acids or by calculation of ratios of their respective quantities. Obviously, this distinguishing criterium is no longer valid in the case of mixed fats /4,5/.

Vegetal Oils: The two areas in which analytical checks on vegetal oils are most important are in determining when one oil has been adulterated with a less expensive one and in ascertaining the quality of processed produce. In general, the composition of fatty acids and of some components of unsaponifiable substances (e.g., sterols, triterpene alcohols) enable the identity of an oil to be determined and the presence of oils from a different source to be signaled /6-8/. From this point of view, no analytical technique provides such accurate and immediate information as GC does. In the space of an hour, vegetal oil glycerides can be transformed into the methyl esters of their respective fatty acids (a method that uses BF_3 as a catalyst /9/) and a gas chromatogram be obtained that identifies the oil examined and immediately signals the presence of possibly fraudulent products. The GC study of fatty acids in glycerides subjected to partial action by pancreatic lipase, or the differentiation of the cis- and trans- isomers of some acids using capillary columns, are further criteria in evaluating oil quality—olive oil particularly. Special attention has been paid to testing for esterification oils, a type which is forbidden by law. It should, however, be noted that oil quality is not necessarily ascertainable by GC analysis alone but may also be determined on the basis of chemical features such as acidity, or by controlling relative technological processes such as refining. With respect to this, UV spectrophotometry has been seen to be as equally important as GC in classifying olive oils as far as some European countries are concerned.

Fish Oils: Marine animals oils contain high quantities of long-chain unsaturated fatty acids and consequently, particular GC conditions are called for /10/ (e.g., the use of high temperature stationary phases such as OV 75, Silar 10, etc.) to identify

TABLE 1. Papers on GC in Food Analysis: 1970-1979. (from GC Literature – Abstracts and Index – Preston Publ. Inc. (*))

COMPOUNDS	'70	'71	'72	'73	'74	'75	'76	'77	'78	'79
Fats and Oils	31	23	55	30	44	36	30	38	10	24
Dairy Products	13	18	9	12	22	15	13	11	–	8
Meat	2	4	7	8	10	9	8	3	2	6
Fish	2	2	3	6	5	7	5	10	3	11
Eggs	3	9	4	–	1	2	2	2	–	–
Fruits	5	17	15	26	19	13	10	12	3	11
Vegetables	5	15	9	11	13	12	11	8	2	7
Alcoholic Beverages	17	17	16	11	13	10	8	8	6	8
Sugar Products	2	3	9	5	3	4	6	8	2	10
Cereals	–	7	3	4	9	8	10	9	–	12
Additives (Preservatives, antioxidants, stabilizers, emulsifiers, artificial sweeteners) and Volatiles(**)										
Nitrosoamines	1	1	7	8	12	8	11	4	5	5
Miscellaneous	6	9	8	5	9	7	2	14	6	18
Total	123	186	195	175	203	176	163	150	58	134

(*) – Papers on GC specifically for contaminants, pesticides, PCBs, methyl-mercury, aromatic polycyclic hydrocarbons and essential oils, all in foodstuffs, have not been tabulated due to the quantity of papers on these topics and the importance these items have acquired in the field of foodstuffs over recent years. These items call for further bibliographic research, which the authors intend to report in a later review.

(**) –Papers on the assessment of volatile compound content in various foodstuffs (oils, meat, fruit, vegetables, dairy products etc.) have been classified under the heading of "flavourings" and "volatiles".

and separate them out.

Classes of compounds in Fats and Oils	Corresponding GC stationary phases
fatty acids (as methyl esters)	DEGS, PEGA, OV 275
triterpene alcohols	SE 30
sterols (as TMS derivatives)	SE 30, OV 17
lactones	QF 1

DAIRY PRODUCTS

Milk: GC technique is used to determine fatty acid composition and contents of milk. Fatty acids are analysed as methyl, ethyl, or butyl esters. Volatile constituents of milk are due either to milk age or rancidness. Constituents may be aldehydes or ketones, both of which are analysable by GC methods /11,12/. Milk sugar composition may also be ascertained by GC analysis after the milk has been dialysed and sugars obtained in the form of TMS derivatives /13/.

Butter: The biggest analytical problem concerning butter is detecting its adulteration with hydrogenated vegetal fats. This possibility may be checked by analysing the fatty acids as methyl or ethyl esters and calculating the content ratio of some fatty acids /14,15/. Alternatively, the sterols, which only comprise cholesterol in the case of butter, may be analysed. In the case of a mixture of vegetal fats, phytosterols are also present and their content is ascertained by GC after having first saponified extracted and derivatized them as TMSs /16,17/.

Cheese: The analytical interest in cheese is not only in its fat composition and content (cheese is rarely adulterated by adding fats or oils), but rather in its flavour constituents which are important in differentiating cheese varieties /18/. Cheese flavour constituents may be detected by GC, keeping isolation temperature as low as possible to prevent the formation of any artefacts.

Classes of compounds in Dairy Products	Corresponding GC stationary phases
fatty acids (as methyl, ethyl or buthyl esters)	DEGS, BDS
biacetyl	TCEP, Carbowax 20M
aldehydes	Apiezon L, capillary
lactones	" "
tocopherols	SE 52
sterols (as TMS derivatives)	SE 30, OV 17
sugars (as TMS derivatives)	SE 30, QF 1
vitamines	OV 1
flavour compounds	Carbowax 20M

MEAT, FISH, EGGS

Meat: In meat analysis, GC determination is not only concerned with lipid composition but also with the volatile flavour components. It is, however, important to isolate flavour volatiles beforehand by distillation or extration and then to proceed to the GC determination of aldehydes, ketones, sulphur-containing compounds and aromatic and aliphatic hydrocarbons. MS should be used for the positive identification of the complex array of peaks /19/.

Fish: Likewise, in the field of fish, besides ascertaining lipid composition, GC technique is also important to determine flavour volatile content to differentiate the various species and, in particular, to assess aliphatic amine content which is caused by fish spoilage. To isolate the volatiles, the sample is first subject to either low-temperature vacuum distillation or extraction. GC and MS then positively identify the compounds which may include amines, aldehydes, ketones, hydrocarbons or sulphur-containing compounds /20/.

Eggs: Ascertaining simple aliphatic acid content such as that of lactic or succinic acids, which are associated with the deterioration of eggs, is the most important use of GC in this sector /21/. To GC-ascertain the cholesterol content of eggs used as ingredients in baked foods, the sample must first be saponified and then a TMS derivative obtained /22/. This method is only applicable if the operator is certain that the ingredients contain no animal fats, i.e. no butter or lard because they contain cholesterol.

Classes of compounds in Meat, Fish, Eggs	Corresponding GC stationary phases
fatty acids (as methyl esters)	DEGS, BDS
amines	THEED-TEP
carbonyl compounds	PEGA, Porapak
sulphur-containing compounds	Carbowax 1500
sterols (as TMS derivatives)	SE 30, OV 17,

FRUITS, VEGETABLES, NON-ALCOHOLIC BEVERAGES

GC is used to identify the natural flavours and aromas that are typical of individual fruit species and in the quantitative determination of polybasic and hydroxy acid content (e.g., citric, malic, etc. acid) in fruit and foodstuffs derived therefrom (juices, jams etc.) /23/. Even amino acids have often been analysed by gas chromatography but, due to the difficulties in preparing suitable volatile derivatives, more easy analytical techniques, such as using ion exchange resins, are normally preferred. To analyse flavours and volatile substances that are normally contained in fruits and vegetables, head space analysis /24/ is generally used and specific detectors based on S emission spectrophotometry measurements are used to look for sulphur-containing compounds. As for non- alcoholic beverages, we may mention the determination of natural aromas (flavanols) in tea or the caffeine in tea or coffee /25,26/.

Classes of compounds in Fruits, Vegetables and Non-alcoholic Beverages	Corresponding GC stationary phases
polybasic and hydroxy acids (as TMS derivatives)	OV 17, OV 61
sugars (as TMS derivatives)	OV 17, NPGS
amino acids (as butylesters-trifluoro-acetates)	EGA

caffeine	SE 30
capsaicin (as TMS derivative)	SE 52, JXR
vanillin, ethylvanillin	Carbowax 20M
sulphur-containing compounds	Carbowax 20M

ALCOHOLIC BEVERAGES

More than determining ethyl alcohol, for which satisfactory chemical methods exist, GC may be used to a certain advantage in assessing methyl alcohol content, important for its toxicology implications. Among the various GC methods that exist, the most highly recommended is the official British method for analysing potable spirits /27/: methanol content is determined separately from acetaldehyde content (after having first added methyl formate used as an internal standard) and then injecting into the chromatograph with a Porapak Q column at temperatures of about 100 °C. GC is also useful in ascertaining fusel oil content which influences the aroma of alcoholic beverages, and the 2-phenyl-ethanol in beer, cider and sherry in particular. GC is a far better tool than any chemical method to determine this component after the sample has first been distilled and the distillate extracted with methylene chloride /28/. Other aroma components (esters, alcohols, aldehydes, ketones, etc.) may be identified and quantified by head space GC analysis or, alternatively, by extraction with pentane or another light solvent and then, if necessary, results may be confirmed by GC-MS using polar capillary columns. The fixed component content of alcoholic beverages, such as organic acids, sugars, and amino acids may be ascertained by GC applying methods such as those mentioned for fruit juices.

Classes of compounds in Alcoholic Beverages	Corresponding GC stationary phases
methanol, ethanol	Porapak Q
2-phenylethanol	Apiezon L
fusel oil	Carbowax 400, 20M
aldehydes	Carbowax 4000
ketones	Carbowax 20M
esters	UCON or FFAP capillary
sulphur-containing compounds	Carbowax 20M
organic acids (as TMS derivatives)	SE 30, OV 17, OV 61
sugars (as TMS derivatives)	SE 52, OV 17
carbon dioxide	Charcoal

SUGAR PRODUCTS

The introduction of trimethylsilyl derivatives (TMS) /29/ has facilitated the GC determination of sugar content in unprocessed produce and foodstuffs that contain such produce. It is now possible to separate the various monose (glucose, fructose) and mono-, di- and tri- saccharides using silicone stationary phases such as OV 1, OV 17, SE 52 either at programmed temperature or under isothermal conditions /30/. Because sugar-containing syrups are often present in flavouring agents, organic acids or amino acids, GC analysis of these components is in line with comments made for "Fruits etc." above.

Classes of compounds in sugars	Corresponding GC stationary phases
sugars (as TMS derivatives)	SE 52, OV 1, OV 17
aconitic acid (as TMS derivative)	SE 30
flavouring agents	Carbowax 20M

CEREALS

GC has enabled fat composition and content of wheat, flour and bread to be ascertained. In this way the age of flour may be determined on the basis of variations in unsaturated fatty acid composition as a result of oxydation phenomena /31/. Using extraction methods that utilise different polarity solvents and TLC separation, it is now possible to differentiate between free and bonded lipids /32/. Normally, fatty acids are examined in the form of methyl esters after lipids have been hydrolysed. GC technique also enables the determining of aldehyde content in flour and bread after they have been transformed into semicarbazones, extracted with methylene chloride, and regenerated with phosphoric acid /33/.

Classes of compounds in cereals	Corresponding GC stationary phases
fatty acids (as methyl esters)	DEGS, BDS
acetic acid	FFAP
amino acids (as butylesters-trifluoro-acetates)	NPGS, QF 1
sugars (as TMS derivatives)	SE 30, OV 17
carbonyl compounds	TCEP

ADDITIVES

Preservatives and Antioxidants: Preservatives and antioxidants are widely used in food technology. GC in many cases is the most reliable analytical method for determining compounds such as benzoic acid, methyl, ethyl and propyl esters of 4-hydroxybenzoic acid and sorbic acid which are generally used as preservatives in foods /34, 35/. Preservatives such as diphenyl and o-phenyl phenol, used to prevent mould growth in citrus fruit, may also be determined by GC methods /36/. BHA, BHT and gallate esters are used to prevent oxidation of fats and oils and they can be found in foods which contain fatty materials. GC methods are available to ascertain these compounds in vegetable oils and margarine /37,38/.

Artificial sweeteners: Artificial sweeteners such as saccharin, dulcin, sorbitol and cyclamates can be determined in food by GC analytical methods /39,40/. Cyclohexylamine, that can be present or produced by cyclamates, is suspected to have some adverse toxicological effects, and its analysis has consequently received attention. After steam distillation, this compound has been determined in artificially sweetened foods by GC, using columns of 10% Carbowax 20M plus 2.5% sodium hydroxide /41/. Sorbitol is especially sought in diabetic foods, and can be determined after extraction with methanol and derivatization as TMS /42/.

Emulsifier and Stabilisers: A number of foodstuffs contain natural or synthetic emulsifiers such as mono- and di- glycerides or the sorbitan esters of fatty acids. Some products possess both emulsifier and stabiliser properties: e.g. brominated vegetable oils, sucrose diacetate hexaisobutyrate, triethyl citrate or triacetin. GC is a valid method of analysis but it must be preceded by separation steps to isolate the emulsifiers and stabilisers from the other various food constituents: column chromatography or TLC are normally used. Detecting emulsifiers and stabilisers is often a complex problem that cannot be summarised in a generalised table. We suggest the reader refer to specific works in the literature /43-47/.

Classes of compounds in Additives	Corresponding GC stationary phases
benzoic acid, sorbic acid, esters of hydroxybenzoic acid	Porapak Q
sorbic, propionic acid (as TMS derivatives)	SE 30
diphenyl, o-phenyl phenol	Silicon Oil
BHA, BHT	Carbowax 20M
BHA, BHT (as TMS derivatives)	JXR
cyclamates	SE 30
saccharin, dulcin	SE 52
cyclohexylamine	Carbowax 20M + NaOH
sorbitol (as TMS derivative)	SE 30
benzaldehyde	Apiezon M
polyols	Carbowax 20M
polyols (as TMS derivatives)	JXR
monoglycerides (as TMS derivatives)	SE 30
brominated vegetable oils (as methyl esters)	JXR, OV 61
triacetin, triethylcitrate	SE 30
sucrose diacetate hexaisobutyrate	JXR
decyl acetate	SE 30

Since the time it was first experimented, GC has possibly become the most widely used modern analytical technique in food analysis because of its suitability to the field and to its capacity to separate organic compounds in general.

It may therefore be considered as routine technology in academic, government and industrial laboratories. However, though the main advantages of GC are to be found in its high sensitivity, selectivity and quantitative accuracy, it is also true that in recent years other techniques, such as HPLC, have also shown a versatility and qualitative specificity to surpass GC. Absolute certainty about some analytical matters both in food analysis (for example, identification and quantification of toxic contaminants) and in other fields of research and application calls for analytical techniques that are extremely specific and sensitive; this is because of the sometimes extremely small quantities involved and the difficulties in separating substances gathered during extraction, concentration and purification operations.

In these cases, GC does have its limits due to possible interferences that may prejudice the validity and selectivity of even some of the most common detector system, e.g., electron capture detectors (ECD), alkali flame ionisation detectors (AFID), or flame photometric detectors (FPD). As a result, over recent years, workers have been "coupling" analytical techniques such as LC-UV, LC-IR, GC-IR, TLC-UV or GC-MS. Because MS analysis is highly specific as well as sensitive and therefore of considerable importance in the structural identification of organic compounds, pairing GC-MS techniques has considerably extended the range of its applications, enabling complex mixtures to be analysed without the laborious preliminary work of separation and purification.

Investigation queries can now be irrefutably answered as mentioned above. HPLC-MS coupling also promises to be a valid and reliable technique. However, the problems to be overcome with this pairing are far more complex than for GC-MS coupling. Though HPLC alone has several advantages over GC in the separation and analysis of substances which are high-boiling and thermally unstable, eliminating the mobile phase prior to entering the MS source is still a difficult problem to solve, even if some devices that show promise for the practical application of HPLC-MS coupling have already been studied. To conclude, there is good reason to state that GC-MS coupling which enables the identification of compounds such as organochlorinated compounds, polycyclic hydrocarbons, pollutants and organic contaminants in general, the toxicity of which have already been demonstrated, constitutes the most effective and sensitive analytical technique currently available not only in the field of foodstuff but also applied to environmental pollution problems and quality checks on chemical products too. So, with this fact in mind, we may truthfully say that "In some specific or advisable instances in which GC food analysis results are either unattainable or uncertain, irrefutable evidence is to be had with GC-MS coupling".

REFERENCES

1 A.T. James and A.J.P. Martin, Biochem. J. 50 (1952), 679
2 A.T. James and A.J.P. Martin, Biochem. J. 52 (1952), 238
3 S.A. Castledine and D.R.A. Davies, J. Ass. Publ. Anal. 6 (1968), 39

4   A.W. Hubbard and W. Pocklington, J. Sci. Food Agric. 19 (1968), 571
5   H.J. Laugner, Fette-Seifen Anstr. 71 (1969), 393
6   D. Grieco and G. Piepoli, Riv. Ital. Sostanze Grasse 41 (1964), 283
7   E. Fedeli et al., J. Amer. Oil Chem. Soc. 43 (1966), 254
8   A. Amati et al., Riv. Ital. Sostanze Grasse 48 (1971), 39
9   L.D. Metcalfe and A.A. Schmitz, Anal. Chem. 33 (1961), 363
10 R.G. Ackman, J. Gas Chromat. 4 (1966), 256
11 R.L. Glass, L.W. Lohse and R. Jenness, J. Dairy Sci. 51 (1968), 1847
12 R.G. Arnold, L.M. Libbey and E.A. Day, J. Food Sci. 31 (1966), 566
13 G.A. Reineccius et al., J. Dairy Sci. 53 (1970), 1018
14 D.F. Withington, Analyst 92 (1967), 705
15 L. Boniforti, Ann. Fals. Exp. Chim. 55 (1962), 255
16 J. Eisner et al., J. Ass. Off. Agric. Chem. 45 (1962), 337
17 L. Boniforti and A.M. Manzone, Boll. Lab. Chim. Prov. XX (1969), 279
18 J.E. Langler, L.M. Libbey and E. Day, J. Agric. Food Chem. 15 (1967), 386
19 C.K. Cross and P. Ziegler, J. Food Sci. 30 (1965), 610
20 N.P. Wong et al., J. Ass. Off. Anal. Chem. 50 (1967), 8
21 H. Salwin and J.F. Bond, J. Ass. Off. Anal. Chem. 52 (1969), 41
22 R.G. Johansen and S.S. Voris, Cereal Chem. 48 (1971), 576
23 E. Fernandez-Flores et al., J. Ass. Off. Anal. Chem. 53 (1970), 17
24 R.J. Romani and L. Ku, J. Food Sci. 31 (1966), 558
25 A.R. Pierce et al., Anal. Chem. 41 (1969), 298
26 J.M. Newton, J. Ass. Off. Anal. Chem. 52 (1969), 653
27 Research Comm. Analysis Potable Spirits, J. Ass. Publ. Anal. 10 (1972), 49
28 M.E. Kieser et al., Nature 204 (1964), 887
29 C.C. Sweeley et al., J. Amer. Chem. Soc. 85 (1963), 2497
30 K.M. Brobst and C. Lott, Proc. Amer. Soc. Brew. Chem. 71 (1966)
31 M.P. Burkwall and R.L. Glass, Cereal Chem. 42 (1965), 236
32 A. Graveland, J. Amer. Oil Chem. Soc. 45 (1968), 834
33 I.R. Hunter and M.K. Walden, J. Gas Chromat. 4 (1966), 246
34 E. Fogden, M. Fryer, S. Urry, J. Ass. Publ. Anal. 12 (1974), 93
35 A. Graveland, J. Ass. Off. Anal. Chem. 55 (1972), 1024
36 R. Thomas, Analyst 85 (1960), 551
37 K.T. Hartman and L.C. Rose, J. Amer. Oil Chem. Soc. 47 (1970), 7
38 E.E. Stoddard, J. Ass. Off. Anal. Chem. 55 (1972), 1081
39 M.L. Richardson and P.R. Luton, Analyst 91 (1966), 520

40 H.B. Conacher and R.C. O'Brien, J. Ass. Off. Anal. Chem. 54 (1971), 1135
41 J.W. Howard et al., J. Ass. Off. Anal. Chem. 52 (1969), 492
42 E. Fernandez-Flores, V.H. Blomquist, J. Ass. Off. Anal. Chem. 56 (1973), 1267
43 M.R. Sahasrabudhe and J. Legari, J. Amer. Oil Chem. Soc. 44 (1967), 379
44 M.R. Sahasrabudhe, J. Amer. Oil Chem. Soc. 44 (1967), 376
45 M.R. Sahasrabudhe and R.K. Chadha, J. Amer. Oil Chem. Soc. 46 (1969), 8
46 H.B.S. Conacher, R.K. Chadha and J.R. Lyengar, J. Ass. Off. Anal. Chem. 56 (1973), 1264
47 L. Kogan and S. Strezlek, Cereal Chem. 43 (1966), 470

Methodological Problems in Vitamin B₆ Estimation

Elisabeth KIENZL[1], P. RIEDERER[1] and J. WASHÜTTL[2]

1) Ludwig Boltzmann Institute of Clin.Neurobiology, Lainz-Hospital, A-1130 Vienna, and 2) Institute Food Chemistry and -Technology, Technical University, A-1060 Vienna, Austria

SUMMARY

In order to assess the B_6 content of foodstuff, we have developed a gas chromatographic method for the separation and identification of these vitamins modified after Imanari and Tamura (1,7). Formation of trifluoroacetyl-derivates and chromatography on SE 30 or OV 1 with electron capture detection were found to provide a simple and adequately sensitive method to determine these compounds. Thus with the method described pyridoxol can be determined at the picogram level with a simple and rapid GC-method.

INTRODUCTION

The three biological active forms of vitamin B_6, pyridoxol, pyridoxal and pyridoxamine are analogues of 3-hydroxy-2-methyl-pyridines different only in their substituents in the 4-position. However, the methods described in the literature are available exclusively to assess the portion of the aldehyde compound in overdosed pharmaceutical preparations or foodstuff enriched with synthetical vitamin B_6. The similarity of these substances has limited the ability of classical physico-chemical techniques such as spectrometry to determine the individual compounds. Thus chromatographic approaches have been made (8) which have limitations in their sensitivity (9,11) or

their applicability to biological samples (10).
Pyridoxal and pyridoxamine are metabolically and functionally interrelated. In living cells vitamin B_6 is transformed mainly to pyridoxal-5'-phosphate, the prosthetic group of a number of enzymes involved in amino-acid metabolism (5). Notable is the characteristic distribution of the three vitamin-compounds in foodstuff: A high concentration of pyridoxol is only found in vegetable material whereas pyridoxamine is enriched in animal tissue. The difficulties in the analysis and separation of natural amounts of the B_6 compounds in foodstuff arise from their extremely low levels (in the range of 1 - 30 µg/g dry substance) and in the necessity of hydrolysis of the bound phosphate group. We tried to overcome these losses by using different preparation steps and preventing destruction in light.

METHODS

PREPARATION OF DERIVATIVES

Vitamin B_6 analogues were obtained commercially by MERCK, Darmstadt, GFR. A method for the rapid and simultaneous quantitative determination was provided. An appropriate quantity of the vitamin standard, pyridoxol (POL) and pyridoxamine (PAM) was diluted and 25 µl were freezedried in 1 ml reactivials (Pierce). For acetylation a mixture of 200 µl tetrahydrofurane (THF) and 100 µl of trifluoroacetylanhydride (TFAA) was added.

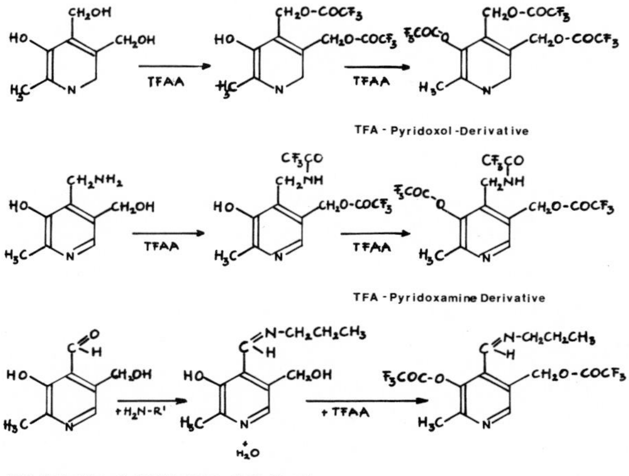

TFA derivative of Schiff's base of Pyridoxal with n-propylamine

FIG. 1

After 30 minutes the reaction mixture was quickly evaporated to
dryness under nitrogen to avoid hydrolysis. The residue was again
dissolved in THF, dried over calcium-hydride.
Pyridoxal has to be treated with n-propylamine prior to TFA - derivatisation. Pyridoxal (PAL), or a mixture of all B_6 compounds was
dissolved in 500 µl n-propylamine and the solution was kept at $40^{\circ}C$
for 20 minutes and subsequently evaporated to dryness under nitrogen.
The residue was then converted to TFA-derivatives and subjected to
gas-chromatography. A 95%-ethanol-solution of α-naphtylamine was
used as standards.

SEPARATION

To determine the samples we have investigated a Perkin-Elmer-Gas-
Chromatograph 3920 equipped with a ^{63}Ni electron-capture-detector
(ECD) and with a glass column 6 ft x 2 mm ID, packed with 3% OV 22
on Chromosorb W-HP 80/100 Mesh. The measurement required a constant
column temperature of $160^{\circ}C$ and a detector temperature of $200^{\circ}C$.
As carrier gas 10% methane in argon (25-50 ml/min) was applied.

The proposed method was also found to be satisfactory for analysis
in foodstuff. The extraction of the samples was performed with
1,8 M H_2SO_4 and heating for 30 minutes at $100^{\circ}C$ as well as by
digestion by diastase (vegetable) or papain (meat) after neutralisation.
In addition, this procedure involves separation on ion exchange resin
(Permutit T) which was performed according to the method described
by Hennessy et al. (6).

RESULTS AND DISCUSSION

The TFA-derivatives of POL, PAL and PAM thus prepared were stable and
volatile enough for gas-chromatographic separation (4). The responses
to different amounts are linear in the range from 100 pg to 100 ng.
(Fig. 2). Fig. 3 and 4 are typical chromatograms. However, PAM
consistently gave one major and three minor peaks. These results
correspond with findings of Dabre and Blau with tyrosine (2,3).
Although there is a possibility of a progressive hydroysis of the
O-TFA-group it was previously observed that the cause of these problems
lies in the use of a polar stationary phase. It is thought that
this catalyses an intramolecular N→ O-acyl shift from the N-TFA

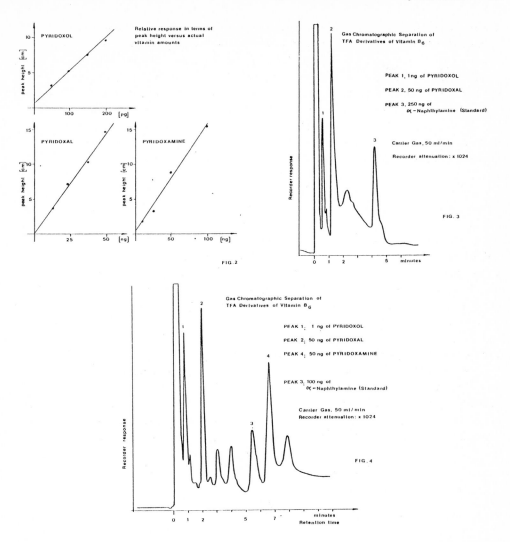

FIG. 2

FIG. 3

FIG. 4

to the O-TFA derivative and the loss of O-TFA; N-TFA may also be lost. This limits the range of stationary phases that may be used for gas chromatographic analysis of B_6. Our investigations have shown that the less polar phases OV 7 and OV 17 minimise this effect yet still provide adequate chromatographic separation.

The evaluation of the chromatograms in picogram-range is of importance for clinical examinations and estimation of bioavailability. Although vitamin B_6 is widely distributed in foods it is of interest to determine the loss in converted nutrients and so avoid the potential risk of deficiency of B_6 which may for example lead to neuromuscular irritability, neuropathy, depression of immune responses and anaemia.

ACKNOWLEDGEMENT

This project was supported by "Österr. Forschungsförderungsfonds", project Nr. 2512.
We thank Dr. G.P. Reynolds for valuable discussion of this manuscript.

REFERENCES

1 K. Blau, G.S., King, Acylation; in: Handbook of derivatives for chromatography; Ed.: K.Blau and G.King,Heyden and Son Ltd.,(1977, 1978), 127
2 A.Dabre, K.Blau, Biochim.Biophys.Acta 100 (1965), 298
3 A.Dabre, K.Blau, Biochim.Biophys.Acta 126 (1966), 591
4 H.Ehrson, T.Walle, H.Buotell, Acta Pharmaceutica Succica 8 (1971),27
5 L.Fieser, M.Fieser, Vitamine; in: Organische Chemie, Ed.: L.Fieser, Verlag Chemie (1975), 1674
6 D.J.Hennessy, A.M. Steinberg, G.S.Wolson, W.P.Keaveney, J.Assoc. agric.Chemists 43 (1960), 765
7 T.Imanari, Z.Tamuka, Chem.Pharm.Bull 15 (1967), 896
8 W.Korytnyk, G.Fricke, B.Paul, Anal.Biochem. 17 (1966), 66
9 L.T.Senello, C.J.Argoudelis, Anal.Chem. 41 (1969), 171
10 R.Strohecker, H.M.Henning Vitamin Assay - Tested Methods, Verlag Chemie GmbH, Weinheim (1965)
11 E.M.Patzer, D.M.Hilker, J.Chromatogr. 135 (1977), 489

Considerations and Remarks about Honey Volatile Components

C.BICCHI, C.FRATTINI, F.BELLIARDO and G.M.NANO

Laboratorio di N.M.R. e Spettroscopie applicate alla Tossicologia, Facoltà di Farmacia, Turin University, Turin (Italy)

SUMMARY

A comparison between four different techniques of honey volatiles extraction is reported. Of the four methods, two gave unsatisfactory results, another one - the direct extraction by heating in a modified Likens-Nickerson apparatus - led to an extract with many furan derivatives, probably arising from pyrolysis and condensation reactions between sugars during heating, the last method, consisting in a preliminary extraction of volatile components from sugar matrix with acetone, followed by extraction of the acetone phase in the Likens-Nickerson apparatus, gave the best results. Sixty components were identified by GLC and GLC/MS and are here reported.

INTRODUCTION

Honey is a complete food well known and appreciated since antiquity. It is produced in large amounts all over the world both by small and large industries and it has acquired great economic importance. The identification of the geographical and botanical origins of honeys is important as honeys from different sources differ widely in aroma and taste, and an effective quality control should be carried out, with obvious commercial implications. Chemical classification becomes necessary because botanical analysis (known as melissopalynologic analysis) which is based on the examination under a microscope of pollen granules and other vegetable residues, such as yeasts, conidia and fungi spores, and analysis of physical properties, such

as colour, specific gravity, viscosity, refractive index and so on, are not
sufficient to characterize different honeys. For chemical classification, several
studies (1-10) have been carried out on the volatile components of honey aroma, a
distinctive feature which would allow a very fine classification of honeys, but, at
least to our knowledge, the problem has not yet been solved. We also have been working on this subject for the last two years, in order to classify honeys of different
origin from the chromatographic pattern of their volatile fractions and to identify
as many volatiles as possible.

EXPERIMENTAL
When we started working with honey we employed two classical extraction techniques:
extraction with dichloromethane in vapour phase, by heating, in a countercurrent
Likens-Nickerson modified apparatus and extraction with dichloromethane in a Soxhlet
apparatus, almost without heating since the b.p. of CH_2Cl_2 is 40.1°C. The gaschromatograms of the two extracts appeared immediately to be different:the first
analyses by GLC/MS showed the presence of a certain amount of furan derivatives in
the sample extracted by heating in the modified Likens-Nickerson, while the quantity
of the Soxhlet extract was decidedly small and furan derivatives were completely
lacking. These first results decided us to examine the problem of honey extraction
thoroughly, and to this end we compared four different methods of extraction, one
of them already reported in the literature by other workers (9):

1) extraction with modified Likens-Nickerson apparatus: 1Kg of honey was suspended
in 1 liter of water, heated to boiling and extracted in vapour phase with 250 ml of
dichloromethane for 5 hours. In this case we had the formation of foam which was
removed hydraulically. The extract was cold-concentrated in rotovap to a 2 ml
volume. After the first extraction, the same honey sample was again extracted in
the same conditions for 4 more hours, and again cold-concentrated to a 2 ml volume.

2) Preliminary extraction of volatile components from sugar matrix with acetone,
followed by extraction of the acetone phase in the Likens-Nickerson modified
apparatus (with this preliminary operation a large part of the sugars was eliminated): 1 Kg of honey was extracted for 2 hours by cold-stirring with 100 ml of acetone
(repeated three times). The acetone phase, containing the volatile components, was
cold-concentrated in rotovap to a 10 ml volume, which was then diluted with 1 liter
of water, heated to boiling and extracted in vapour phase with 250 ml of dichloromethane for 5 hours. Methylene chloride extract was cold-concentrated in rotovap
to a 2 ml volume. Extraction of the sample was continued for 12 more hours in the
same conditions in the Likens-Nickerson apparatus, and the second extract was cold-
concentrated to a 2 ml volume.

3) Continuous extraction in Soxhlet apparatus: pentane (b.p.36.2°C) and dichloromethane (b.p. 40.1°C) were employed as extracting solvents. The quantity of pentane
extracts, however, was decidedly too low, and dichloromethane was employed in all
the successive experiments. In this case too, we had the formation of large amounts

of foam, which caused the emulsion of the two extraction phases. This problem was solved by mixing the honey with silica sand: 1 Kg of honey was mixed with 1 Kg of sand and extracted with 2.5 l of dichloromethane continuously for 7 days. The extract was cold-concentrated in rotovap to a 2 ml volume.

4) Extraction with ethyl acetate (without heating): this method was reported in the literature by Wootton and colleagues (9). 100 g of honey were extracted by cold-stirring with 30 ml of ethyl acetate for one hour (repeated three times). The ethyl acetate combined extracts were concentrated in rotovap under vacuum, at low temperature, to a 2 ml volume.

These four extraction methods were applied to several honey samples. We report here the results for a homogeneous sample of a multifloral honey, produced in 1980 in the mountains of Pragelato (Sestriere) near Turin. Melissopalynological analysis was carried out according to the methods of International Commission for Bee Botany. The honey examined was multifloral: pollens of *Poligonum* spp., *Helianthemum* group, *Rhododendron ferrugineum*, *Hippocrepis comosa* L., *Lotus corniculatus*, *Myosotis* spp. were present in its sediment. Also spores, conidia and fungi hyphas were present in small amounts.

The gaschromatographic conditions were the following: gaschromatograph Carlo Erba 2900, glass capillary column OV 1 50 mt, injection splitless 40", film thickness 0.5 microns, injector temperature 200°C, detector temperature 225°C, initial temperature 40°C for 1 min., programmed to 100°C at 1°C/min, then programmed to 180°C at 5°C/min.

Gaschromatographic-mass spectral analyses were carried out on a Varian Mat CH7A spectrometer, coupled with a Varian Aerograph 1400 gaschromatograph, by means of a direct coupling system (made by ourselves): capillary column-restrictor-line off sight inlet-ion source.

RESULTS AND DISCUSSION

In fig. 1 - 4 are reported the gaschromatograms of the samples obtained with the four methods. In Table I are reported the components identified.

From the results reported in the Table and the chromatograms we can make the following considerations:

1) the ethyl acetate extraction method resulted unproductive for us. Probably the differences between our results and those of Wootton and coll. are a consequence of having operated on very different samples: for instance, they report that most of the components possessing honey aroma had high retention times (above 80 minutes) while we found no traces of them in our samples.

2) The modified Likens-Nickerson extraction method led to an extract with many furan derivatives, probably arising from pyrolysis and condensation reactions which may occur between sugars at the operating temperature (100°C).

3) The extraction in Soxhlet apparatus led to a very small overall quantity of extract, expecially in the more volatile components, probably responsible for

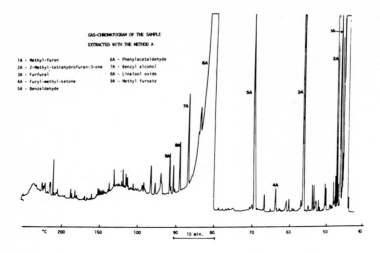

Fig. 1

TABLE 1

COMPOUND	MW	METHOD A	METHOD B	METHOD C	IDENTIFIED IN PREVIOUS SAMPLES
1) 2-methylfuran	82	*			
2) 3-pentan-2-one	84			*	
3) crotonic acid	86			*	
4) phenol	94		*	*	
5) 2-furfural	96	*			
6) pyran-2-one	96	*			
7) 5-methylenfuran-2-one	96	*			
8) 5-methyl-3-hydrofuran-2-one	98	*			
9) furfuryl alcohol	98				*
10) 4-methyl-3-penten-2-one	98		*		*
11) 3-hexan-2-one	98			*	
12) 2-methyl-2-tetrahydrofuran-3-one	100	*		*	
13) 2,3 pentandione	100			*	
14) styrene	104	*			
15) dimethylbenzene	106			*	
16) benzaldehyde	106	*	*	*	
17) benzyl alcohol	108	*	*	*	*
18) 2-acetylfuran	110	*	*	*	
19) 5-methyl-2-furfural	110	*		*	
20) 4-hydroxy-4-methyl-2-pentanone	116		*	*	
21) phenylacetaldehyde	120	*	*	*	
22) trimethylbenzene	120			*	
23) phenylethylalcohol	122			*	*
24) 5-hydroxymethylfurfural	126	*		*	
25) methyl furoate	126	*			
26) 2-methyl benzofuran	132	*			*
27) 1-methyl-4-isopropenyl-benzene	132			*	
28) M.W. 132 unknown	132	*	*		
29) 3-phenyl-2-propenal	132	*	*		
30) 2-etoxyethyl acetate	132	*		*	
31) p-cymene	134		*		*
32) 3,4,5 trimethyl phenol	136				
33) 2,3,5 trimethyl phenol	136		*		*
34) β-pinene	136				
35) 3-hexin-1-ol acetate	140				*
36) p-cymen-α-ol	150				
37) 1-p-menthen-9-ol	152	*	*		*
38) M.W. 152 unknown	152		*		*
39) 1,8 cineole	154		*		
40) eugenol	164				
41) cis-3-hexenyl butyrate	170				
42) linalool oxide (1)	170	*	*		*
43) linalool oxide (2)	170	*	*		*
44) chrysantenyl acetate	194				
45) 4-terpinenyl acetate	196				
46) carvomenthyl acetate	198				
47) sesquiterpene hydrocarbon	204				
48) 3-methylbutyl octanoate	214				
49) n-hexadecane	226				*
50) ethyl laurate	228				
51) n-heptadecane	240				
52) n-octadecane	254		*		*
53) 3-octanol	270				*
54) n-eicosane	282				
55) ethyl palmitate	284				*
56) n-heneicosane	296				
57) ethyl octadecanoate	312				*
58) n-tricosane	324				
59) n-pentacosane	332				*
60) n-heptacosane	340				*

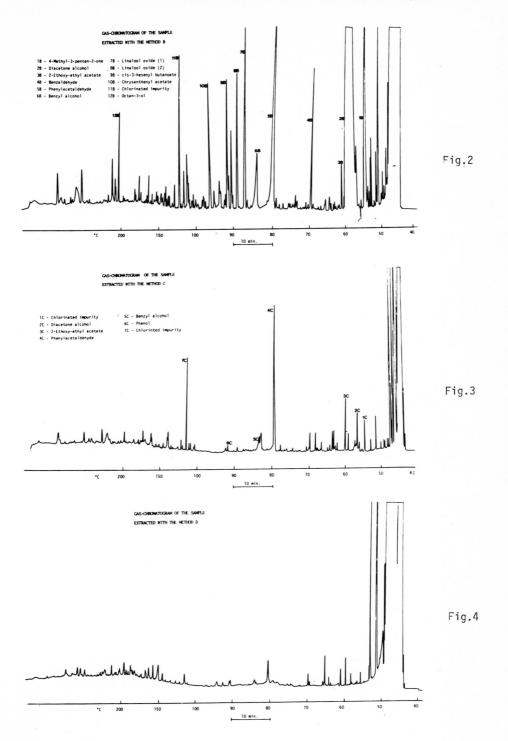

Fig.2

Fig.3

Fig.4

honey flavor. Moreover, this method is time and solvent consuming.

4) The method which gave the best results is the separation of the volatiles from the sugar matrix by cold-stirring with acetone, followed by extraction in the modified Likens-Nickerson. The peak of diaceton alcohol is probably an artifact, coming from acetone itself. In the end we must point out that three extract out of four possessed a strong honey-like aroma, extremely similar to that of the starting product.

The aim of this study was to find a proper method for extracting honey aroma, reducing to a minimum the formation of artifacts; we are still far from having identified the components responsible for honey flavor and from being able to characterize honeys of different origin from their chromatographic patterns. We can only say that, in the case of sample reported here, at least in our opinion and in the opinion of experts, the characteristic and floral aroma is given by phenylacetaldehyde, which is also the most abundant component.

REFERENCES

1   W.Dorrscheidt and K.Friedrich - J. Chromatogr., 7 (1962) 13
2   H.J.G. Ten Hoopen - Z.Lebensm.Unters.Forsh., 119 (1963) 478
3   J.H.Merz - J.Apic.Res., 2 (1962) 55
4   E.Cremer and M.Riedmann - Z.Naturforsh., 119b (1964) 76
5   E.Cremer and M.Riedmann - Monatsh. Chem., 96 (1965) 364
6   E.Cremer and M.Riedmann - Z.Anal.Chem., 212 (1973) 31
7   Tschogowadse and coll. Lebens. Industr., 20 (1973) 224
8   C.E.M.Ferber and H.E.Nursten - J.Sci.Food Agric., 28 (1977) 511-518
9   M.Wootton, R.A.Edwards and R.Faraji-Haremi - J.Apic.Res., 17(3)(1978) 167
10   D.Graddon, J.D.Morrison and J.F.Smith - J.Agric.Food Chem., 27 (1979) 4.

Use of Glass Capillary Columns for the Determination of Fusarium Mycotoxins in Food

Á.BATA, R.LÁSZTITY, J.GALÁCZ

Department of Biochemistry and Food Technology, Technical University
H-1521 Budapest, Pf. 19. Hungary

SUMMARY

A glass capillary column gaschromatographic method has been developed for the determination of three mycotoxyns /zearalenone, T-2 toxin and deoxynivalenol/ produced by Fusarium species. The toxins were extracted by ethylacetate resp. metanol-water and cleaned by column chromatography. The recovery and standard deviation values were 73-87% resp. 10,3-13,7% at a 0,1 mg/kg concentration level.

INTRODUCTION

As the initial step of our work, a method has been developed for the detection of some metabolites produced by Fusarium fungi. The reason for our selection is that in Hungary and in other countries of Middle Europe the toxins of Fusarium fungi cause severe veterinary problems in certain years of adverse weather conditions.

DISCUSSION

Of the compounds investigated first zearalenone and its derivatives will be discussed /Table 1./ [1]

Table 1. Naturally Occuring Derivatives of Zearalenone.

	R_1	R_2	R_3
Zearalenone	=O	=H_2	=H_2
Zearalenol	-OH	=H_2	=H_2
8'-hydroxyzearalenone	=O	=H_2	=OH
7'-dehydrozearalenone	=O	-H	-H
6'-8'-dihydrozearalene	-OH	=H_2	-OH

This oestrogenous substance is produced in large quantities by F. graminearium, F. culmorum and F. moniliforme.

Zearalenone causes in animals dysorexia, oedemic swelling of the mammary glands, of the pera-, and vulva vaginitis. In the case of pregnant animals it may cause pseudo-rutting, abortion.

The other two compounds to be determined belong to the compounds with trichothecene skeleton. /Table 2./. The common feature of these compounds is to contain the 12,13-epoxy group, more over, the LD_{50} value of the toxins is relatively low, 0.1 - 10 mg/kg body weight, so that they justly belong to the group of strongly toxic compounds [2].

Table 2. Structures of trichothecenes with simple hydroxyl or acyl substituents.

	R_1	R_2	R_3	R_4	R_5
T-2 toxin	-OH	-OAc	-OAc	-H	-OOCH$_2$CH/CH$_3$/$_2$
HT-2 toxin	-OH	-OH	-OAc	-H	-OOCH$_2$CH/CH$_3$/$_2$
Diacetoxyscirpenol	-OH	-OAc	-OAc	-H	-H
Trichothecene	-H	-H	-H	-H	-H

F. oxysporum, F. sporotrichoides, F. moniliforme, F. nivale, F. poea all can synthesize one or more compounds with trichothecene skeleton.

Of the compounds with trichotecene skeleton T-2 toxin and deoxyvalenol /Table 3./ are first investigated.

Table 3. Structures of 8-keto trichothecenes.

	R_1	R_2	R_3	R_4
Nivalenol	-OH	-OH	-OH	-OH
Deoxynivalenol	-OH	-H	-OH	-OH
Fusarenon-X	-OH	-OAc	-OH	-OH

On consuming T-2 toxin the whole alimentary tract of the animal is injured. Injury begins from the mouth, the mucous membrane becomes excoriated, and as the toxin is passing through the alimentary tract, haematoma, oedema appear in the stomach and intestines and after absorption the oedeams apeear in the cavities, the bone-marrow necrotizes, the blood picture changes, the red blood count decreases, and finally the animal dies.

Deoxynivalenol, called by some of the authors vomitoxin, is responsible in accordance with this later name for vomiting, but at higher doses it produces the general trichotecene toxic effect described above.

They cause in humans vertigo, vomiting, drowsiness, hausea, and in more severe cases bone atrophy and skin irritation. Compounds of trichotecene type are responsible for the disease called in the literature ATA /Alternary Toxic Aleukia/ [3].

Having dealt with the physiological effects of the compounds investigated, results of our work will be reported. Several methods are published in the literature for the detection of the compounds mentioned. For example for the TLC determination of zearalenone Sarudy proposed 4-methoxybenzene diazonium fluoroborate reagent, further Japonese workers proposed a colour reagent nitrobenzene piridine and tetraetilenpentamin for the measurement of trichotecene toxins proved to be good. The disadvantages of the TLC method are its relatively low sensitivity /o.l mg/kg/, and the non-specificity of the colour reagents used. This latter makes the reliability of the determination questionable.

In addition to the TLC method gas chromatographic methods are also known for the determination of the said compounds. In the gas chromatographic methods described earlier packed columns have been used, but the great number of interferring components makes the determination below a concentration of 0.1 mg/kg uncertain.

The HPLC technique can be used only in the case of zearalenone. The trichothecene compounds mentioned have no u.v. absorption, and so, their detection is not yet solved.

The analytical method developed will be demonstrated first for zearalenone.

In the literature organic solvents of medium polarity are recommended for the extraction of zearalenone from cereal samples, for example chloroform, acetonitrile, ethyl acetate. Of the solvents recommended ethyl acetate was found by us the most suitable, because extraction with ethyl acetate was the most efficient, and the ethyl acetate extract contained according to our experiences the fewest contaminants.

The infected sample was extracted at room temperature. The solvent needed was taken as 15-20-fold of the quantity of the sample /Fig.1./

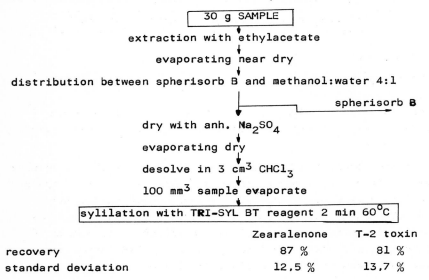

Fig.1. Cleaning procedure for determination of zearalenone and T-2 toxin.

A Packard instrument was used for the gas chromatographic measurements. The capillary column has been prepared in our own laboratory. An effective plate number above 2000 per meter was achieved with the columns, both in the case of the OV 17 and the SE 52 columns /4,5,6/.

For the determination of T-2 toxin a preparation process identical with that of zearalenone has been used.

The recovery and standard deviation values were calculated from 11 parallel measurements after the admixture of 0.1 mg/kg standard substance to healthy cereal.

In the case of deoxynivalenol the sample, was extracted with a methanol-water mixture. /Fig.2./.

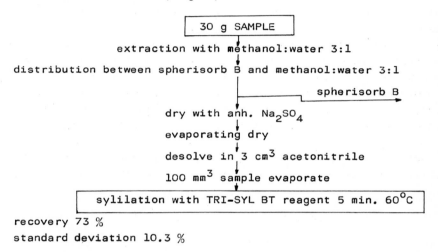

recovery 73 %
standard deviation 10.3 %

Fig.2. Cleaning procedure for determination of deoxynivalenol.

For the determination of recovery and standard deviation values 7 parallel determinations were carried out at a toxin concentration level of 0.1 mg/kg.

With the methods presented 9 samples from different mills in Hungary have been investigated. Seven of these were corn and 2 rice samples. 2 samples from 9 were out of sorted. The sorting out was ordered because of high mould number. In the two sorted out wheat samples zearalenone was found at a level of 90 and 110 ppb. The sample of higher zearalenone content contained also 50 ppb T-2 toxin. The samples investigated did not contain deoxynivalenol.

In summary, it can be established that capillary gas chromatography offers a possibility for increasing the sensitivity of mycotoxin detection, and the reliability of the determination method increases.

REFERENCES

1. Mirocha, C.J., Christensen, C.M. and Nelson, G.H.: Appl. Microbiol. 15. /1967/ 497.
2. Pathre, S. and Mirocha, C.J., in Rodricks, J.V., Hesseltime, C.W. and Mehlman, M.A.
   Mycotoxins in Human and Animal Health.
   Health Pathotex Publ. Park Forest South, III. 1977. p. 238.
3. Bamburg J.R., in Rodricks, J.V., Hesseltime C.W. and Mehlman M.A.: sea above.
4. Grob K. and Grob G.: J.Chromatogr. 125 /1976/ 471.
5. Grob K., Grob G. and Grob Jr.: Chromatographia 4 /1977/ 181.
6. Grob, Jr., Grob, G. and Grob, K.: Chromatogr. 156 /1978/ 1.

Propylene Carbonate, a Superior Solvent for the Extraction of Polycyclic Aromatic Hydrocarbons

K. POTTHAST

Federal Center for Meat Research, Institute of Chemistry and Physics
Kulmbach, Federal Republic of Germany.

SUMMARY

A method is described for the extraction of polycyclic aromatic hydrocarbons (PAH) from foodstuffs and other materials. In this procedure propylenecarbonate has proved to be an excellent solvent to produce a quantitative extraction of PAH. The method requires short time and is simple applied.

INTRODUCTION

Some of the PAH have been proven to be carcinogenic in test animals. PAH are found in curing smoke, in exhaust from motor vehicles, in emissions from industries and from other sources as well. The contamination of many foodstuffs is mainly dependent upon the above mentioned sources of contamination. Because of the hazardous effect of different PAH there is a great interest in analytical determination of these components in foodstuffs in order to estimate the health hazard in relation to their amount. In general PAH are found in foodstuffs only in quantities of parts per billion. To determine these samll amounts one can use different chromatographic methods. For instance, liquid and thin layer chromatography in connection with optical methods as well as gaschromatography have been applied successfully in identification and in quantitative determination. The extraction of PAH from foodstuffs presents on the other hand great difficulties. These difficulties may be diminisked by using special solvents for the extraction procedure.

In this paper propylenecarbonate is described as an eluent which has been successfully applied in many experiments. Propylenecarbonate will not mix with linear and cyclic hydrocarbons or with fat or water. Therefore, it can be employed in liquid-liquid chromatography to great advantage. Furthermore, propylenecarbonate causes no known hazards to health as do most of other solvents which are commenly used in trace analysis of residues. Therefore it is very surprising that propylenecarbonate is not more frequently used.

MATERIALS AND REAGENTS

Unless otherwise specified, the materials used were obtained from Merck.
Celite 545 (Macherey and Nagel, Düren, G.F.R.), heated at 500°C for 3 h
Sodium sulphate p.a., heated at 500°C for 3 h.
Florisil, 60-100 mesh (ASTM) for column chromatography, aged in a solution containing 10 % sulphuric acid for 4 h at a temperature of 60°, washed free from acid using distilled water, dried at 110°, heated at 500° for 3 h and, after cooling in a dessiccator, activated with 15 % of water.
Calcium chloride p.a., powder form, water-free.
sea sand p.a., heated at 500° for 3 h.
Propylenecarbonate residue analyses grade (Merck or Riedel de Häen).
Light petroleum (b.p. 40°) purified by distillation at 20° with a rotating vacuum evaporator.
Cyclohexane, Uvasol.
Chloroform p.a.
Sodium hydroxide, 20 % (w/v) solution.
Chromatography columns (60 cm x 6 cm) with glass frit (pore diam. D0) and PTFE stopcocks (Normag, Hofheim, G.F.R.)
Separatory funnels (2 litre) with glass stopcocks and long stems (Schott, Mainz, G.F.R.)

METHODS

Propylenecarbonate has been used for the extraction of pesticides or PAH from vegetable and animal food products (1,2,3,4), as well as from food additives (5) and inorganic or organic matter. In these investigations propylenecarbonate demonstrated a special advantage in the extraction of fat layers which covered inert anorganic materials as a thin film. For the extraction of PAH from fat containing foodstuffs it is recommended to solublize the fat in a proper solvent, to dry it with sodium sulphate and to mix it thoroughly with suitable adsorbents to obtain a thin layer. With this method one proceeds as follows: Foodstuffs are thoroughly mixed with chloroform and sodium sulphate using a kitchen mixer in order to bind the water of the product to the sodium sulphate and thus to improve the extraction of the lipid fraction. To this mixture celite is added and carefully mixed so that the chloroform extract is uniformly distributed on celite surface. To evaporate the chloroform the mixture is dried at $40^{\circ}C$ in a vacuum drying chamber until the vacuum remains constant at a pressure lower than 10 Torr. Only if the vacuum remains constant can one be sure that the chloroform has been quantitatively evaporized. Otherwise residual chloroform introduces fat into the extract disturbing further analysis. The dry powder mixture is placed in column with a glass-frit containing a layer of approximately 5 cm celite. This layer at the same time protects the glass-frit and adsorbs colloidal particles which become soluable in propylenecarbonate. The food/celite mixture is packed densely in a column, and 2 cm of sea sand layered over this mixture. For the elution process propylenecarbonate is utilized in an amount sufficient to fully wet the dry mixture contained in the column. The propylenecarbonate is displaced with water from the column. The first few millilitres of eluent contain the main portion of PAH. One avoids elution loss by trapping about 50 - 200 ml of the eluent, depending on the amount of foodstuff. The total extract is further analyzed.

The propylenecarbonate extract is saponified in a separatory funnel with a solution containing 20 % sodium hydroxide and water. For every hundred ml of propylenecarbonate extract, 150 ml sodium hydroxide and 300 ml of water are required. The saponification process causes a warming of the eluent and therefore must be cooled to room temperature (20 - $22^{\circ}C$) with running water. During cooling one must observe that no Na_2CO_3 crystallizes because a crystallization makes the elution of the PAH from the alkaline water solution more difficult. To separate PAH from the water solution one utilizes petrolether with a boiling point of less than $40^{\circ}C$. The petrolether quickly separates from the water phase after shaking. The elution of PAH from the water phase must to be repeated three times with low boiling petrolether. The different petrolether extracts are collected in a separatory funnel. A mixture of calcium chloride and celite is poured into this solution. It is briskly shaken and then left undisturbed for several minutes until the calcium chloride mixture settles out. The valve of the separatory funnel is carefully opened, so that none of the now clear petrolether extract can pass through the opening. By pumping air into the funnel (thus slightly racing the pressure), the speed with which the extract leaves the separatory funnel is increased. A further clean up of the petrolether extract is achieved by filtering the extract through several absorbent layers. As absorbent layers one can use: aluminium oxid acid or basic as well as silica gel with 15 % water respectively, or florisil with 15 % water which makes separation of isocyclic PAH from heterocyclic PAH possible. The petrolether extract, prepared as described, can be separated by thin layer chromatography without further clean up procedures (5). The quantity of some well separated PAH can be determined by evaluation of the fluorescence with certain spectrofluorometers using specific excitation- and emission maxima. With the method described by Tóth et al. (6) good results have been achieved particularly with benzo-a-pyrene. Fractionation of the raw extract on a sephadex LH 20 column (7), or on a preparative thin layer plate, should be carried out before gaschromatography if one wants to separate as many PAH as possible. For the preparative thin layer chromatography (TC) the following system has been developed. TC-plates, 0.5 mm thick, have been prepared from silica gel RP 2 (Fa. Merck, Darmstadt, G.F.R.). After the plates are air-dried they are cleaned by immersion in a chromatographic tank with acetonitrile. Using a syringe, the petrolether extract is spread on the plates producing a line about 5 cm long. The chromatogram is first developed with a methanol:water mixture in a ratio 3:1 and after drying with a fan, a second time with pure acetonitrile. The acetonitrile must be free of water. In the methanol-water mixture the PAH remain at the start position. With acetonitrile the PAH migrate only as one small zone with a RF-value of about 0.9. This zone is outlined under UV-light and collected from the plate by a zone collector (Desaga, catalog-nr. 120930, Heidelberg, G.F.R.). This zone collector works like a vacuum cleaner. The layer from the outlined zone is collected in a tube closed with a glass-frit. The PAH are separated from the silica gel RP 2 phase with 1 ml acetonitrile. This extract is concentrated using a nitrogen stream to about 50 µl and analyzed by gaschromatography. ß,ß´-Dinaphthyl has proved a good internal standard in these investigations. Because of good separation during thin layer chromatography and because of the low quantities of solvent used for the extraction of PAH from the zone collector the PAH fraction is free from interfering substances.

One can see the advantageous properties of propylenecarbonate by comparing the distribution coefficients of propylenecarbonate and other commonly used eluents for the extraction of PAH from cyclohexane, iso-octane and petrolether. Distribution of benzo-a-pyrene between cyclohexane and propylenecarbonate leads to a 90 % gain of benzo-a-pyrene in the propylenecarbonate phase. Using acetonitrile - as a common solvent for the elution of PAH - only about 55 % of the PAH are transfered from the cyclohexane phase. The distribution coefficients between propylenecarbonate or acetonitrile and iso-octane and petrolether are similar.

The distribution of PAH within the described polar and unpolar solvents does not reflects at extractability of these components from foodstuffs. A comparison of propylenecarbonate with acetonitrile, dimethylformamide and dimethylsulphoxide, used for the elution of benzo-a-pyrene and other PAH from country type ham, elucidates the use of propylenecarbonate as a solvent for these components.

Table 1: Benzo-a-pyrene Extraction from Country Type Ham with Different Solvents

solvent	B-a-P (ppb)	remarks
propylenecarbonate	1.84	free of fat, clear, good separation on TC
dimethylformamide	0.5	unclear, bad separation on TC
acetonitrile	0.5	unclear, bad separation on TC
dimethylsulphoxide	0.51	slightly muddy, separation of some PAH indicated

As table 1 shows, the extractable benzo-a-pyrene from the same sample of country type ham is 3.7 times higher by using propylenecarbonate instead of the other solvents. In addition, the extracts produced by propylenecarbonate showed significantly better chromatographic behaviour when separated on thin layer plates or by capillary gas liquid chromatography.

Propylenecarbonate has been used for the extraction of PAH from different foodstuffs, for instance from smoked meat products, smoked fish, smoked cheese, natural and smoked spices, liquid smoke and smoke condensed on organic matter. From the preliminary investigations, this solvent appears very suitable for the extraction of PAH with subsequent detection by gas or thin layer chromatography. The extraction procedure takes about four hours time.

REFERENCES

1) Hadorn, H. and K. Zürcher: Mitteilung aus dem Gebiete der Lebensmitteluntersuchung und Hygiene. Veröffentlicht vom Eidg.-Gesundheitsamt in Bern. Band 61, Heft 2, 141-169 (1970)

2) Potthast, K. and G. Eigner: J. Chromatogr. 103, 173-176 (1975)

3) Potthast, K., G. Eigner and R. Eichner: Fleischwirtschaft 57, 1294 (1977)

4) Potthast, K.: Fleischwirtschaft 60, 1941 (1980)

5) Potthast, K.: Fleischwirtschaft 54, 183 (1974)

6) Tóth, L.: J. Chromatogr. 50, 72 (1970)

7) Grimmer, G. and H. Böhnke: J. AOAC 58, 725 (1975)

Theory and Special Methods of Mass Spectrometry

S. ABRAHAMSSON

Department of Structural Chemistry, University of Göteborg,
Box 33031, S-400 44 Göteborg, Sweden

I.   INTRODUCTION

Food science is truly multidisciplinary. Not only does the external form which you meet in food vary considerably but also its physical and chemical behaviour show a diversity which require the application of a number of analytical techniques for objective assessment of various properties. So far in this Congress we have heard reports on the successful use of HPLC and GC and today we shall try to illustrate the possibilities of mass spectrometry (MS) in food analysis.

Mass spectrometry has been used for more than 25 years for structure determination of organic molecules. It represents in itself a large scientific discipline which has produced spectacular results in various fields including food chemistry. During the time allotted for my talk it is, of course, impossible to cover all of this wide field so I have tried to limit myself to some basic principles and added outlooks into nearest future developments.

II. INSTRUMENTATION

In a mass spectrometer the substance under investigation is ionized into positive or negative fragments, which are analyzed as to their mass over charge ratio. The basic features of a mass spectrometer is thus

        (a) the ion source
        (b) the mass analyzer
        (c) the ion detector

Let us look at these in turn.

(a) In most instruments today ionization is performed by letting the sample interact with an electron beam with an energy of 10-100 eV. This usually leads to extensive fragmentation. In spite of this electron ionization is widely used as no other source has a similar sensitivity in relation to reliability and ease of operation.

In another fragmentation procedure the ionization chamber is kept very tight so that an externally added gas can be held at several torr. This high pressure gas is transformed into a plasma by an electron beam and the sample is ionized by the plasma in a way depending on the externally added gas (chemical ionization). The method represents a convenient way of determining fragmentation pathways as the ionization pattern can be influenced by the added gas. Chemical ionization for instance usually gives molecular ions - which helps the interpretation of the spectrum - when electron impact fragmentation often does not.

Molecular ions are formed as a rule when field ionization is used. Here the ions are formed in the high electric fields produced at very fine points of less than a micrometer kept at about 10 kV.

There are a number of other less used ionization methods as well but they must for time reasons be left out from this lecture.

(b) The earliest instruments were all of the so-called magnetic type. The ions are accelerated into a magnetic field where they are deflected with a radius depending of the accelerating voltage, magnetic field strength and their m/z ratio. Normally a stationary slit system with a photomultiplier detector is utilized and the magnetic field or the accelerating voltage must be varied if heavier and heavier ions should be recorded by the detector. Though the magnet has some focussing properties and allows input of a slightly divergent ion beam the slits must be kept rather narrow and the useful resolution lies in the order of 1000-2000.

This means for instance that if you can record masses 1000 and 1001 with a valley of 10% between them you have a resolution of 1000. For higher resolution an electrostatic lens has to be added to give a double focussing instrument. Here you can reach resolutions over 100 000. The cost for such an instrument is, of course, much higher than that of a low resolution one.

The concept of tandem mass spectrometry (MS/MS) was introduced in the later half of the sixties. The output of ions with the same m/z-value from one mass spectrometer is brought into a collision chamber of a second instrument where the ions undergo further fragmentation to a new "ion mass spectrum". This technique has been proven powerful for specific component determination in complex mixtures.

A very versatile low-cost and consequently popular mass analyzer is the quadropole one. It consists of four pole pieces of hyperbolic cross section arranged symmetrically along the direction of the ion beam. Mass discrimination is achieved by a DC voltage and a ratio signal component that is variable in both frequency and amplitude. The exact physical relationships are, however, rather complicated and cannot be described during this lecture.

III.   SAMPLE INPUT

What about the samples? In GC/MS applications they must be brought into the gas phase. In order to achieve this the GC column can be heated up to several hundred degrees centigrade. The volatility of a substance can furthermore be increased by blocking polar groups for instance by methyl, trimethylsilyl esterification and amino group acylation. The choice of derivatives is often very important. For instance in the case of complex fatty acids the methyl esters give spectra that are often difficult to interpret. The corresponding pyrrolidides on the other hand give spectra that are more easily evaluated.

Both in the case of studies of structure and fragmentation pathways it may be of advantage to use isotope labelled compounds.
As stable isotopes are easy to handle in contrast to radiactive ones most MS studies of organic compounds use deuterium and C^{13}. At this department very efficient methods have been developed to prepare compounds which contain these isotopes either at specific locations or throughout the whole molecule.

In the beginning of the sixties direct coupling between a gas chromatograph and MS (GS/MS) was achieved resulting in an explosion of new analytical applications. The famous jet separator was introduced to remove most of the carrier gas. Nowadays the output from modern glass capillary columns is such that it can be brought directly into the ion source without disturbing MS-performance. With the increased use of HPLC it became natural to arrange interfaces to the MS (HPLC/MS). One type which has been used contains a moving belt which transfers the LC effluent to the ion source. There are also inlet systems using lasers for use in quadropole instruments. Though commercial versions are available further research is required in this area.

IV. USE OF MASS SPECTRA

The main purpose with MS was originally to determine the molecular structure of isolated compounds. As few references were available the early works to determine the chemical structure from low resolution data required great ingenuity. Nowadays the situation is quite different and spectra of e.g. hydrocarbons represent straightforward interpretation cases.

Spectra of glycolipids on the other hand gives a very complicated fragmentation pattern. In spite of this, glycolipids have been analyzed with success with mass spectrometry and the carbohydrate sequence determined in species with more than ten sugar residues. Here though the use of a high resolution instrument was a necessary prerequisite.

The output from the mass detector is drastically different now from the chart papers originally presented where you had to measure the relative intensities at different amplifications for every single mass numbers. Now computers have taken over and digitalize the analogue signal, determine peak height and m/z and finally present the spectra in the usual bar diagram form. With high resolution instruments the m/z value can be measured with five decimal points. This allows determination of the exact composition of the ions, e.g. C_2H_4 can easily be distinguished from N_2.

In cases where the various chromatographic separation stages have not been efficient enough, e.g. in the analysis of body fluids the computer can assemble consecutive spectra to a threedimensional mass chromatogram which might be interpretable.

As most spectra of resolved compounds are unique the mass spectrometer constitutes a very efficient chromatograph detector. Optical isomers and sometimes <u>cis</u>-<u>trans</u> isomers offer exceptions in giving identical mass spectra.

Many laboratories work with a rather narrow substance range of mass spectra. They therefore usually cope easily with structure determination of species in a conventional way. However, when a large number of different substance categories are present, it might be necessary to consult the various volumes of reference spectra that are now available. These are also to a large extent stored as data banks (more than 30.000 substances), that can be accessed at low cost over public data communication networks such as TYMNET/TELENET and SCANNET. The search systems are often such that if the unknown substance is not found in the data band - which is still likely due to the enormous number of organic compounds - at least the most similar spectra are retrieved.

V. SELECTED ION DETECTION (SID)

As was mentioned earlier present instruments have to be scanned to bring ions with different m/z-values through the narrow slit of the stationary detector. If for instance the mass range 100-500 amu is

scanned in 1 second the detector "sees" each m/z-value only during
2,5 ms whereas the total ions produced during the one second are
mostly lost. In order to improve the sensitivity SID was introduced.
This technique is useful when one or a few characteristic ions are
looked for in a complicated sample. The instrument is set to focus
on these few m/z-values only by repeated switching between the corresponding accelerating voltages rather then scanning the full spectrum.
Hereby the sensitivity is drastically increased and detection of compounds down to the femtogram level has been possible.

The main drawback with SID is the fact that one has to know beforehand
what m/z-values have to be looked for. Therefore new instruments are
under development in several laboratories to record the full mass
spectrum simultaneously. This means that SID sensitivity can be
achieved for all ions of the spectrum.

The ions leaving the source are focussed on to a line of detectors
usually multichannel plates (1 or 2 depending on the required sensitivity). The electrons leaving each channel is accelerated towards
a fiber optic faceplate covered with a light conversion layer on the
vacuum side. The other side of the faceplate, which is at atmospheric
pressure, is in optical contact with a flexible fiber optics bundles.
The latter is divided into parts that fit the input window a photo-
diode array. The signal from the array is preprocessed by a micro-
computer which then transfers its data to a larger computer for final
presentation. Repeated tests in our laboratory with a low-cost permanent magnet prototype covering simultaneously 50-500 amu have shown
the expected performance and the instrument is now used to study its
versatility in various application areas.

VI.   CONCLUSIONS

Mass spectrometry is a highly efficient tool in the structure analysis
of organic compounds. Several types of instruments in various price
classes are available from manufacturers, the number of which seem
to decrease due to the present economical situation in the world. Inno-
vations in both hard- and software are ready to be incorporated into
commercially available spectrometers further enhancing the versatility
of mass spectrometry in various fields.

spectra that are often difficult to interpret. The corresponding pyrrolidides on the other hand give spectra that are more easily evaluated.

Both in the case of studies of structure and fragmentation pathways it may be of advantage to use isotope labelled compounds.
As stable isotopes are easy to handle in contrast to radiactive ones most MS studies of organic compounds use deuterium and C^{13}. At this department very efficient methods have been developed to prepare compounds which contains these isotopes either at specific locations or throughout the whole molecule.

In the beginning of the sixties direct coupling between a gas chromatograph and MS (GS/MS) was achieved resulting in an explosion of new analytical applications. The famous jet separator was introduced to remove most of the carrier gas. Nowadays the output from modern glass capillary columns is such that it can be brought directly into the ion source without disturbing MS-performance. With the increased use of HPLC it became natural to arrange interfaces to the MS (HPLC/MS). One type which has been used contains a moving belt which transfers the LC effluent to the ion source. There are also inlet systems using lasers for use in quadropole instruments. Though commercial versions are available further research is required in this area.

IV.   USE OF MASS SPECTRA

The main purpose with MS was originally to determine the molecular structure of isolated compounds. As few references were available the early works to determine the chemical structure from low resolution data required great ingenuity. Nowadays the situation is quite different and spectra of e.g. hydrocarbons represent straightforward interpretation cases.

Spectra of glycolipids on the other hand gives a very complicated fragmentation pattern. In spite of this, glycolipids have been analyzed with success with mass spectrometry and the carbohydrate sequence determined in species with more than ten sugar residues. Here though the use of a high resolution instrument was a necessary prerequisite.

A version of this lecture has also been published in Ernährung 5 (1981) by agreement of the publishers.

Selectivity of Reagent Gases as used in the Analysis of Flavour Mixtures

U. RAPP, M. HOEHN, C. KAPITZKE and G. DIELMANN

Finnigan MAT GmbH, Barkhausenstraße 2, D-2800 Bremen 14, W-Germany

SUMMARY

Together with high performance capillary columns the chemical ionization technique has been used to gain selectivity during the analysis of complex mixtures. This is achieved by applying different reagent gases such as isobutane, ammonia, water, heavy water and others. The selectivity is obtained because the proton affinities of the substrates as well as of the reagent gases are different. It results in discrimination or enhancement of the component signals in the total ion current trace as is shown by means of an arrak extract and an artificial flavour component mixture.
On the other hand, the mass spectra are presenting molecule specific information when different reagent gases are used. Results of aliphatic aldehydes, terpene alcohols and acetates are discussed.

INTRODUCTION

Chemical Ionization (CI) is now well introduced as a supplementary ionization method in mass spectrometry [1]. In the ion source of the mass spectrometer an ion molecule reaction takes place. In contrast to electron impact ionization (EI), here a pressure of about 0.3-1 torr has to be maintained to yield optimal reagent gas cluster formation. Chemistry is carried out in the mass spectrometer because the choice of the reagent gas will strongly influence the reactions with the substrate molecules. A reagent gas dependant selectivity can be

achieved which is to be documented in two different ways:
1. Selectivity as presented in the total ion current trace (TIC trace or quasi GC trace)
2. Selectivity as presented in the mass spectra

Details on acid-base reactions or redox reactions through charge exchange and the mechanisms involved should not be discussed here. Richter and Schwarz have published recently a paper dealing with this topic with instructive examples [2].

RESULTS AND DISCUSSION

1. TIC trace selectivity

Figure 1 shows the TIC traces of an arrak extract measured with different reagent gases. It is easily recognisable that the EI and the isobutane CI traces are very similar. But there are occuring significant discriminations of distinct compounds when using ammonia as reagent gas. Discriminations are indicated by arrows whereas an enhancement (meaning a better response to the reagent gas) is marked with an asterisk.

An exceptional selectivity is observed for compound 12 (3-methyl-1-butanol) which has nearly completely disappeared in the ammonia CI trace whereas it is the main component under EI conditions.

The compounds discriminated to a lower or higher extent in the ammonia CI measurement are mainly belonging to aliphatic alcohols and aldehydes together with aliphatic and aromatic hydrocarbons. In few cases a higher response of a component towards ammonia is found in this particular example. Here we mainly deal with ethoxy-group containing aliphatic compounds and most probably N-heterocycles such as pyrazine derivatives.

Figure 2 displays the TIC traces of another mixture of components, also yielding discrimination effects for certain substance classes in the ammonia CI mode.

The compounds involved are given within the figure, where the isobutane CI and ammonia CI are to compare with the electron impact trace as a "standard". The discrimination of the aliphatic aldehydes $\underline{2}$, $\underline{7}$, and $\underline{8}$ as well as of the glycol $\underline{6}$ in the ammonia CI trace is in accordance with the data of figure 1.

2. Mass spectral selectivity

Besides the general response of a compound to a particular reagent gas which was discussed in the first chapter, the mass spectrum itself may contain different characteristics depending also on the reagent gases used. Some typical substance classes are selected and explained according to their behaviour.

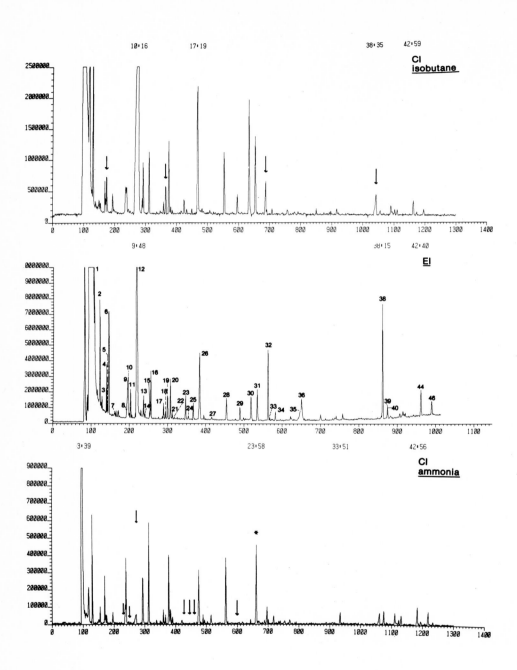

Figure 1. Total ion current (TIC) traces of arrak extract.

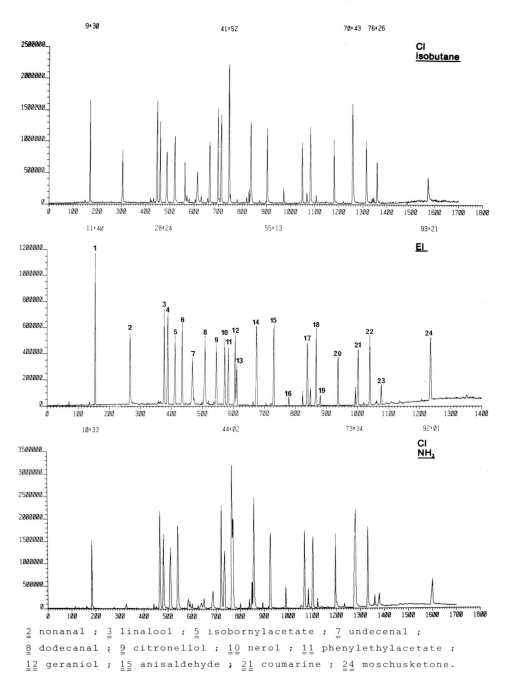

Figure 2. Total ion current (TIC) traces of an artificial mixture.

2 nonanal ; 3 linalool ; 5 isobornylacetate ; 7 undecenal ;
8 dodecanal ; 9 citronellol ; 10 nerol ; 11 phenylethylacetate ;
12 geraniol ; 15 anisaldehyde ; 21 coumarine ; 24 moschusketone.

a. Aliphatic Aldehydes

The aliphatic aldehydes did not receive too much attention from the MS analysts. The EI spectra are not very characteristic; molecular ions are only detectable for the short chain compounds [3]. Figure 3 and 4 show a comparison of spectra of the nonanal $\underline{2}$ and the undecenal ($\underline{7}$). The isobutane CI and the ammonia CI spectra show unequivocally the molecular weight information: $(M+H)^+$ and $(M+57)^+$ in the case isobutane and $(M+NH_4)^+$ for ammonia respectively. It seems to be a typical behaviour of this substance class that the $(M+57)^+$ ion appears as the base peak since other homologues investigated show similar results.

b. Terpene Alcohols

The terpene alcohols belong under EI conditions to the more difficult to characterize compound classes because very frequently the molecular ion is weak or even absent. Hence one key characteristic is missing. The chemical ionization produces in many cases a quasi molecular ion whereby the molecular weight can be determined. Ammonia turned out to be the best suited reagent gas.

Figure 5 and figure 6 show the mass spectra of citronellol and linalool respectively. The EI spectrum of citronellol exhibits already a small molecular ion but in the CI spectra the quasi molecular ions m/z 157 isobutane: (M+H) and m/z 174 ammonia: $(M+NH_4)$ are undoubtedly present as the base peaks. The linalool turns out to be more labile. Under isobutane CI conditions the water elimination from the quasi molecular ion $M+H-H_2O$ (m/z 137) is the most significant peak. Ammonia CI also shows only weak quasi molecular ion intensities, whereas fragmentation processes are yielding the dominating ions.

Very similar behaviour has been found for other terpene alcohols such as nerol and geraniol. Further investigations with other reagent gases are under preparation for this class of compounds.

c. Acetates

Acetates and also longer chain esters - of high importance in flavour research - show frequently no molecular ion information in their mass spectra. Figure 7 shows an instructive example, where under EI conditions the two characteristic moieties of the molecule are clearly detectable; m/z 43: acetate group and m/z 104: phenylethyl part. No molecular ion is detectable. Ammonia CI easily yields the supplementary information with an $M+NH_4$ quasi molecular ion (m/z 182) as base peak. Isobornylacetate $\underline{5}$ is more critical; the quasi molecular ion under ammonia CI conditions is present (about 5 % Irel) but m/z 137 (M-59) shows up as the base peak.

Selectivity of Reagent Gases in the Analysis of Flavour Mixtures   167

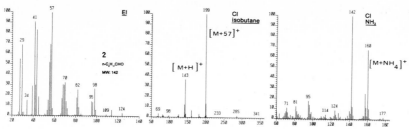

Figure 3. EI and CI mass spectra of nonanal.

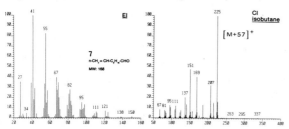

Figure 4. EI and CI mass spectra of undecenal.

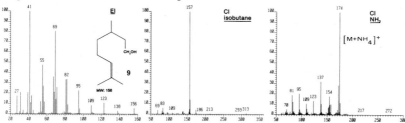

Figure 5. EI and CI mass spectra of citronellol.

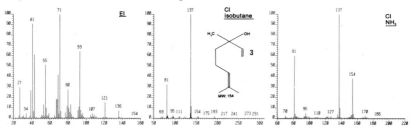

Figure 6. EI and CI mass spectra of linalool.

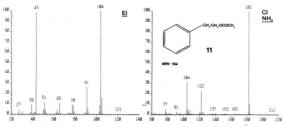

Figure 7. EI and CI mass spectra of phenylethylacetate.

Some Examples for the Identification of Artificial Flavours in Food

W. PFANNHAUSER, R. EBERHARDT, H. WOIDICH

Forschungsinstitut der Ernährungswirtschaft, A 1190 Wien
Blaasstraße 29

SUMMARY

Artificial flavouring and its proof is demonstrated presenting some practical examples from a food analysis laboratory. Intensifying flavour by addition of natural substances, identification of solvents used for aroma compounds and proof of added natural or artificial aromamixture are the three categories of interest.

INTRODUCTION

At first some remarks on the analytical techniques used.
The concentration range reaching from about 1 ppb up to some ppm needs efficient methods of separation, concentration and determination of the aroma compounds under consideration.
Extraction was carried out on the rotationsperforator according to Ludwig. Separation by steam destillation was combined with an extraction in an apparatus introduced by Likens-Nickerson (1,2).
As the separation method of choice capillary chromatography was used coupled with a mass spectrometer for identification of unknown compounds. The identification was carried out using mass fragmentography and identification of separated compounds by their mass spectrum.

INTENSIFYING OF NATURAL FLAVOUR

As an example the addition of benzaldehyde to roasted hazelnuts is given. In Fig. 1 two chromatograms of extracts from hazelnut products are shown. Chromatogram A has been identified from original roasted nuts. B represents a chromatogram of a sample of unknown history. B shows an excess amount of benzaldehyde. In addition A shows some pyrazines, identified by GC-MS techniques, whereas B due to extrem low content of roasted hazelnuts did not show pyrazines in the chromatogram.

Fig. 1: Extract from hazelnut products
A: original roasted hazelnuts
B: benzaldehyde added

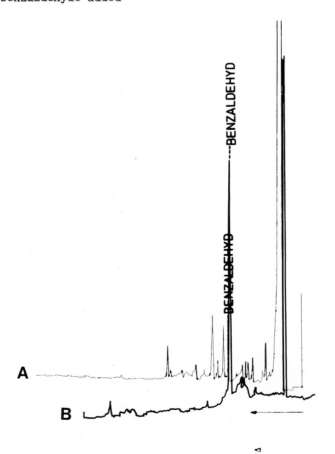

The second example deals with "raspberry ketone" - 1-(4'-hydroxyphenyl)-butanone-3 - which is responsible for the well known typical raspberry aroma. In raspberries the concentration does not exceed 1 mg/kg.
Fig. 2 shows two chromatograms. Chromatogram A originates from an extract of raspberries and B from an artificial enhanced product. It can be seen, that the "raspberry ketone" was added excessively.

Fig. 2: Raspberry ketones
      A: natural raspberry extract
      B: artificial enhanced product

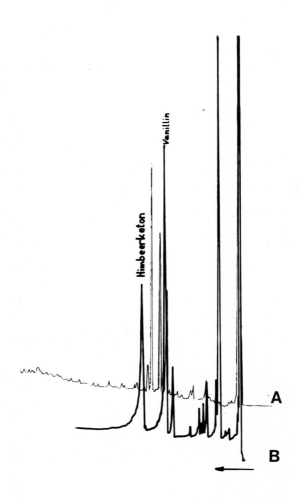

Some Examples for the Identification of Artificial Flavours in Food   171

IDENTIFICATION OF SOLVENTS USED FOR FLAVOURS

The proof of addition of natural or artificial flavour can be done
by identification of solvents, not of original origin, like triacetine, triethylcitrat and triethylacetylcitrate.
We identified triacetine, which is glyceroltriacetate, e.g. in products from banana and coconuts by GC/MS analysis.
In Fig. 3 strawberry aroma of a unknown sample is compared with the
GC of a natural one.

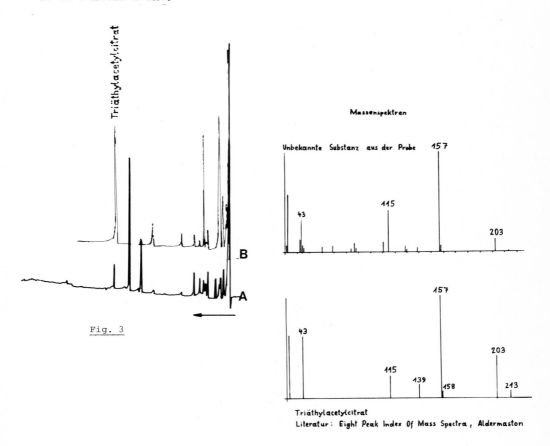

Fig. 3

Fig. 4: Mass spectrum of triethylacetylacetate compared with the
unknown peak from Fig. 3

A peak, identified as triethylacetylcitrate by mass spectrometry
was found (Fig. 4). This finding confirmes addition of strawberry
aroma to the product.

An interesting case in the determination of added aroma compounds to food were samples of fruit joghurt. Aroma of cherries and apricot type joghurt was enhanced by addition of alcoholic aroma concentrates. The proof was done by finding ethanol in the range of about 0,1 % in samples.

ADDITION OF ARTIFICIAL FLAVOURS

The presence of natural compounds not present in the fruit may be of value for confirmation of addition of flavouring substances to the product.
Pear aroma extracts contain decadienesters, which are not present in apple juices. In a sample of Jonathan apple juice decadienester was found an addition of pear juice was therefore assumed.
Coconut products contain as intensive flavour compounds δ - lactons. Artificial flavouring is done by using the coconutlike odour of γ - lactons. These compounds are synthetised from γ - hydroxyacids. For instance γ - nonalactone, which is called "coconutaldehyde" is no aldehyde and its origin is not from coconut, but it is a very intensive used compound in simulating coconut flavour.
We found as well mixture of γ and δ - lactones and γ - lactons alone in coconut products like alcoholic drinks and sweets. Also in these case the solvent triacetine was often found (3).
Fig. 5 shows the massfragmentogram of an artificial enhanced product containing δ - and γ - lactones (m/e = 99 und 85). Vanilline (m/e = 152) was also identified. The ultimate confirmation is given by the spectrum of the separated and first fragmentographically identified compound.

Fig. 5: Mass fragmentogram of artificial enhanced coconut flavour
S = total ion current chromatogram
m/e 152, 99, 85, 43 characteristic mass fragments of
vanillin, δ - lactones,
γ - lactones and acetates.

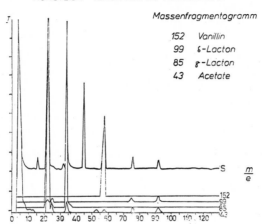

REFERENCES

1 S.T. Likens, G.B. Nickerson
   Am. Soc. Brewing Chemists Proc. (1964), 5

2 H. Pyysalo, A. Kiviranta, S. Lahtinen
   J. Chromatogr. <u>168</u>, (1979), 512

3 W. Pfannhauser, H. Woidich
   Mikrochim. Acta (in press)

Study on the Composition of Cruciferae Oils by Gas Chromatography and Mass Spectrometry

S. LORUSSO [1], L. BONIFORTI [2], A. SELVA [3], E. CHIACCHIERINI [1]

[1] Istituto di Merceologia, Università, Rome, Italy
[2] Istituto Superiore di Sanità, Rome, Italy
[3] Politecnico, Istituto di Chimica, Centro C.N.R. per lo Studio delle Sostanze Organiche Naturali, Milan, Italy

SUMMARY

Rapeseed oils most commonly used in European Community countries, some for forage, others for oil production, are investigated. Chemical and physico-chemical characteristics and quantitative analysis of positional and geometric isomers (in particular mono-unsaturated long chain $C_{20:1}$ and $C_{22:1}$ fatty acids) of oils extracted from the above-mentioned seeds are reported. Capillary column GLC and GLC-MS were used.

INTRODUCTION

It is well-known that Cruciferae oils, in particular rapeseed oils, induce biochemical, functional and anatomical lesions in many animal species (1-2). In former papers (3-4) we have studied the different aspects connected with the rapeseed oil problem, such as its stereochemical, analytical and nutritional aspects in connection with its declared toxicity. In this paper, we have examined the various most widely used rapeseed oils in Common Market countries, some oils for forage, others for further oil manufacturing (*). Many authors have investigated rapeseed oils without specifying the varieties examined. In recent years, however, many papers (6-7) have been published on the influence of variety, environment and seeding time on rapeseed oil composition and on its fatty acid composition in particular. Geneticists have differentiated many rapeseed varieties on the basis of erucic acid ($C_{22:1}$, 13-cis-docosenoic) toxicity or different content of erucic acid (for industrial use) or its total absence (for nutritional use). Other than the determination of some chemical and chemico-physical features of the oils extracted from some of the aforementioned seeds, our study is concerned with acid composition and the qualitative and quantitative detection of geometric and positional isomers of $C_{20:1}$ and $C_{22:1}$ long-chain mono-unsaturated fatty acids. Initially our analytical investigation was performed by GLC-FID. A packed column was used for the analysis of the rapeseed oils. Then we used a capillary column to separate and quantify the geometric isomers of the eicosenoic acids ($C_{20:1}$, $\Delta 11$; gadoleic = cis, gondoic = trans) and docosenoic acids ($C_{22:1}$, $\Delta 13$; erucic = cis, brassidic = trans). Some researchers have shown the possibility to determine the positional isomers of these mono-unsaturated acids by GLC (8) and combined GLC-MS (7) analysis. Then, in order to confirm the positional isomers, we have analysed by GC-MS trimethylsilyl derivatives (TMS) of the diols obtained by oxydation with performic acid of the double bound of the rapeseed oil fatty acids. The TMS mass spectra have shown the diagnostic fragments that indicate the presence of $C_{20:1}$ and $C_{22:1}$ acid positional isomers.

EXPERIMENTAL

The extracted and commercial oils were transesterified with 4% W/W H_2SO_4/CH_3OH and then subjected to GC analysis with the following equipment set up: Perkin Elmer Sigma 1 gas chromatograph equipped with FID; column: stainless steel (3.60 m x 3 mm i.d.) packed with 10% BDS on 80-100 mesh Chromosorb W HP; temperatures: oven 205 °C, injector 290 °C; carrier gas flow: nitrogen 30 ml/min. These samples were also injected into a DANI 3600 gas chromatograph fitted with a capillary column so as to dose the geometric isomers of the mono-unsaturated $C_{20:1}$ and $C_{22:1}$ acids. The analytical set

(*) There are over 80 varieties of "Brassica napus oleifera", i.e. rape cultivars, on the O.C.D.E. (Organisation de Coopération et de Développement Economiques) - monitoring market. However, because there are several countries that are not O.C.D.E. member states, it may be estimated that at least 250 varieties of rape cultivars exist in the world (from an extensive study by Dr. B. Kokeny, ex-consultant at F.A.O. (5)).

up was as follows: DANI 3600 gas chromatograph equipped with FID; capillary column: stainless steel (90 m x 0,25 mm i.d.) coated with Apiezon L and with a number of theoretical plates based on the methyl palmitate peak of 200,000; temperatures: oven 200 °C isot., injector 290 °C, detector 250 °C; carrier gas flow: helium 1.2 ml/min; auxiliary gas flow: helium 15 ml/min; inlet slit ratio 50:1. Gas chromatograms for the oils extracted show that the aforesaid isomers are present, even if only in relatively small amounts. At this stage it became important to establish whether the isomers were geometric, positional or both. Consequently, the gas chromatograms for the methylesters of the oils were compared with those for oils to which erucic, brassidic, gadoleic and gondoic acid had been added and transesterified. The gas chromatograms for a sample of oil (Petranova) were also compared.

The Petranova oil was previously trans-isomerized according to the following method: 5 g of oil were placed in a frosted cap test tube. 2.5 ml of HNO_3 (d = 1.40) were added. The test tube was shaken for a few minutes. 0.5 g of Hg were added and dissolved by energetic shaking. The mixture solution was left to rest at a temperature no lower than 25 °C. A pasty, yellow-brown liquid resulted. The liquid was diluted with 10 ml of distilled H_2O and extracted with approx. 40 ml of petroleum ether thereby removing the entire isomerised product from the test tube.

After a few hours, the underlying aqueous phase was separated out. Approx. 0.3 g of transisomerised oil contained in the ether fraction was treated as described above, i.e. transesterified first for GC-FID analysis to detect the new fatty acid composition and then for GC-FID analysis with a capillary column to detect trans isomers.

To confirm that the geometric and/or possible positional isomers of the aforesaid fatty acids were, in fact, present, rapeseed oils were subjected to GC-MS analysis, after having derivatised the samples as follows (9): 10 mg of the methyl esters of the fatty acids were hydroxylated with performic acid and then treated with 100 µl of HCHO and 50 µl of H_2O_2 30% for 2 hours at 40 °C, shaking occasionally. The mixture was then diluted with 2 ml of distilled water and extracted three times with 5 ml aliquots of ethyl ether. The ether extracts were evaporated in an N_2 gas flow. The product obtained was saponified for ½ an hour at 100 °C with 0.2 ml of 3N NaOH and then acidified to pH 2 or 3 by adding diluted H_2SO_4 after the alkaline solution was cooled. The dihydroxy derivatives obtained were extracted three times with 5 ml of ethyl ether. The ether extracts were pooled, dried with anhydrous Na_2SO_4 and then evaporated. Dioil residues were methylated with CH_2N_2 and converted to corresponding TMS's with HMDS (hexamethyldisylazane) and TMCS (trimethylchlorosylane) in pyridine (3/1) for half an hour at 60 °C. The derivate was then ready for GC-MS analysis, with the following equipment:

LKB 2991 gas chromatograph - mass spectrometer interfaced with a Digital PDP 11 computer system.

<u>Gas chromatographic conditions</u>: quartz capillary column (30 m x 0.3 mm i.d.) coated with OV 101; Carrier gas flow: helium 1.5 ml/min ; LKB solvent-less injection system; temperatures: oven programmed from 150 to 250 °C (with increments of 4 °C/min.), injector 280 °C.

<u>Mass spectrometric conditions</u>: Becker-Rhjiage separator heated to 270 °C; electron-impact ion source operated at an electron energy of 70 eV; ion source temperature: 250 °C; mass range 4 to 650 m/z; scan time: 1.2 sec; scan period: 4 sec.

RESULTS AND DISCUSSION

Although only 12 rapeseed oils, either extracted or taken from market, were examined, we believe that this constitute a sufficiently representative preliminary stock for the purpose of a chemical and commodity science knowledge of several of the varieties of rapeseed.

As shown in Table 1, the varieties of rapeseeds chosen were suitably selected among those used for different purposes, i.e. for fodder, oil production and, in the case of the Olimpiade variety we distinguished between original and reproduced seed oils. A variety of turnip rapeseed used for fodder was also investigated. Lastly, three samples of commercial oils were also examined i.e. crude rapeseed oil, refined, and oil from several different seed types.

Table 2 gives the fatty acid composition of the oils examined. It may be noted that 13-docosenoic acid is present in large quantities (approx 50%) in the extracted rapeseed oils for fodder and industrial use. Oils for nutritional use (Europa, Zephir and those in commerce) have a low 13-docosenoic acid content. On the other hand, 11-eicosenoic acid, was found to constitute approx 10% of the composition of the oils with a high $C_{22:1}$ percentage. The percentage drops to 1 or 2% in those oils with $C_{22:1}$ acid amount under 1%.

Figures 1, 2, 3 show the interesting sections of gas chromatograms obtained with a capillary column for mono-unsaturated and saturated acids with 20 and 22 atoms of carbon. Figure 1 refers to crude rapeseed oil. Peaks 1-2-3 represent 11-cis-eicosenoic acid, its 13-cis-eicosenoic positional isomer and saturated docosenoic acid (arachidic acid), respectively. Peaks 4-5-6 represent 13 cis-docosenoic acid, its 11-cis-docosenoic positional isomer, and saturated docosenoic acid (behenic acid). It must be noted that Peak 5 indicates the presence of at least two components.

Table 1. Rape seeds examined and common characteristics of extracted oils.

RAPE SEEDS	GENUS: Brassica		SPECIES: Brassica napus var. oleifera					
	Observations	Breeder	Country	% extracted oil	density at 22 °C	degree of refraction at 22 °C	refraction index at 25 °C	iodine number
Varieties for forage								
Petranova (original seed)	forma aestiva	Lochow Petkus	Germany	34.8	0.9108	69.3	1.4718	105.0
Torrazzo (" ")	" "	C.A.P. Cremona	Italy	41.7	0.9109	69.2	1.4718	102.7
Varieties for oil production (for industrial or nutritional use according to high or low content of 13-docosenoic acid)								
Matador (reproduced seed)	forma ibernalis	Institute of Research of Smalof	Sweden	36.0	0.8958	69.8	1.4715	98.1
Leonessa (original seed)	" "	Sis-foraggers (Bologna)	Italy	38.2	0.9282	74.0	1.4748	92.2
Olimpiade (reproduced seed)	" "	Institute of Plant Breeding (Lonigo-Vicenza)	Italy	41.3	0.9107	70.3	1.4725	105.8
Olimpiade (original seed)	" "	Institute of Plant Breeding (Lonigo-Vicenza)	Italy	42.4	0.9169	70.6	1.4726	100.3
Eurora	" "	Roffi Sem. (Bologna)	Italy	39.6	0.9178	69.3	1.4718	113.9
Zephir	" "	Sis-foraggers (Bologna)	Italy	42.1	0.9146	68.2	1.4711	114.8
TURNIP RAPE SEED	GENUS: Brassica		SPECIES: Brassica campestris					
Variety for forage								
Perko (hybrid interspecific)	forma ibernalis	Kleinwanzlebener Saatzucht AG vorm. Rabethge und Giesecke, Pa.	Germany	39.1	0.9142	70.5	1.4726	98.1

Table 2. Fatty acid components of rape seed oil extracts and rape seed - containing commercial oils.

No.	Varieties	$C_{16:0}$	$C_{16:1}$	$C_{18:0}$	$C_{18:1}$	$C_{18:2}$	$C_{18:3}$	$C_{20:0}$	$C_{20:1}$	$C_{21:0}$	$C_{22:0}$	$C_{22:1}$
1	Petranova	3.87	0.17	1.07	11.87	12.41	7.16	0.47	9.79	0.69	0.41	50.08
2	Torrazzo	4.43	0.19	1.03	13.62	12.05	7.17	0.36	10.10	0.51	0.81	48.50
3	Matador	2.78	0.21	0.75	11.18	10.49	5.73	0.64	7.63	0.38	0.43	58.82
4	Leonessa	5.72	0.16	1.30	15.42	10.53	3.29	0.24	8.86	tr.	0.21	53.78
5	Olimpiade original	3.87	0.24	1.04	11.97	11.50	6.97	0.52	9.90	0.47	0.74	51.78
6	Olimpiade reproduced	2.96	0.22	0.86	11.25	9.81	5.67	0.50	7.44	0.33	0.67	52.69
7	Eurora	4.57	0.25	1.79	68.25	15.25	4.99	0.47	2.09	0.04	0.46	1.24
8	Zephir	5.84	0.12	1.65	64.50	17.81	6.30	0.42	2.15	0.05	0.11	0.61
9	Perko	2.78	0.14	0.91	19.21	10.47	4.08	0.57	11.91	0.11	0.48	48.96
10	Coarse rape seed oil (commercial) (Chiari e Forti, Italy)	5.81	0.23	1.37	60.74	21.31	6.96	0.38	1.54	0.07	0.30	0.76
11	Refined rape seed oil (commercial) (Chiari e Forti, Italy)	6.65	0.16	1.49	54.48	27.33	6.74	0.34	1.39	tr.	0.25	0.79
12	Oil of different seeds (commercial) (Chiari e Forti, Italy)	8.40	0.08	2.68	37.86	43.41	5.00	0.30	0.86	0.14	0.26	0.55

Figure 2, which refers to the same oil + 3% standard gondoic acid ($C_{20:1}$ trans) and to the oil + 3% standard brassidic acid ($C_{22:1}$ trans), shows the Peaks 1-2'-3 that correspond to 20-carbon-atom acids, i.e. cis, trans and saturated acids. Peaks 4-5'-6 correspond to 22-carbon-atom acids, i.e. cis, trans and saturated acids. The retention times of peaks 2' and 5' taken at their tops, and respectively corresponding to eicosenoic and docosenoic acids, indicate the presence of positional isomers of the same acids, in spite of poor resolution. In Figure 3, which refers to the isomerised oil, Peaks 1-2'-3 and 4-5'-6 have the same sequence as that mentioned before. In Figures 2 and 3 it is interesting to note that the retention times for the trans isomers do not correspond to those of the isomers present in the crude oil. This excludes the presence of detectable amounts of trans isomers in the rapeseed oils investigated. Finally Figure 4 shows the fragmentogram of each pair of fragments a, b. After identification of the double bond position of isomeric $C_{20:1}$ and $C_{22:1}$ acids we could determine the percentage amount of each isomer present in the oils which contain high percentages of these acids. These values are around 9% for $C_{20:1} \Delta 13$ (the other 91% consisting

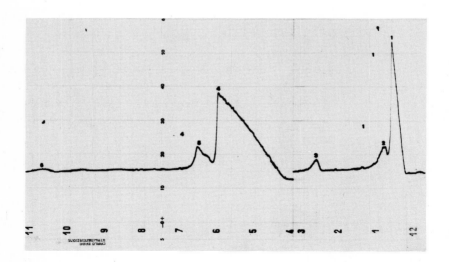

Fig.1. GLC analysis (capillary column) of crude rapeseed oil extracted by us.

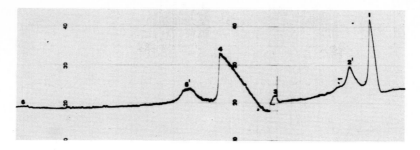

Fig.2. GLC analysis (capillary column) of crude rapeseed oil extracted by us added with standards.

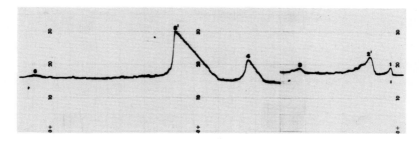

Fig.3. GLC analysis (capillary column) of crude rapeseed oil extracted by us after isomerisation.

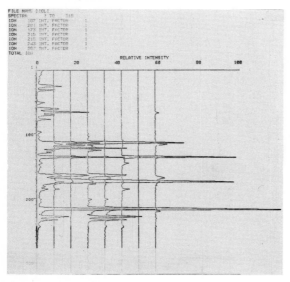

Fig.4. Mass fragmentogram of pair of fragments <u>a</u>, <u>b</u> of crude rapeseed oil extracted by us.

of gadoleic acid $C_{20:1} \Delta 11$) and 6% for $C_{22:1} \Delta 11$ (the other 94% consisting of erucic acid $C_{22:1} \Delta 13$). In a forth-coming paper we shall extend our research of positional or geometric isomers of $C_{20:1}$ and $C_{22:1}$ acids also in order to correlate the analytical results with the age/or type of seeds. Such a study will be concerned also with mustard, wallflower, nasturtium seeds. Even margarine obtained by hydrogenation of rapeseed oils will be examined.

REFERENCES

1. M. Cocchi, P.L. Biagi, M. Cipolla, G. Tavernese, S. Lorusso and E. Turchetto, Boll. Soc. It. Biol. Sper. $\underline{50}$ (1973), 294
2. E. Turchetto, S. Lorusso, V. Barzanti and W. Ciusa, Boll. Soc. Biol. Sper. $\underline{52}$ (1976), 1464
3. S. Lorusso, Atti dell'VIII Convegno sulla Qualità, Perugia, ottobre 1976
4. W. Ciusa and S. Lorusso, Oli Grassi Derivati $\underline{13}$ (1977), 39
5. B. Kokeny, Sementi Elette $\underline{5-6}$ (1976), 65
6. R. Izzo, C. Paradossi and G. Vicentini, Riv. It. Scienza Alim. $\underline{1-2}$ (1976), 9
7. G. Lercker, P. Capella, L.S. Conte and P. Pasini, Riv. It. Sost. Grasse, $\underline{55}$ (1978), 130
8. R.G. Ackman, J. Am. Oil Chem. Soc. $\underline{43}$ (1966), 483
9. S. Passi, M. Nazzaro, L. Boniforti and M. Merli, Giorn. e Min. Derm. $\underline{112}$ (1977), 463

Analysis of Organophosphoric Pesticide Residues in Food by GC/MS Using Positive and Negative Chemical Ionisation

H.-J. STAN and G. KELLNER

Institute of Food Chemistry, Technical University
D-1000 Berlin 12, Müller-Breslau-Straße 10 (FRG)

SUMMARY

The fragmentation patterns of selected organophosphorus pesticides for electron impact ionization and for negative and positive ions with chemical ionization are compared. The fragmentation pathways of negative ions generated by electron resonance capture are discussed in more detail. Information obtained by positive and by negative ion recording is complementary allowing the detection and identification of organophosphorus pesticide in food samples by GC/MS in a single run using selected ion monitoring of both the positive and the negative ions.

INTRODUCTION

The analysis of pesticide residues in food samples is performed nowadays commonly by gas chromatography using selective detectors. Organophosphorus pesticides (OP's) are detected either with the thermionic or the photometric detector. Due to the large number of OP's in use the introduction of glass capillary and fused silica capillary columns is to be considered as a major progress for the determination of these substances by means of retention data. The unequivocal identification of any compound at trace level, however, can be attained only by the direct combination of mass spectrometry and gaschromatography (GC/MS)

The application of GC/MS on OP's with glass capillary columns and electron impact as well as chemical ionization with positive ion record was demonstrated (1,2). Detailed discussions of the fragmentation pattern of OP's were also given (3,4,5).

We wish to present preliminary data which show that the simultaneous record of positive and negative ions produced by chemical ionization with methane is to be considered as a very valuable tool for the identification and determination of OP residues in food at the ppb-level.

IONIZATION OF SAMPLE MOLECULES

Electron impact ionization in the high vacuum of the ion source produces in a primary process positive molecule ions of high energy with an odd electron number. This primary radical ions frequently undergo subsequent fragmentation into a multitude of smaller ions. For the trace analysis in biological samples, however, it is essential to generate ions in the high mass region with high levels of intensity in order to avoid interferences with the biological matrix. This can be achieved by chemical ionization with reactant gases generating Brønstedt acids which are able to transfer protons in a gas phase reaction to the sample molecule without a simultaneous high energy exchange. The resulting protonated cation has an even electron number and a low energy content and therefore frequently exhibits high stability.

The reactions occuring during the chemical ionization process are outlined in figure 1 for the reactant gas methane. The electron impact ionization (eq. 1) yields a variety of primary positive ions which undergo gas phase reactions with the reactant gas generating secondary ions. The formation of the three thermodynamic most favoured reactant gas ions from methane is demonstrated in equations 2 - 5. These secondary ions are the effective reaction partners for the formation of the positive sample ions (eq. 6 - 9). The electron impact ionization (eq. 1) also leads to the generation of a population of low energy electrons (in eq. 1 marked with an asterix). Formation of negative ions by interaction of these electrons and samples can occur by three different mechanisms (eq. 10 - 12), each of which is dependent on the electron energy. An ion-molecule reaction applying a Brønstedt base can be used to generate anions by proton abstraction (eq. 13). With high energy electrons negative ions usually are produced with low efficiency and the fragments occur at the low mass end of the spectrum. In contrast to this situation many sample molecules capture near thermal energy electrons and are converted to either stable molecular anions, M^-, or high molecular weight fragment ions.

$$CH_4 + e^- \longrightarrow [CH_4]^{+\cdot}, [CH_3]^+, [CH_2]^{+\cdot}, [CH]^+, [C]^{+\cdot}, [H_2]^{+\cdot}, [H]^+, e^*, e^- \quad (1)$$

$$[CH_4]^{+\cdot} + CH_4 \longrightarrow [CH_5]^+ + [CH_3]\cdot \quad\quad 17 \quad\quad (2)$$

$$[CH_3]^+ + CH_4 \longrightarrow [C_2H_5]^+ + H_2 \quad\quad 29 \quad\quad (3)$$

$$[CH_2]^{+\cdot} + CH_4 \longrightarrow [C_2H_3]^+ + H_2 + H\cdot \quad\quad (4)$$

$$[C_2H_3]^+ + CH_4 \longrightarrow [C_3H_5]^+ + H_2 \quad\quad 41 \quad\quad (5)$$

$$[CH_5]^+ + M \longrightarrow [MH]^+ + CH_4 \quad M+1 \quad\quad (6)$$

$$[C_2H_5]^+ + M \longrightarrow [MH]^+ + C_2H_4 \quad M+1 \quad\text{BRØNSTEDT ACID}\quad (7)$$

$$[C_2H_5]^+ + M \longrightarrow [M(C_2H_5)]^+ \quad M+29 \quad\quad (8)$$

$$[C_3H_5]^+ + M \longrightarrow [M(C_3H_5)]^+ \quad M+41 \quad\text{LEWIS ACID}\quad (9)$$

$$AB + e^- \xrightarrow{\sim 0\text{ eV}} AB^{-\cdot} \quad\quad \text{RESONANCE CAPTURE} \quad\quad (10)$$

$$AB + e^- \xrightarrow{0\text{ -15 eV}} A\cdot + B^- \quad\quad \text{DISSOCIATIVE RESONANCE CAPTURE} \quad (11)$$

$$AB + e^- \xrightarrow{>10\text{ eV}} A^+ + B^- \quad\quad \text{ION PAIR PRODUCTION} \quad\quad (12)$$

$$B_1H + B_2^- \longrightarrow B_1^- + B_2H \quad\quad \text{ION - MOLECULE REACTION} \quad\quad (13)$$

Fig. 1: Chemical ionization with methane
Origin of primary ions

A more detailled discussion of the formation and fragmentation of negative ions was given by Hunt (6,7).

MASS SPECTRA OF ORGANOPHOSPHORUS PESTICIDES

For the application in the residue analysis it is a great advantage that the protonated positive ions and the anions formed by resonance capture of electrons arise both under the same experimental conditions of chemical ionization with reactant gases as methane. Using GC/MS both positive and negative ions can be detected nearly simultaneously with

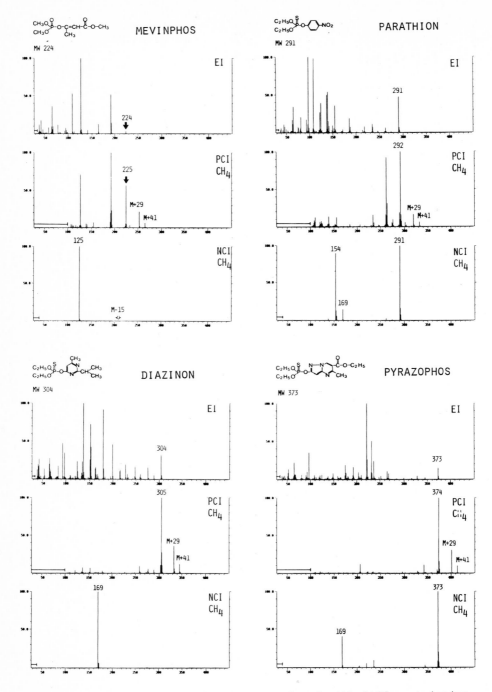

Fig. 2: Comparison of mass spectra produced with different ionization processes

the technique of pulsed positive and negative ion record.
In figure 2 a comparison of mass spectra obtained by electron impact (EI) and chemical ionization (CI) with positive and negative ion record is given. Four samples were selected in order to represent the different structural types of OP's. The EI mass spectra of all four compounds exhibit a higher degree of fragmentation in comparison to the positive and negative CI mass spectra. The well known parathion containing an aromatic moiety with an electron-withdrawing nitro group in para position is able to stabilize both positive and negative molecule radical ions by charge delocalization. Therefore, intense molecular ions can be observed in the spectra together with considerable fragmentation in the case of EI. Pyrazophos and diazinon contain very different heterocyclic moieties in their molecules. Pyrazophos demonstrates a remarkable stable molecular anion whereas diazinon shows fragmentation into one single anion. Fragmentation under EI conditions leads in both samples to less characteristic fragment patterns. Mevinphos is a typical example for OP's with aliphatic moieties. They undergo extensive fragmentations with EI as well as with negative CI conditions, molecular ions are scarcely to be observed because of the lack of any stabilizing charge delocalization in the molecule. It is evident from figure 2 that CI with methane produces in all samples protonated abundant quasimolecular cations which are frequently the base peak in the mass spectra (4). In contrast stable radical molecular anions are formed only with a minority of the more than 50 investigated OP's. Most of them undergo dissoziation leading to stable anions of the various phosphate groups because these are obviously in most cases stronger gas phase acids than the moieties which are eliminated as neutral radicals. The general fragmentation scheme of OP's is summarized in figure 3. In the anion mass spectra the majority of the OP's exhibit as base peak the fragments indicative for the type of phosphorus group.

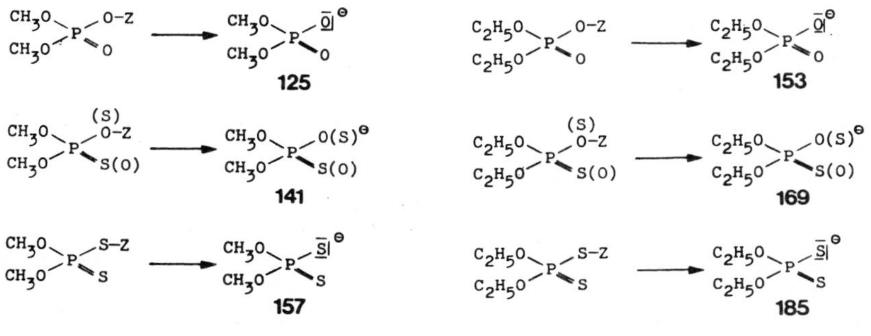

Fig. 3: Fragmentation pathway of negative ions of organophosphorus pesticides

The exceptions of this rule are the substances of the parathion group. In the negative CI mass spectrum of parathion (figure 2) the nitrothiophenoxide ion is the second most abundant ion. It results from a rearrangement with a transfer of the aromatic moiety from the oxygen to the sulfur. From gas phase acidity measurements it is known that thiophenoxide exhibits the greater electron affinity compared to phenoxide and therefore forms the more stable anion (8).
The fragmentation scheme for parathion methyl is outlined in figure 4.

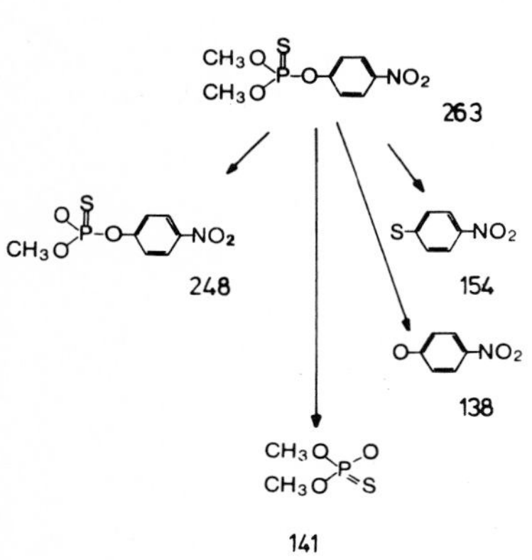

Fig. 4: Negative ion fragmentation of parathion-methyl after electron resonance capture in the methane reactant gas plasma

APPLICATION TO RESIDUE ANALYSIS

From the view point of residue analysis it is important that all OP's including the parathion group yield negative CI mass spectra with the corresponding phosphate group ion as listed in figure 3.
The information about the phosphate type from negative CI is complementary to the information about the molecular weight obtained from the positive CI mass spectrum. Together with the retention data on capillary columns there are three independent parameters allowing an unequivocal identification of the OP's at trace concentrations in the food matrix.
Another application of negative CI for trace analysis of OP's in food is the screening with multiple ion monitoring. In this method the six characteristic ions described in figure 3 are selected. The background from the biological matrix is very low so that ppb concentrations of OP's in food samples can be detected after a simple extraction without any clean up.

In figure 5 the detection of 10 pg (= 2 ppb) of parathion and chlorpyrifos in an ethylacetate extract from leek is demonstrated. With the thermionic as well as with the photometric detector the chromatogram was dominated by peaks from the biological matrix making any identification impossible. Screening in the NCI mode showed m/z 169 pointing to a diethylthiono or -thiolphosphate, which because of the gaschromatographic retention time could be parathion or chlorpyrifos only. Additional scanning of m/z 291 (M^-, parathion) and m/z 313 ($[M-HCl]^-$, chlorpyrifos) verified the presence of both OP's.

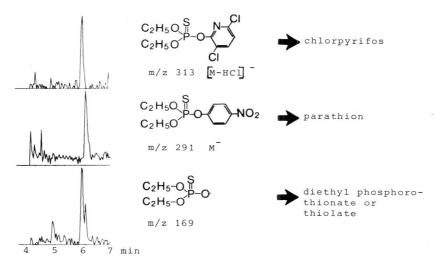

Fig. 5: Detection of parathion and chlorpyrifos in leek, 2 ppb, selected ion monitoring, NCI

REFERENCES

1) H.-J. Stan, Chromatographia, 10 (1977), 233
2) H.-J. Stan, Z. Lebensm. Unters.-Forsch. 164 (1977), 153
3) J.N. Damico, J. Assoc. Off. Anal. Chem. 49 (1966), 1027
4) H.-J. Stan, B. Abraham, J. Jung, M. Kellert and K. Steinland, Fresenius Z. Anal. Chem. 287 (1977), 271
5) H.-J. Stan, Fresenius Z. Anal. Chem. 287 (1977), 104
6) D.F. Hunt, G.C. Stafford,Jr., F.W. Crow and J.W. Russell, Anal. Chem. 48 (1976), 2098
7) D.F. Hunt and S.K. Sethi in ACS Symposium Series No. 70, "High Performance Mass Spectrometry: Chemical Applications", ed. by M.L.Gross p. 150, Amer. Chem. Soc. (1978)
8) B.K. Janousek and J.I. Brauman in "Gas Phase Ion Chemistry", ed. by M.T. Bowers, p. 53, New York (1979)

Use of Positive and Negative Chemical Ionisation Techniques to Determine and Characterize Free Amino Acids by Mass Spectrometry

D. FRAISSE[1], F. MAQUIN[2], J.C. TABET[3] and H. CHAVERON[2]

[1] C.N.R.S.- S.C.A. BP 22   69390 VERNAISON - FRANCE
[2] Laboratoire d'Analyses Alimentaires. Université de Compiègne
    BP 233 - 60206 COMPIEGNE - FRANCE
[3] Laboratoire de Synthèse Organique - Ecole Polytechnique
    91120 PALAISEAU - FRANCE

SUMMARY

Positive and negative chemical ionization techniques have been utilized to determine amino acids without derivatization, i.e., free amino acids by mass spectrometry. Electron impact ionization mode produces mass spectra showing large fragmentation so that molecular ions are small and most of the time absent. The positive and negative chemical ionization mass spectra give very little or, no fragmentation and can be used to determine free amino acids in raw materials by direct introduction. The CID-MIKE mass spectra of the quasi-molecular ions $[MH^+, (M-H)^-]$ have been carried out to characterize the amino acids.

INTRODUCTION

Proteins, a word meaning "of first importance" are as commun as hair, wool, skin or finger nails (all of which are mostly proteins). Proteins are responsible for much of the structure and most of the functions of a living cell. As proteins are actually polymers of amino acids, we can say amino acids are the building blocks of life.

Some amino acids are biosynthesized and others are to be considered as nutrients for life.

Although determination of amino acids has been investigated by many workers, we have found that mass spectrometry has not often been used for this purpose. One of the reasons may be due to the fact that the electron impact ionization mode was used.

On table 1 the different methods of derivatization for the determination of amino acids by gas chromatography are summarized.

TABLE 1 : Different techniques of derivatization.

DERIVATIZATION TECHNIQUE	REFERENCES	REMARKS
1 - SILYLATION (Silylation of the aminol and carboxylic acid groups)	GERHRBE et al. J. of Chromatogr. 45, 24, 1969	Stable derivatives not obtained for all amino acids depending on reagents and experimental conditions
2 - SILYLATION AND ESTERIFICATION (Silylation of the amine group, esterification of the carboxylic acid group).	H. IWASE et al. Anal. Biochem, 78, 340, 1977 - Chel. Pharma Bull, 27, 6, 1307, 1979	Used for fragmentation studie and quantification of amino acids by GC-MS M.I.D.
3 - ESTERIFICATION	RACHELE et al. Org. Chem, 28, 2898, 1963 PEGON et al. J. Of chrom, 151, 163, 1978	Technique now utilized only in combination with acylation.
4 - ACYLATION (Acylation + esterification).	FRANCK et al. J. Chromatogr. 67, 187, 1978 FRANCK et al. J. Chromatogr. Sci. 15, 174 1977 POOLE et al. J. Chromatogr. 150, 439, 1978 H. IWASE et al. chem Pharm Bull 25, 285, 1977 JJ. RAFTER et al. Acta Biol med. germ. 38, 321 1979	N-perfluoro acyl alkyl esters drivatues have been proposed. The interest of this technique lies in the absence of degradation of the compounds (i.e. specificity of the reaction). Quantitative analysis by GC-MS-MID has been performed for 46 N-TFA-L-propyl amino acid n-butyl esters.

Several types of derivatives are currently in use, the choice depending very much on the subsequent separation technique, that is wether gas chromatography or liquid chromatography is to be used.

Before derivatization the amino acids under investigation must be extracted from raw materials (bloods, foods ...) these clean-up techniques are time-consuming and some times difficult. One experimental procedure discribed by Abramson et Coll (1) is reported on table 2.

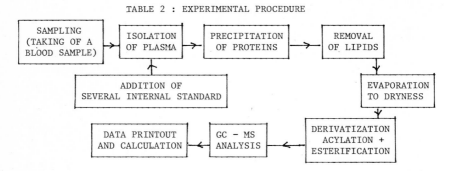

TABLE 2 : EXPERIMENTAL PROCEDURE

Some derivatization techniques require high quality GC resolution to separate a large number of amino acids derivatives formed, and despite the use of chirasil-val capillary columns the separation of enantionmeric compounds remains quite difficult. Artefacts due to the derivatization may also occur.

During the past decade new mass spectrometry techniques have been developed. First of all positive chemical ionization, introduced by MUNSON and FIEL (2) ; is now widely utilized to get molecular weight of the compounds. More recently negative chemical ionization mass spectrometry has proved to be a useful analytical tool.

Others technique to study metastable transitions have been developed (B/E, B^2/E linked scanning techniques give interesting information on the daughter or precursor ions). One the newest techniques is collision induced dissociation mass analyzed ion kinetic energy (CID-MIKE) spectrometry. This method involves using a reverse geometry mass spectrometer (the magnet preceding the electric analyser (3).

All amino acids investigated were introduced into the source by using the direct introduction solid probe.

POSITIVE AND NEGATIVE CHEMICAL IONIZATION MASS SPECTRA

On table 3 we have compiled all the amino acids we studied. Mass spectra we acquired with a VG Micromass 70-70. In EI mode mass spectra show very weak molecular peaks, thus extensive fragmentation is taking place. As Vetter (4) reported $(M+1)^+$ ions are also observed even at low pressure (10^{-5} - 10^{-6} Torr) for some amino acids (ex : aspartic acid). Thus EI ionization mode is quite unsatisfactory for identification of amino acids in mixtures. In PCI ionization mode with NH_3 as reactant gas all investigated amino acids yield very intense $(M+H)^+$ ions. This appears to be very interesting from an analytical point of view, because the amino acids can be easily determined by their protonated molecular ions. Apart from one case, i.e. Glutamic acid for which the base peak is $(MH-18)^+$, the protonated ion MH^+ is base peak for the other amino acids investigated and very little fragmentation is induced.

The negative chemical ionization mass spectra were obtained using the hydroxyl ion OH^- as reactant ion. OH^- is formed by electron impact bombardment of a mixture of N_2O and CH_4 at a total pressure of 0,5-1 Torr. Thus an intense OH^- ion beam which reacts with the amino acids to form $(M-H)^-$ ions, is obtained.

As can be seen on table 3 all amino acids investigated give $(M-H)^-$ ions which are base peaks and very little fragmentation occurs.

TABLE 3 : AMINO ACIDS

	ELECTRON IMPACT	MOLECULAR PEAK INTENSITY	POSITIVE CI	BASE PEAK	NEGATIVE CI	BASE PEAK
A-MONOAMINE, MONOCARBOXYLIC						
Glycine (75) Gly			+	$MH^+= 76$		
Alanine (89) Ala	+	0	+	$MH^+= 90$	+	$(M-H)^-= 88$
Valine (117) Val	+	0	+	$MH^+=118$	+	$(M-H)^-=166$
Leucine (131) Leu	+	1 %	+	$MH^+=132$	−	$(M-H)^-=130$
Isoleucine (131) Ile	+	$<0,5$ %	+	$MH^+=132$	+	$(M-H)^-=130$
B-HYDROXY MONOAMINO MONOCARBOXYLIC						
Serine (105) Ser			+	$MH^+=106$		
Threonine (119) Thr	+	1 %	+	$MH^+=120$	+	$(M-H)^-=118$
C-MONOAMINO, DICARBOXYLIC AND AMIDOCARBOXYLIC						
Glutamic acid (147) Glu	+	<1 %	+	$(MH-18)^+=130$	+	$(M-H)^-=146$
Aspartic acid (133) Asp	+	MH^+ 1 %	+	$MH^+=134$	+	$(M-H)^-=133$
Glutamine (146) Gln						
Asparagine (132) Asn	+		+	$MH^+=133$		
D-DIAMINO-MONOCARBOXYLIC						
Lysine (146) Lys			+	$MH^+=147$		
Arginine (174) Arg						
E-SULFUR CONTAINING						
Cysteine (121) Cys			+	$MH^+=122$		
Methionine (149) Met	+	24 %	+	$MH^+=150$	+	$(M-H)^-=148$
F-CYCLIC (INCLUCLING AROMATIC)						
Phenylalaline (165) Phe	+	≤ 1 %	+	$MH^+=166$	+	$(M-H)^-=164$
Tyrosine (181) Tyr	+	≤ 3 %	+	$MH^+=182$	+	$(M-H)^-=180$
Tryptophane (204) Trp	+	≤ 4 %	+	$MH^+=132$	+	$(M-H)^-=203$
Histidine (155) His	+	1 %	+	$MH^+=156$	+	$(M-H)^-=154$
Proline (115) Pro	+	1 %	+	$MH^+=116$	+	$(M-H)^-=114$

CID-MIKE MASS SPECTRA

Although these PCI and NCI are very interesting from an analytical point of view as they give molecular weight information and thus can suggest the presence of some amino acids in mixture, they do not prove their presence with certainty. Thus before determining quantitatively the amino acids the qualitative analysis has to be completed by performing the CID-MIKE mass spectra of the protonated molecular ions MH^+ formed in positive chemical ionization or $(M-H)^-$ ions formed in negative chemical ionization.

TABLE 4 : CID-MIKE spectra of amino acids

Amino acids	$(MH-NH_3)^+$	$(MH-H_2O)^+$	$(MH-NH_3-H_2O)^+$	$(MH-2H_2O)^+$	$(MH-HCOOH)^+$ or $(MH-H_2O-CO_2)^+$	$(MH-HCOOH-NH_3)^+$	$(MH-H_2O-HCOOH)^+$
Glycine	−	−	−	+	+++	−	−
Alanine	+	−	+	−	+++	++	−
Valine	−	+	−	+	+++	++	−
Leucine	−	+	−	+	+++	−	−
Isoleucine	−	−	−	−	+++	−	−
Serine	−	++	−	−	+++	−	+
Threonine	−	++	−	−	+++	−	++
Glutamic acid	−	−	−	−	+++	−	−
Aspartic acid	++	++	+	−	+++	−	−
Glutamine	−	−	−	−	−	−	−
Asparagine	−	−	−	−	−	−	−
Lysine	+++	−	+	+	+	++	−
Arginine	−	−	−	−	−	−	−
Cysteine	+++	−	+	−	++	+	−
Methionine	−	−	−	−	−	−	−
Phenylalaline	−	++	−	−	+++	−	−
Tyrosine	+++	−	−	−	++	−	−
Tryptophane	+++	−	−	−	++	−	−
Histidine	+	+	−	−	++	−	−
Proline	−	+	−	−	+++	−	−

+++ very intense
++ intense
+ weak
− absent

CID-MIKE mass spectra were acquired using a VG-MICROMASS ZAB-2F with he as collision gas. The pressure in the collision cell located between the two sectors was adjusted so that the parent ion beam (MH)$^+$ was reduced to one-third of its value with no gas in the cell.

Table 4 shows results obtained using the CID-MIKE spectrometry technique to get ions formed by unimolecular decomposition of the protonated molecular ions MH$^+$ of the amino acids in NH_3 PCI.

CONCLUSION

Thus in the analysis of a mixture of different materials the NH_3 PCI or OH$^-$ NCI mass spectra provide useful information on the presence of amino acids CID-MIKE spectra confirm their presence.
The characterization of free amino acids by NCI and PCI ionization modes in raw materials is under investigation and we expect to use this technique for quantitative analysis as well.

REFERENCES

1 - M.S. SCHULMAN, F.P. ABRAMSON, Biomed. mass spect. 2, 9, 1975
2 - M.B. MUNSON, F.H. FIELD, J. Am. Chem. Soc. 88, 8621 (1966)
3 - J.H. BEYNON, R.G. COOKS, Res. Dev. 22, 26 (1971)
4 - W. VETTER, Biochem, Applications of mass spectrometry, John Wiley
    NEW YORK, 14, 439, (1980).

Qualitative and Quantitative Aspects of the Analysis of Trace Contaminants in Foods by Selected Ion Monitoring

J.R. STARTIN and J. GILBERT

Ministry of Agriculture, Fisheries and Food, Food Science Division, Haldin House, Queen Street, Norwich NR2 4SX.

SUMMARY

Some examples of the application of selected ion monitoring to the analysis of trace contaminants in foods by gas chromatography - mass spectrometry are given. The advantages of the technique in terms of sensitivity and selectivity in the headspace analysis of monomers in plastics packaged foods, and in the analysis of deoxynivalenol in cereals are illustrated. Conversely the difficulties encountered with interfering peaks and the means of overcoming these limitations in the analysis of nitrosodimethylamine are discussed.

INTRODUCTION

In recent years increased attention has been given to the analysis of foods for a variety of organic contaminants at concentrations below 1 part per million. The object of this paper is to illustrate the role that the mass spectrometer can play in such analyses when used in the selected ion monitoring mode (SIM) using examples from our recent experience.

In the SIM technique the mass spectrometer is adjusted to focus ions of a selected mass/charge ratio continuously onto the detector. By an appropriate choice of ion SIM can be made responsive to a vast range of compounds, in marked contrast to most other selective GC detectors. The versatile nature of the selectivity, coupled with high sensitivity makes SIM a very powerful tool for trace analysis.

DISCUSSION

ANALYSIS OF MONOMERS IN FOODS

The versatile selectivity of SIM is illustrated in the analysis of foods for styrene (1). Polystyrene packaging is now extensively used for food retailing and may contain substantial styrene residues which can migrate into the contained foods. A particularly rapid and convenient technique for the analysis of volatile compounds such as styrene is headspace GC which requires little sample preparation, gives good reproducibility, and for which automated equipment is available.

Figure 1 compares headspace chromatograms obtained with FID and SIM (m/z 104) detection, and demonstrates the need for selectivity; apart from SIM suitable detectors are not readily available. Table 1 compares detection limits obtained for various foods with FID and SIM detection

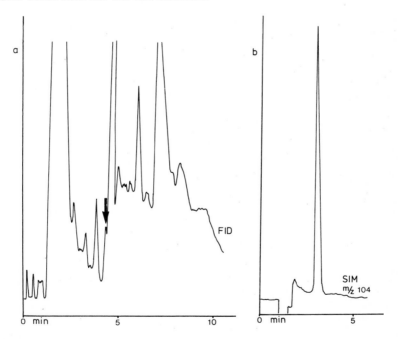

Figure 1. Headspace chromatograms from fruit flavoured yoghurt. 2 m x 2 mm ID column packed with 5% Carbowax 20M on Diatomite C-AW (a) 30 ppb styrene with FID detection. (b) 20 ppb styrene with low resolution SIM.

	Dectection Limit ppb	
	FID	SIM
Double Cream	60	5
Yoghurt	20	1
Cottage Cheese	35	1
Soft Margarine	350	15
Chocolate Spread	200	1
Honey	15	1

Table 1. Detection limits for styrene in foods.

The figures for yoghurt refer to the unflavoured product; the presence of fruit flavours caused additional interference with FID but not with SIM.

In the analysis of styrene using SIM at a resolution just sufficient to separate unit masses at m/z 104 the nominal mass was sufficiently characteristic of styrene to provide a high degree of selectivity. This was not the case in similar experiments for the analysis of acrylonitrile (AN) in soft margarine packaged in ABS containers when the molecular ion of AN was monitored at m/z 53. In this case operation of the mass spectrometer at 3000 resolution (10% valley) provided an improvement in selectivity and despite the consequent decrease in absolute sensitivity gave a detection limit of 1 ppb.

ANALYSIS OF DEOXYNIVALENOL IN CEREALS

Deoxynivalenol (I) is a mycotoxin generated by <u>Fusarium</u> species which often infect cereal crops (3). The presence of I in grain has been implicated in inducing

vomiting, emesis and feed refusal in swine (3). Analysis by capillary GC with FID detection has been reported (4) with a detection limit of 5 ng per injection (50 ppb), but in order to obtain lower detection limits from a method which was uniformly successful with a variety of cereals we have found selective detection to be necessary. Chromatograms obtained by SIM of trimethylsilylated extracts are shown in figure 2. The selectivity and sensitivity are such that an increase in amplification provided a detection limit of 30 pg (1 ppb).

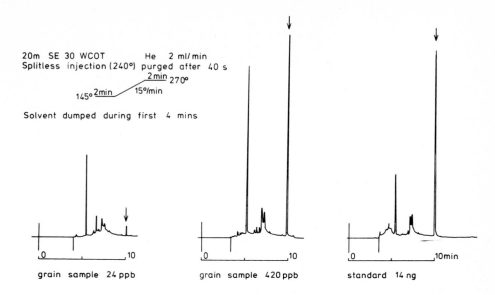

Figure 2. Capillary column low resolution SIM chromatograms of deoxynivalenol TMS (m/z 512).

ANALYSIS OF NITROSODIMETHYLAMINE IN MALTED BARLEY PRODUCTS

The analysis of volatile nitrosamines in food has been extensively studied and the application of MS is reviewed in several publications (5, 6). Webb et al (7) have indicated that the MS should achieve at least 7000 resolution. In the capillary column analysis of NDMA in products derived from malted barley by SIM under EI conditions a coeluting component gave rise to abundant ions requiring a resolution of 8500 to preserve selectivity when monitoring the molecular ion at 74.048. This component also produced a massive temporary decrease in sensitivity (presumably due to source defocussing) and relatively large amounts of NDMA, when coinjected with a sample, were not detected. We have observed that such suppression effects are largely absent when chemical ionisation (CI) is used, and figure 3 demonstrates the detection of subnanogram quantities of NDMA under high resolution CI conditions.

EXPERIMENTAL

Throughout this work a VG 70-70 H double focusing mass spectrometer interfaced to a Carlo Erba 4160 GC was used. In EI measurements the MS source was operated at 200°C with 70eV electron energy and a trap current of 200 μA. For NH_3 CI the source was operated at 180° with 70eV electron energy and 1 mA emission. The reagent gas pressure was adjusted to maximise the yield of $(M + 1)^+$ ions from pyridine.

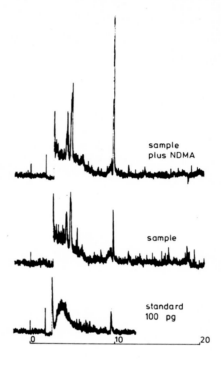

Figure 3. Capillary column CI SIM chromatogram of NDMA (m/z 74.048; 8500 resolution). 25 m x 0.32 mm UCON 5100 programmed from 37° at 5°/min.

For headspace gas chromatography a Perkin-Elmer F42 automated headspace analyser was interfaced and operated as described elsewhere (1, 2).

Deoxynivalenol was extracted from grain samples by the method of Scott (8) and silylated with TBT reagent (Pierce Chemical Co., Rockford, Illinois, USA). Chromatography used spitless injection on a 20 m x 0.3 mm SE 30 WCOT column connected to the MS via a single stage jet separator after addition of make-up gas.

Nitrosodimethylamine was extracted by a distillation method and chromatographed on a 25 m x 0.32 mm UCON 5100 WCOT column. On column injections were made using a cooled septumless device and the column was directly coupled to the mass spectrometer via a glass restrictor. The temperature was programmed from 37°C at 5°/min.

REFERENCES

1 J. Gilbert and J.R. Startin, J. Chromatogr., 205 (1981), 434.
2 J. Gilbert and J.R. Startin, Food Chemistry, (in press).
3 Yoshio Ueno, in Mycotoxins in Human and Animal Health, Ed. J.V. Rodricks et al, Pathotox Publishers Inc., 1977, p197.
4 Cs. Szathmary, J. Galacz, L. Vida and G. Alexander, J. Chromatogr., 191 (1980), 327.
5 T.A. Gough, Analyst, 103 (1978), 785.
6 J. Gilbert and R. Self, Chem. Revs., (in press).
7 K.S. Webb, T.A. Gough, A. Carrick, D. Hazelby, Analyt. Chem., 51 (1979), 989.
8 P.M. Scott, J. Assoc. Off. Anal. Chem., (in press).

2 Bioassay and Enzymatic Methods

Bioassays in Food Analysis

J. LÜTHY

Institut für Toxikologie der Eidg. Technischen Hochschule und der Universität Zürich, CH-8603 Schwerzenbach, Switzerland

SUMMARY

One of the most useful applications of bioassays is the possible detection of yet unknown toxic compounds in foodstuffs. Recent successful examples are the isolation and identification of mutagens in fried meat and the recognition of mycotoxins. Advantages and disadvantages of biological methods in comparison to chemical analytical methods are discussed for aflatoxins and for paralytic shellfish poison.

INTRODUCTION

Historically, bioassays have been of great importance in the history of nutrition research. Practically all known essential factors in food were detected initially by means of biological methods before the chemical characterization and the elucidation of their structures was possible. As an example, one may remember the exciting and admirable work which led to the detection of the most important vitamins between about 1910 and 1940 (1). However, what is the meaning of bioassays in modern food analysis, in nutrition research and in food toxicology?

A bioassay may be most generally defined as "the determination of the potency of a physical, chemical or biological agent by means of a biological indicator" (2). In the following discussion I will concentrate on chemical agents only and their interaction with living systems such as whole animals, plants, microorganisms or cell cultures. Two further biochemical methods are closely related to bioassays: The enzymatic methods use the interaction of chemical compounds with enzymes and the immunological methods employ antibodies for analytical purposes. These two methods will be discussed in two later plenary lectures and I propose to say nothing further about them here.

Generally three applications of bioassays in food analysis are important:
1. The detection of (toxic) contaminants with known chemical structure, e.g. mycotoxins (3, 4, 5), musseltoxins (6), pesticides (7, 8, 9, 10), antibiotica (11, 12), hormones (13) or bacterial toxins (14, 15).
2. The application of bioassays in nutrition research and for the determination of food quality, e.g. the analysis of vitamins (16, 17, 18), trace elements (19, 20), protein quality (21, 22) and the detection of antinutritive factors (23).
3. The detection of yet unknown toxic compounds in foodstuffs, e.g. "new" mycotoxins or toxic constituents occurring naturally or which are formed during processing of food.

In this latter field the application of bioassays is often the most successful method, in contrast to application 1 where chemical analytical methods are available also and so the analyst has to evaluate carefully which method is the most suitable. In the following some examples of bioassays will be discussed to demonstrate the advantages or disadvantages of the biological methods in comparison to the generally better known chemical methods.

AFLATOXINS

Aflatoxins are, from the toxicological viewpoint, the most important mycotoxins. In the last ten years highly sensitive chemical methods using thinlayer chromatography or HPLC in combination with fluorimetric detection were developed allowing the detection of even less than 1 ng of aflatoxin B1 (24, 25). For the confirmation of aflatoxin-positive samples several biological test systems (26, 27, 28) were proposed, but only one of them has still some importance, the chicken embryo bioassay which was introduced as early as 1964 by Verrett et al. (29, 30). Aliquots of the chromatographically purified aflatoxin fraction have to be injected into the air cell of fertile eggs. 25 - 100 ng of aflatoxin B1 are lethal to the egg embryo. The U.S. FDA tested more than 500 aflatoxin containing food samples with the chicken embryo bioassay; the agreement with chemical methods was 100 %. However, the disadvantages of the chicken embryo bioassay cannot be overlooked; the procedure is time consuming and not cheap. For statistical reasons the AOAC-protocol requires more than 200 eggs for testing one sample. For the same reason relatively large amounts of aflatoxin B1, more than 2 μg, are necessary to perform the test. The chemical confirmation methods such as derivatisation (31, 32) or especially mass spectrometry (33, 34) are much more sensitive and give the same security of identification with less expenditure.

PARALYTIC SHELLFISH POISON

The paralytic shellfish poison is a typical example where a bioassay was and still is highly useful for the detection of toxic mussel samples. The most prominent of these potent neurotoxins of natural origin is saxitoxin. The structure was elucidated by Schantz et al. (35) and Bordner et al. (36) as late as 1975 and this might be one of the reasons why the mouse bioassay (6) was for a long time the only possible method of detection of this group of toxic compounds. The principle of the mouse bioassay is simply a measurement of the death time after intraperitoneal toxin administration. Saxitoxin specifically blocks the sodium channels in the nerve membranes, and the mouse dies after a few minutes by respiratory failure. 0.2 μg of STX (40 μg/100 g mussel sample) are still detectable with a normally sensitive mouse strain (37). Most countries have adopted a value of 80 μg STX per 100 g of shellfish tissue as a basis of control activities. The assay has been tested collaboratively and a standard error of 20 % has been observed. In the lower concentration range, however, systematic errors may occur, because normal food constituents (e.g. sodium chloride) can influence the toxicity of STX (38).

Bates et al. (39) proposed a more sensitive fluorimetric method for the determination of STX with 0.4 μg/100 g as a limit of detection. An essential disadvantage of this method is that the structurally related and also toxic gonyautoxins are missed by this method. A complete analysis of toxic shellfish originating from Spain which were responsible for a massive epidemic of paralytic shellfish poisoning in Western Europe, 1976, possessed a STX / GTX ratio of about 1 : 1 (40). The application of the fluorimetric method would result in an underestimation of the toxicity in this case (41).

PESTICIDES

Between 1950 and 1970 several bioassays (10) were described in the literature and also applied in pesticide analysis. The test organisms most frequently used were Daphnia magna (7), Drosophila melanogaster (8) and Zebrafish larvae (9). These tests were mainly used in a qualitative manner or as sorting tests, giving indications of the chemical type of a pesticide. Today the physico-chemical methods are clearly dominating in routine pesticide analysis, especially for the organochlorine and organophosphorous pesticides, mainly because the technique of chromatographic separation, the detection methods and the possibilities for automatization are so highly sophisticated now that there is not much need for the relatively slow, unspecific, and generally less sensitive bioassays. An exception may be the highly sensitive cholinesterase-inhibition technique for the detection of organophosphorous pesticides (42, 43).

PROTEIN QUALITIY

The evaluation of the protein quality in foodstuffs is a basic and important task of nutrition research. Terms like "protein quality" (44) or "relative nutritive value" (45) are biological characterizations by definition. Analytical indices such as "chemical score" or "essential amino acid index" or "available lysine" may be helpful but any true assessment of the nutritive value in proteins of human interest may finally rely on biological evaluation; employing laboratory animals (46), insects (47), protozoa (48), bacteria or fungi (49, 50). The different bioassay procedures have been recently discussed in an excellent review by Pellett (22).

DETECTION OF YET UNKNOWN TOXIC COMPOUNDS IN FOODSTUFFS

At least the same attention is deserved by the third of the three mentioned applications of bioassays in food analysis i.e. the detection of yet unknwon toxic compounds in foodstuffs. Today the role of nutrition in carcinogenesis and mutagenesis is most often discussed. Weisburger and Williams (51) evaluated recently the causes of human cancer using mainly epidemiological data. These authors estimate that about 50 % of human cancer is food-related. It is not the object of this contribution to fully discuss these complex problems but this figure may give an idea how important the search for genotoxic compounds in foodstuffs really is. In fact, the greatest efforts have been made during the most recent years to obtain simple and reliable short-term bioassays for the detection of mutagenic and carcinogenic compounds (52, 53). Presently, the most popular procedures employ microorganisms, together with microsomes for activation, to measure the mutagenicity of a compound (54).

Since the discovery of the prominent toxic and carcinogenic properties of aflatoxins 20 years ago the search for other mycotoxins has been intensified. Besides the above mentioned tests for mutagenicity, screening tests most often used in mycotoxicology are performed with (55):

- lower organisms such as Zebrafish larvae or Artemia salina (56, 57)

- cultured mammalian cells (58)

- rabbit skin tests (59)

- chicken embryos (30, 60)

- ducklings (61)

- laboratory rodents (mice and rats) (62, 63)

Although the tests on lower organisms or on rabbit skin perform very well with certain classes, e.g. with Fusarium toxins, they do not respond to several other important compounds such as luteoskyrin and griseofulvine and only moderately to other mycotoxins such as citrinin, patulin and penicillic acid. Since the physiology and biochemistry of the lower organisms differ greatly from those of mammals, a lack of conformity is not surprising. A disadvantage of the Artemia salina test is its susceptibility to fatty acids (64). When crude extracts are examined this has to be taken into account. Skin tests on rabbits or other rodents, mammalian cell systems in culture, or chicken embryos are able to detect general cytotoxicity. However, effects due to metabolism, absorption, distribution and elimination are not predictable by these systems.

As an example of screening fungi cultures for mycotoxins the work of Bachmann et al. (65) and Blaser et al. (66) will be shortly discussed here. A representative selection of 25 strains of the ubiquitous mold Aspergillus glaucus were tested on their ability to produce mycotoxins. As bioassays three tests were used: (1) The already discussed chicken embryo test, (2) the acute toxicity test in mice and (3) the salmonella/microsomes test according to Ames (54). The organic solvent extracts of most strains were found to be toxic in chicken embryos and produced lethal local irritations after intraperitoneal injection into mice. However, no toxicity was found after oral administration to mice. A positive correlation between toxicity and mutagenicity was observed, but mutations were induced in only one of the five normally used Salmonella strains, in TA 1537. One of most toxic mold strains were then selected for further investigations and the i.p. toxicity in mice used as bioassay for fractionation. This finally led to the isolation and identification of four different anthraquinone derivatives which had similar toxic properties to the crude mold extracts.

One of the most exciting discoveries in food toxicology recently was the detection of mutagens in fried meat (67, 68, 69). Sugimura et al. (70) found that the crude extract of fried beef (and other food rich in protein as well) shows a very weak mutagenicity in Salmonella typhimurium TA 98. Using this test as bioassay and starting with more than 2 kg crude beef extract about 100 μg of a highly mutagenic compound were isolated. The elucidation of the chemical structure of this compound has not yet been fully achieved. The structure seems to be different from the earlier found mutagens isolated from the pyrolysate of amino acids (71, 72). The compounds with the highest biological activity were derivatives of tryptophan, Trp-P-1 and Trp-P-2 (73). The most recent result of a classical long-term study for carcinogenicity performed on mice indicate that both compounds are carcinogenic (74). Trp-P-1 and Trp-P-2 have been detected in trace amounts in broiled fish using GC-MS as analytical method (75).

Not enough is known yet concerning the potency, the organ specificity and the occurrence in foodstuffs of this new group of mutagens and probably carcinogens and it is too early for a risk assessment. However, this last example clearly points out how important short-term bioassays for the recognition and subsequent isolation of hazardous compounds in foodstuffs are.

REFERENCES

1. R.A. Morton, Intern. Z. Vitaminforsch. 38 (1968), 5
2. J.R. Dipalma, Basic Pharmacology in Medicine, p. 8, McGraw-Hill, Inc. (1976)
3. R.F. Brown, Proc. 1st U.S.-Japan Conf. on Toxic Microorganisms, p. 12, Ed. by M. Herzberg (1965)
4. B. Jarvis and M.O. Moss, Proc. 8th International Symposium on Food Microbiology, p. 293 (1973)
5. J. Harwig and P.M. Scott, Appl. Microbiol. 21 (1971), 1011
6. Assoc. of Offic. Anal. Chemists: Paralytic shellfish poison. In: Official methods of analysis, 12th edition, AOAC 1975, 305
7. D.E.H. Frear and N.S. Kawar, J. Econ. Entomol. 60 (1967), 1236
8. H. Rothert, Deut. Lebensm. Rundschau 63 (1967), 81
9. S.S.H. Qadri and K.P. Kashi, J. Food Sci. & Technol. 7 (1970), 8
10. P.S. Hall, J. Assoc. of Public Analysts 1 (1963), 5
11. T.M. Cogan and P.F. Fox, J. Dairy Res. 37 (1970), 165
12. F.V. Kosikowski, Science 126 (1957), 844
13. B. Hoffmann, H. Karg, K.H. Heinritzi, H. Behr and E. Rattenberger, Mitt. Gebiete Lebensm. Hyg. 66 (1975), 20
14. J. de Waart, F. van Aken and H. Pouw, Zbl. Bakt. Hyg. I. Abt. Orig. 222 (1972), 96
15. M. Kienitz and G. Schmelter, Zbl. Bakt., I Abt. Orig. 193 (1964), 447
16. J.F. Gregory and J.R. Kirk, J. Agric. Food Chem. 26 (1978), 338
17. F.J. Mulder and R. van Strik, JAOAC 61 (1978), 117
18. L.J. Harris and Y.L. Wang, Biochem. J. 35 (1941), 1050
19. R.A. Anderson, J.H. Brantner and M.M. Polansky, J. Agric. Food Chem. 26 (1978), 1219
20. G.W. Pla, J.C. Fritz and C.L. Rollinson, JAOAC 59 (1976), 582
21. P.A. Lachance, R. Bressani and L.G. Elias, Nutrition Rep. Int. 16 (1977), 179
22. P.L. Pellett, Food Technol. 32 (1978), 60
23. I. Gontzea and P. Sutzescu: Natural Antinutritive Substances in Foodstuffs and Forages, S. Karger-Verlag, Basel (1968)

24 J. Lüthy, Mitt. Gebiete Lebensm. Hyg. 69 (1978), 200
25 W.A. Pons jr. and L.A. Goldblatt, in: Aflatoxin, ed. by L.A. Goldblatt, Academic Press, New York (1969)
26 M.S. Legator, in: Aflatoxin, ed. by L.A. Goldblatt, Academic Press, New York (1969)
27 M.R. Daniel, Brit. J. Exptl. Pathol. 46 (1965), 183
28 T. Arai, T. Ito and Y. Koyama, J. Bacteriol. 93 (1967), 59
29 M.J. Verrett, J.-P. Marliac and J. McLaughlin jr., JAOAC 47 (1964), 1003
30 Assoc. of Offic. Anal. Chemists: AOAC methods (1980), 426
31 W. Przybylski, JAOAC 58 (1975), 163
32 A.E. Pohland, L. Yin and J.G. Dantzman, JAOAC 53 (1970), 101
33 W.F. Haddon, M.J. Masri, V.G. Randall, R.H. Elsken and B.J. Meneghelli, JAOAC 60 (1977), 107
34 F. Friedli and B. Zimmerli, Mitt. Gebiete Lebensm. Hyg. 70 (1979), 464
35 E.J. Schantz, V.E. Ghazarossian, H.K. Schnoes, F.M. Strong, J.D. Springer, J.O. Pezzanite and J. Clardy, J. Am. Chem. Soc. 97 (1975), 1238
36 J. Bordner, W.E. Thiessen, H.A. Bates and H. Rapoport, J. Am. Chem. Soc. 97 (1975), 6008
37 P. Krogh, Nordisk Veterinaermed. 31 (1979), 302
38 E.J. Schantz, E.F. McFarren, M.L. Schafer and K.H. Lewis, JAOAC 41 (1958), 160
39 H.A. Bates, R. Kostriken and H. Rapoport, J. Agric. Food Chem. 26 (1978), 252
40 J. Lüthy, U. Zweifel, Ch. Schlatter, G. Hunyadi, S. Häsler, Ch. Hsu and Y. Shimizu, Mitt. Gebiete Lebensm. Hyg. 69 (1978), 467
41 H.A. Bates, R. Kostriken and H. Rapoport, Toxicon 16 (1978), 595
42 H. Ackermann, J. Chromatog. 36 (1968), 309
43 T. Stijve and E. Cardinale, Mitt. Gebiete Lebensm. Hyg. 62 (1971), 24
44 Y.Y.D. Wang, J. Miller and L.R. Beuchat, J. Food Sci. 44 (1979), 540
45 H.W. Kästner, A.K. Kaul and E.G. Niemann, Qual. Plant.-Pl. Fds. Hum. Nutr. XXV (1976), 361
46 G.M. Evancho, H.D. Hurt, P.A. Devlin, R.E. Landers and D.H. Ashton, J. Food. Sci. 42 (1977), 444
47 D.R. Metcalfe, S.R. Loschiavo and A.J. McGinnis, Canad. Ent. 104 (1972), 1427

48 R.M. Warren and T.P. Labuza, J. Food Sci. 42 (1977), 429
49 M. Mohyuddin, A.K. Kaul, T.R. Sharma and E.-G. Niemann, Qual. Plant. Pl. Fds. Hum. Nutr. XXV (1976), 317
50 Y.-Y.D. Wang, J. Miller and L.R. Beuchat, J. Food Sci. 44 (1979), 1390
51 J.H. Weisburger and G.M. Williams, in: Cancer Medicine, ed. by J.F. Holland and E. Frei, Lea and Febinger, Philadelphia (1979)
52 IARC: Screening Tests in Chemical Carcinogenesis, ed. by R. Montesano, H. Bartsch and L. Tomatis, IARC Scientific Publ. No. 12 (1976)
53 Applied Methods in Oncology, Vol. 3: The Predictive Value of Short-Term Screening Tests in Carcinogenicity Evaluation, ed. by G.M. Williams, R. Kroes, H.W. Waaijers and K.W. van de Poll, Elsevier / North-Holland (1980)
54 B.N. Ames, J. McCann and E. Yamasaki, Mutat. Res. 31 (1975), 347
55 Ch. Schlatter, Pure & Appl. Chem. 52 (1979), 225
56 Z.H. Abedi and P.M. Scott, JAOAC 52 (1969), 963
57 J. Harwig and P.M. Scott, Appl. Microbiol. 21 (1971), 1011
58 M. Umeda, in J.V. Rodricks, C.W. Hesseltine and A. Mehlman, Mycotoxins in Human and Animal Health, p. 713, Pathotox Publishers (1977)
59 B. Gedeck, Zbl. Vet. Med. B 19 (1972), 15
60 J. Boehringer, Inauguraldissertation, Ludwig Maximilians Universität, München (1972)
61 B.H. Armbrecht and O.G. Fitzhugh, Toxicol. Appl. Pharmacol. 6 (1964), 421
62 K. Ohtsubo, M. Enomoto, T. Ishiko, M. Saito, F. Sakabe, S. Udagawa and H. Kurata, Japan. J. Exp. Med. 44 (1974), 477
63 H.R. Burmeister, R.F. Vesonder and W.F. Kwolek, Appl. Environm. Microbiol. 39 (1980), 957
64 R.F. Curti, D.T. Coxon and G. Levett, Fd. Cosmet. Toxicol. 12 (1974), 233
65 M. Bachmann, J. Lüthy and Ch. Schlatter, J. Agric. Food Chem. 27 (1979), 1342
66 P. Blaser, H. Ramstein, W. Schmidt-Lorenz and Ch. Schlatter, Lebensm.-Wiss. Technol. 14 (1980), 66

67 T. Kawachi, M. Nagao, T. Yahagi, Y. Takahashi, T. Sugimura, S. Takayama, T. Kosuge and T. Shudo, in: Advances in Medical Oncology, Research and Education, Vol. 1: Carcinogenesis. Ed. by G.P. Margison, Pergamon Press (1979)
68 T. Sugimura, Mutat. Res. 55 (1978), 149
69 J.S. Felton, S. Healy, D. Stuermer, C. Berry, H. Timourian, F.T. Hatch, M. Morris and L.F. Bjeldanes, Mutat. Res. 88 (1981), 33
70 N.E. Spingarn, H. Kasai, L.L. Vuolo, S. Nishimura, Z. Yamaizumi, T. Sugimura, T. Matsushima and J.H. Weisburger, Cancer Lett. 9 (1980), 177
71 T. Sugimura, T. Kawachi, M. Nagao, T. Yahagi, Y. Seino, T. Okamato, K. Shudo, T. Kosuge, K. Tsuji, K. Wakabayashi, Y. Iitaka and A. Itai, Proc. Japan Acad. 53 (1977), 58
72 T. Sugimura, M. Nagao, T. Kawachi, M. Honda, T. Yahagi, Y. Seino, S. Sato, N. Matsukura, T. Matsushima, A. Shirai, M. Sawamura and H. Matsumoto, in: H.H. Hiatt and J.D. Watson (eds.), Origin of Human Cancer, Cold Spring Harbor, New York (1977), p. 1561
73 T. Sugimura, T. Kawachi, M. Nagao, T. Yahagi, Y. Seino, T. Okamoto, K. Shudo, T. Kosuge, K. Tsuji, K. Wakabayashi, Y. Iitaka and A. Itai, Proc. Japan Acad. 53 (1977), 58
74 T. Sugimura, personal communication
75 Z. Yamaizumi, T. Shiomi, H. Kasai, S. Nishimura, T. Takahashi, M. Nagao and T. Sugimura, Cancer Lett. 9 (1980), 75

Immunochemistry in Protein Analysis

J. DAUSSANT

Laboratoire de Physiologie des Organes Végétaux C.N.R.S., 4 ter, Route des Gardes, 92190 Meudon, France

SUMMARY

This report aims at showing the possibilities of different immunochemical approaches in the identification of the origin of proteins used in food products particularly as concerns the quantitàtive aspect. Some important characteristics are underlined and schematized concerning immune serums, antibodies and antigens as well as the antigen antibody reaction. The principles of some basic methods are described. The determination of denatured proteins is discussed.

INTRODUCTION

Immunochemical studies of proteins involve the use of a biological reactive, the immune serum, the active constituents of which are the antibodies.

This biological reactive is used in a series of methods applied in the fields of biochemistry, physiology, molecular biology, genetics. The reactive is currently applied in medicine, particularly for analytical purposes.

As far as food analysis is concerned, applications of immunochemistry in protein analysis were developped essentially for the controll of quality in order to identify the origin and to evaluate the amount of certain products used in food. For the last two decades, several reviews have already been published dealing with immunochemical methods in the detection of foreign proteins (DEGENKOLB and HINGERLE, 1967,

1969 ; KRÜGER and GROSSKLAUS, 1970, 1971a, 1971b ; BESSEMANS and LAMBION, 1972 ; HERRMANN et al.,1973 a, 1973 b ; Colloquium EFRAC, 1977 ; BAUDNER, 1978 ; LLEWELLYN, 1979 ; SCHECK, 1980).

The large number of publications in this field dealt with the following concerns : Milk products in order to warrant the origin of milk and cheese ; Detection of flour adulteration (barley flour in wheat flour or vulgare flour in Durum flour) ; identification of cereals used for beer ; Use of foreign proteins in meat products, etc.

A major difficulty has been pointed out in all these studies concerning the detection and mainly the quantitative determination of proteins which have undergone denaturation. Denaturation can be caused by the conditions of pH as well as by the conditions of temperature used during the processing and the sterilization of the food products. They can also be caused by the reactions between proteins and other constituents of the food products which can be initiated at certain temperatures (the Maillard reaction for example). Thus, the physico-chemical properties of the proteins can be modified, namely in their solubility and in their antigenical structure. The processes used for solubilizing the proteins for analytical purposes can further be detrimental to the initial antigenic structure of the protein.

This report does not aim at describing immunochemical methods which have largely been dealt with in the litterature. It rather aims at underlining possibilities offered by different immunochemical approaches of the problem of identification, particularly in its quantitative aspect. In this respect, the reaction between the biological reactive and the proteins will be schematized and the description of the principle of some basic techniques will illustrate the topic.

THE BIOLOGICAL REACTIVE

High vertebrates defend their integrity by recognizing and then rejecting foreign constituents which penetrated their organism. Their defense system is well elaborated, both at the cellular and at the molecular levels. The antibodies are one of the expressions of the defense system at the molecular level.

When a foreign constituent of sufficient size, called antigen, is injected in an animal, seric proteins appear in the animal with a new function, these proteins are called antibodies. The antibodies react specifically in vitro with the antigen which induced their formation.

When one antigen is injected, a whole family of antibodies is induced which are specific for different structural features of the antigen (called antigenic determi-

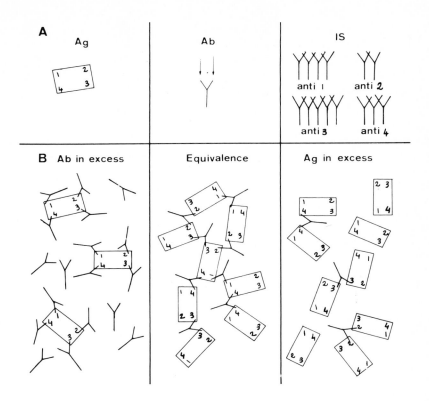

Fig. 1 - Diagramatic representation of antigens (Ag) ; antibodies (Ab), immune serum (IS) and the reaction between antigens and antibodies.
A - Antigen (Ag) - One antigen has several different antigenic determinants (each represented by a different number). No repetitive antigenic determinants on one antigen molecule are represented here.
Antibody (Ab) - The quantitatively most important part of the antibodies circulating in the serum (the immunoglobulins G) are represented here in a simplified form. Each antibody molecule has two antibody sites specific for a same antigenic determinant (arrows) on the N terminal side of the chains.
Immune serum (IS) - In the immune serum of the animal immunized with the antigen represented here there is a family of antibodies directed against the different antigenic determinants.
B - Reaction between antigens and antibodies.
Different situations according to the proportions between antigen and immune serum amounts can be obtained. Three extreme cases are represented.
Large excess of antibodies - The antigenic determinants of all antigens are saturated by antibodies but only one antibody site on the antibodies is occupied by an antigenic determinant. Small complexes without formation of large networks will be obtained.
Large excess of antigens - Only some of the antigens react with the antibodies and in such a case only one antigenic determinant is occupied by an antibody. Both antibody sites are occupied for all antibody molecules. Small complexes without formation of large networks will be obtained.
Equivalence - Between both preceeding situations there is a proportion between antigens and antibodies corresponding to a maximum saturation of antigenic determinants on antigens and of both antibody sites on antibodies. That corresponds to the formation of large networks. The optimum proportion is referred to as equivalence point (The beginning of such a lattice is represented here).

nants). Thus, an immune serum specific for one protein contains markers of several different structural features of the protein.

The quantitatively most important part of the antibodies in the serum of mammals belongs to the immunoglobulin G class. They have two antibody sites which are specific for a same antigenic determinant. The immunoglobulins G are formed by two identical large chains (heavy) bound together by disulfide bridges and by two identical smaller chains (light) each bound to one heavy chain by disulfide bridges. The antibody sites are located in the area comprised between a heavy chain and the corresponding light chain towards the N terminal of these chains. Figure 1 shows the multivalence of the antigens with different antigenic determinants, the divalence of the antibodies, the reaction between antigens and antibodies.

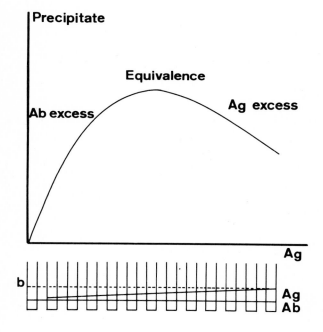

Fig. 2 - Reaction of precipitation between antigens and antibodies.
Ag : antigen ; Ab : antibodies ; b : buffer solution.
In a series of test tubes, increasing amounts of the antigen solution are added to a same amount of the immune serum. The tests are then adjusted to a same volume with a buffer solution. After incubation at 4°C and a preliminary exposure at 37°C (in order to accelarate the formation of the Ag-Ab complexes), precipitates can be observed. The solutions are centrifuged, the precipitates are washed and their amount measured by using a protein determination technique.
The maximum amount of precipitate obtained corresponds to the equivalence zone (see Fig.1).

In protein analysis the immune serum can be used directly or indirectly for detecting and determining proteins. The principles of different types of methods will be briefly described.

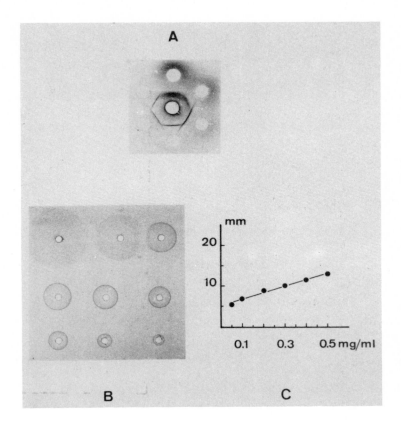

Fig.3 - Basic techniques of immunoprecipitation in gel, immunodiffusions
A - Double diffusion (OUCHTERLONY, 1949)
Principle - Antigens (peripherical wells) and antibodies (central well) diffuse towards each other. After a time of diffusion, at the place in the agarose gel where the proportion between antigens and antibodies correspond to the equivalence zone, the Ag-Ab complexes precipitate.
Remarks - The technique is useful for problems of identification, for antigenical comparison between constituents in different solutions. It can be used for semi quantitative determinations. The test is difficult to read when protein mixtures are analyzed with an immune serum containing antibodies for several proteins.
B, C - Single Diffusion (MANCINI et al., 1965)
Principle - Antigens diffuse in an agarose gel containing an uniform concentration of immune serum. After a time of diffusion where the proportion between Ag and Ab reaches the equivalence zone, the Ag-Ab complexes form a precipitate which appears as a ring surrounding the wells (B).
Remarks - Several relationships were obtained between the Antigen concentration and the diameter, or the surface of the circle delimited by the precipitin ring. For the example shown (B), a linear relationship between the antigen concentration and the diameter of the ring was obtained for the conditions used (immune serum concentration, time and temperature of diffusion) and for concentrations of antigen ranging from 0.1 to 0.5 mg/ml (C).
The technique is widely used for quantitative determination. An immune serum monospecific for the antigen is necessary in order to avoid problems of identification.

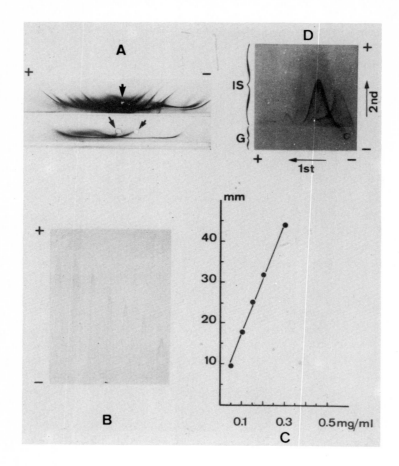

Fig. 4 — Basic techniques of immunoprecipitation in gel, immunoelectrophoreses
A — Immunoelectrophoretic analysis (IEA), (GRABAR and WILLIAMS, 1953)
Principle — Antigens are first separated by agarose gel electrophoresis, (the upper part of the figure shows the single IEA, the lower part the tandem disposition of the IEA ; the arrows indicate the wells where the antigen solutions are deposited before electrophoresis). After electrophoresis a canal is cut in the gel parallely to the migration axe and filled with the immune serum.
Antigens and antibodies diffuse, meet each other and form precipitin arcs.
Remarks — The technique is well adapted to the qualitative analysis of protein mixtures by using immune serums specific for several antigens.
Homologous antigens in different solutions can be antigenically compared by using the tandem disposition.
B, C — Immunoelectrodiffusion, or rocket immunoelectrophoresis (LAURELL, 1966).
Principle — The agarose gel contains an uniform concentration of the immune serum. Antigens are deposited in the wells. Electrophoresis is conducted at a pH value for which antibodies do not move or if so only slightly. The antigens are made to migrate in the gel containing antibodies and when the ratio between Ag and Ab in the complexes corresponds to the equivalence zone, they precipitate in form of peaks (B).
Remarks — For a constant concentration of antibodies in the gel, the surfaces of the peaks, or even their heights, are in linear relationship with the antigen concentrations (CLARKE and FREEMAN, 1967), see part C of the figure.
D — Crossed immunoelectrophoresis (RESSLER, 1960 ; LAURELL, 1965).

Principle — The technique combines a first electrophoretic separation in agarose gel of the antigen (1st) with the electroimmunodiffusion in a second step (2nd).
Remarks — The technique has a high power of resolution, particularly for constituents of close electrophoretic mobilities.
Different types of gel or of electrophoresis can be used in the first dimension and combined with the electroimmunodiffusion in agarose gel.
Quantitative results may be obtained for comparing amounts of homologous antigens in different solutions.

TECHNIQUES OF SPECIFIC PRECIPITATION

Most antibodies, particularly those belonging to the immunoglobulins G, form with the corresponding antigens complexes which have the property of precipitating. A precipitation curve can be obtained by adding increasing amounts of antigens to a same amount of antibodies (Fig.2). The particular form of the curve is due to the divalence of the antibodies for a same antigenic site and to the multivalence of the antigens with different antigenic determinants : in the extreme cases represented on Fig.1, corresponding to large excess in either antibodies or antigens, the complexes are small and isolated, they remain soluble. Between these two extreme situations, the antibody-antigen complexes can form lattices. The lattices increase in size and form aggregates. The complexes become insoluble as a result of the loss of some of their surface charges and of the increase in their size. The ratio between antigen and antibody amounts corresponding to the equivalence zone (Fig. 1) favours the formation of a maximum of precipitating aggregates. For this ratio, the amount of precipitate is maximum.

The reaction of precipitation in solutions is still used for the determination of antigens. That is done by using the ring test technique or better by quantitating the specific precipitate either after its separation by centrifugation (MAURER, 1971) or during its formation by using nephelometric techniques (LI and WILLIAMS, 1971). For these determinations, a reference scale is established by using the immune serum and serial amounts of the purified antigen, generally by using the antibody excess side of the curve.

In order to determine a protein in a protein mixture, an immune serum specific for the protein has to be used. If the serum is specific for several different proteins (which means that it contains several families of antibodies specific for several different proteins), the reaction of specific precipitation in solutions cannot be used. The identification of individual precipitates corresponding to distinct antigen-antibody complexes can be done by carrying out the reaction of precipitation in gels. Some of the techniques of specific precipitation in gel, are very easy to carry out and they do not consume much serum. Thus, several of them are commonly used instead of the reaction of precipitation in solutions. There are many techniques of specific precipitation in gel (OUCHTERLONY and NILSSON, 1973 ; AXELSEN, 1973,

1975 ; KAMINSKI, 1979) and a great number of applications have been reported. Five basic techniques are shown on Fig.3 and 4. In the examples shown on figures 2, 3 and 4, antigens are identified or determined by direct techniques.

PASSIVE AGGLUTINATION INHIBITION

The principle of this classical technique is shown on Fig. 5. It is an indirect technique for determining antigens (STAVITSKY, 1977 ; LITWIN and BOZICEWICH, 1977). As for the techniques of immunoprecipitation, the agglutination techniques involve the formation of a sort of network.

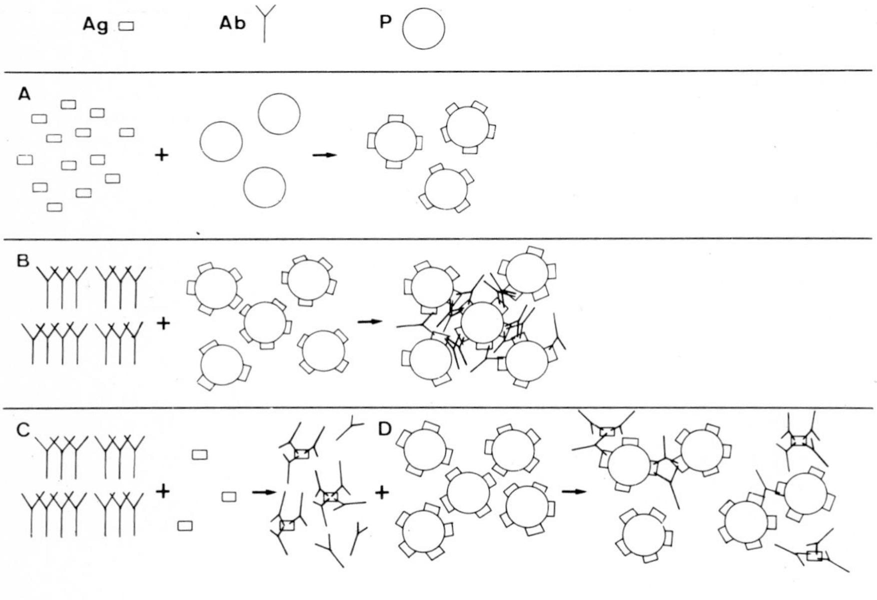

Fig. 5 - Passive agglutination inhibition
Ag : antigen ; Ab : antibodies ; p : particles (for example red cells or latex particles)
A - Sensitization of the particles - The particles are coated with the purified antigen.
B - Agglutination (check sample) - When antibodies specific for the antigen used for coating the particles are added to the sensitized particles the particles agglutinate.
Remarks : several antibodies can be involved in the bridge between two particles.
The maximum dilution of the immune serum for which the agglutination is visible is taken as reference : that is the "titer" of the immune serum.
C - D - Agglutination inhibition (test sample) - C - The immune serum is added to a protein mixture containing the antigen. D - The treated immune serum is added to the sensitized particles. The capacity of the immune serum for agglutination is reduced. Thus, the maximum dilution for which the agglutination is visible is reduced. By using solutions of a known content in the antigen, a scale can be established between the titer of the immune serum and the amount of the antigen added in step C.

ENZYME LINKED IMMUNOSORBENT ASSAY (ELISA)

The labelling of antibodies was used in histological techniques as an alternative to immunofluorescence. The use of enzymes bound either to antibodies or to antigens in the detection and determination of the antigens in unknown solutions gave the start to a new series of methods for antigen or antibody analysis (VOLLER et al., 1979). The principle of one of these techniques for the antigen determination is shown on Fig.6. It constitutes an indirect determination of the antigen by using a reaction of competition as it was the case in the example shown on Fig.5.

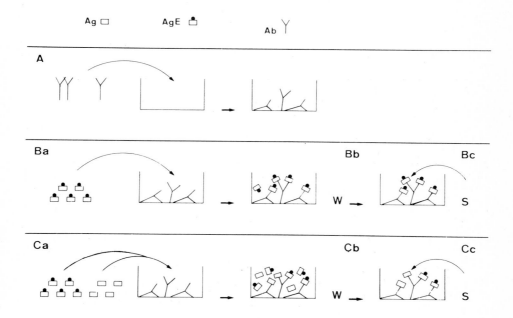

Fig. 6 - Enzyme linked immunosorbent assay (ELISA)
Ag : Antigen ; Ag-E : Enzyme labelled antigen ; Ab : antibodies ; S : substrate
A - Bottom innersurface of microtest tubes or microplates are coated with the antibodies (generally with the immunoglobulin fraction of the immune serum).
B - Check sample
Ba - Enzyme labelled antigens are transfered on the microplates. The antigens become immunoabsorbed after a time of incubation.
Bb - W : washing ; the non immunoabsorbed antigens are eliminated.
B'c - Reaction of enzyme characterization. The reaction can be visualized by eye or measured by using a photometer.
C - Test sample
Ca - Enzyme labelled antigens are transfered on the microplates together with the solution containing unknown antigen amount. Compared with Ba, a reduced number of labelled antigens will be bound to the antibodies.
Cb - W : washing ; the non immunoabsorbed antigens (labelled and unlabelled) are eliminated.
Cc - Colour reaction for enzyme activity. In this case, compared with Bc, the enzymatic staining is reduced. The inhibition is proportional to the amount of unlabelled antigens in the unknown solution.

In the ELISA techniques, the observation of the results does not necessitate the formation of a network as it is the case for the precipitation and the agglutination techniques.

The principle of the ELISA techniques is similar to the one of the radioimmuno assay (RIA) in which one of the reactants is labelled with radioactive markers. The RIA techniques are not commonly used in all laboratories because they involve the use of radioactive elements. Because of its high sensitivity, the ELISA technique may in certain cases be used as a substitute to the radioimmuno assay (VOLLER et al., 1979).

REMARKS

The Complement fixation method has also to be mentioned here (RAPP and BORSOS, 1970). This very classical technique involves the property of a series of seric constituents which bind in a certain order to the antibodies when the antibodies are fixed on the antigens. (As for the hemagglutination technique, the fixation Complement method involves the use of red cells coated with the purified antigen). The fixation of the Complement constituents on the antibodies bound to an antigen coating the red cells results in the lysis of the red cells. The intensity of the lysis is read by a spectrophotometer. The technique is used in its indirect form (competition reaction between free antigens and antigens coating the red cells).

Among the techniques, Complement fixation and ELISA are the most sensitive ones. Agglutination techniques are more sensitive than precipitation techniques.

In all of these techniques, several antigenic determinants are involved in the quantitative determination of one antigen. As concerns the determination of proteins likely to be denaturated, there is a prerequisite : the immune serum has to react in the same way with the protein sample taken as a reference and with the protein extracted from the samples submitted to analysis. If that is not the case, the quantitative result will be erroneous. Generally, denaturated proteins react with only some of the antibodies of the anti "native protein" immune serum, they may even react very poorly with this immune serum. Inducing antibodies against the denaturated protein and using the denaturated protein as a reference for the determination partially solves the problem. However, since the denaturation in different samples is probably not always of the same nature and does not reach the same degree of intensity, the answer may not always be completly satisfactory. A relatively new technique the immunoaffinity chromatography, may provide an alternate answer to this question when used in certain conditions.

IMMUNOAFFINITY CHROMATOGRAPHY

Immunoaffinity chromatography (LIVINGSTON, 1974) is an immunoabsorption technique the principle of which is schematized on Fig. 7. Antibodies are immobilized on a support which is for example packed on a column. The protein solution is passed through the column which specifically retains the antigen. After washing, the immunoabsorbed antigen is desorbed by using appropriate solutions (for example low pH, urea, etc ...). An advantage of this technique is that the immunosorbent can be used many times.

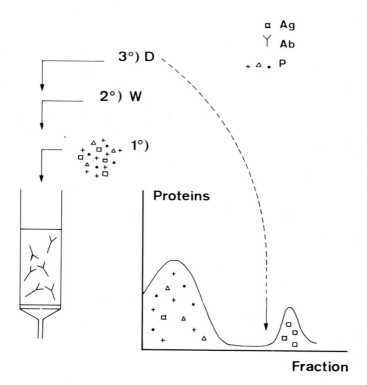

Fig. 7 - Immunoaffinity chromatography
Ag : Antigens ; Ab : antibodies ; P : other proteins
The antibodies are immobilized on an insoluble support which is packed on a column.
1- The protein solution is poured onto the column. Antibodies retain specifically the antigens.
2 - The column is washed (W) with a buffer solution. The excess of antigens and the proteins non specifically recognized by the antibodies are eluted in a first fraction.
3 - A solution for desorbing (D) the specifically attached antigen onto the immunosorbent column is then passed through the column (arrow on the elution curve).

In order to determine the amounts of antigens in a protein solution, the technique would have to be used in conditions of unsaturation, in order to retain all antigen molecules from the protein solution. After desorption, a chemical dosage would have to be used for determining the amount of antigen. In other words, the antibodies would not be used for titrating but for purifying the antigen. Thus, the fact that more or fewer antibodies react with the antigen, according to its degree of denaturation, would not interfere with the titration (if the immunosorbent is used in unsaturated conditions). Problems remain to be solved for such an application. That implies the use of immunosorbent with sufficient capacity. Moreover, the desorption procedure has to desorb all immunoabsorbed antigens without releasing the immobilized immunoglobulins.

CONCLUDING REMARKS

This report aimed at underlining the fact that an anti protein immune serum contains several antibodies each specific for distinct antigenic determinants on the protein. It aimed also at schematizing the reaction between antigens and antibodies. Another concern was to schematize the principle of some main immunochemical techniques for the analysis of proteins. Some of these techniques are extensively used in the analysis of food products in order to identify and to determine some of their components. As for the major difficulty in the use of this analysis, namely the determination of denatured proteins, some of the last developments in immunochemical methods may open a new approach to the problem.

Moreover, immunochemical analysis of proteins may also be used for other applications two of which are mentioned here.

It could be used for evaluating the intrinsic quality of certain major proteins of food products which were submitted to different physicochemical and thermal pressures during their preparation. For example, at a certain degree of the heating procedure, the nutritional value of the soybean meal is dammaged. The immunochemical approach was used in order to recognize which kind of heat treatment (temperature and humidity) was given to soybean meal samples and thus to provide an estimate of the nutritional value of the product (KOIE and DJURTOFT, 1977). The field of application is largely open. In this regard the approach could be used for estimating the specific activity of enzymes by evaluating both their enzymatic activity and the amount of their enzymatic protein.

In the food industry, processes must be adapted to the quality of the raw materials and homogeneity becomes an important aspect of the quality. In the cereal industry, the techniques shown on Fig. 3b and 4b are perfectly adapted to the seed by seed analysis. Thus they could be applied for controlling the homogeneity of a seed sample on the basis of one or several proteins. Derivated techniques of diffusion and immunoabsorption in gel were shown to provide a means for evaluating the homogeneity of a malt sample in a seed by seed analysis taking α and β -amylases as references (DAUSSANT, 1981).

It is to be expected that by its new developments immunochemical methods in protein analysis will contribute to solve the still existing analytical problems. Moreover, the immunochemical analysis of proteins will find other applications for avaluating different aspects of the quality of food products and raw materials used in the food industry.

REFERENCES

1 N.H. AXELSEN, Scand. J. Immunol. 2 (Suppl. N°1) (1973), 1
2 N.H. AXELSEN, Scand. J. Immunol. (Suppl. N°2) (1975), 1
3 S. BAUDNER, Getreide Mehl Brot 32 (1978), 330
4 J. BESSEMANS and R. LAMBION, Revue des fermentations et des industries alimentaires, 27 (1972), 61
5 H.G. CLARKE and T. FREEMANN, in Prot. Biol. Fluids, PEETERS, H. edit. Elsevier, 14 (1967), 503
6 COLLOQUIUM organized by the "Fractionation and Reassembly of Biological Units" working party of the Council of Europe's Committee on Science and Technology on "identification of protein additives or substitutes in meat products". Annales de la Nutrition et de l'Alimentation, 31, (1977), 129
7 J. DAUSSANT, Monograph VII of the European Brewery Convention, (1981), in press
8 E. DEGENKOLB and M. HINGERLE, Archiv für Lebensmittelhygiene, 18, (1967), 24
9 E. DEGENKOLB and M. HINGERLE, Archiv für Lebensmittelhygiene, 20 (1969), 73
10 P. GRABAR and C.A. WILLIAMS, Biochim. Biophys. Acta, 10 (1953), 193
11 C. HERRMANN, C. MERKLE and L. KOTTER, Fleischwirtschaft 1 (1973a), 97
12 C. HERRMANN, C. MERKLE and L. KOTTER, Fleischwirtschaft 2 (1973b), 249
12 M. KAMINSKI, La Pratique de l'immunoélectrophorèse, Masson édit., Paris (1979)
14 B. KOIE and R. DJURTOFT, Annales de la Nutrition et de l'Alimentation, 31 (1977) 183
15 H. KRÜGER and D. GROSSKLAUS, Fleischwirtschaft, 11, (1970), 1529
16 H. KRÜGER and D. GROSSKLAUS, Fleischwirtschaft, 2 (1971), 181
17 C.B. LAURELL, Anal. Biochem. 10 (1965), 358
18 C.B. LAURELL, Anal. Biochem. 50 (1966), 344
19 I.W. LI and C.A. WILLIAMS, in Immunology and Immunochemistry, C.A. WILLIAMS and M.W. CHASE edit., Academic Press, Vol.III (1971), 94

20 S.D. LITWIN and J. BOZICEVICH, in Immunology and Immunochemistry, C.A. WILLIAMS and M.W. CHASE edit., Academic Press, vol.IV (1977), 115
21 D.M. LIVINGSTON, in Methods in Enzymology, Academic Press, New York, Vol. XXXIV (1974), 723
22 J.W. LLEWELLYN, International Flavours Food Additive, 10 (1979), 115
23 G. MANCINI, A.O. CARBONARA and J.F. HEREMANS, in Immunochemistry, Pergamon Press, Oxford (1965), 235
24 P.H. MAURER, in Immunology and Immunochemistry, C.A. WILLIAMS and M.W. CHASE edit., Academic Press, vol. III (1971), 1
25 O. OUCHTERLONY, Acta Pathol. Microbiol. Scand., 26 (1949), 507
26 O. OUCHTERLONY and L.A. NILSSON, in Immunochemistry, second edition, WEIR edit., Blackwell Scientific publications Oxford/London/Edinburgh/Melbourne, Vol.1 (1973), 19.1
27 H.J. RAPP and T. BORSOS, Molecular Basis of Complement Action, Appleton Century-Crofts, Meredith Corporation, New York (1970)
28 N. RESSLER, Clin. Chim. Acta, 5 (1960), 795
29 K. SCHECK, Fleischwirtschaft, 60 (1980), 406
30 A.B. STAVITSCKI, in Immunology and Immunochemistry, C.A. WILLIAMS and M.W. CHASE edit., Academic Press, vol. IV (1977), 30
31 A. VOLLER, D.E. BIDWELL and A. BARTLETT, The enzyme linked immunosorbent assay (ELISA), Dynatech. Europe, Borough House, Guernsey, G.B. (1979).

A version of this lecture has also been published in Ernährung 5 (1981) by agreement of the publishers.

A New Method for the Rapid Detection of Microbial Contamination of Fruit Juices

J.G.H.M. VOSSEN, H.D.K.J. VANSTAEN

LUMAC B.V., 6372 AD Schaesberg, The Netherlands

ABSTRACT

A new method for the detection of microbial contamination of fruit juices was developed using the luminescent ATP-assay. The principle of the measurement involves the removal of somatic and free ATP by incubating the sample in the presence of SOMASE^R. After this step, the microbial cells are measured by extracting their ATP with L-NRB^R and assaying the liberated ATP in an ATP-specific bioluminescent system. The light produced in the bioluminescent reaction is proportional to the ATP concentration and was measured with a BIOCOUNTER^R M2010. For sterility testing of fruit juices, the products have to be incubated until the microbial contamination reaches the sensitivity threshold of the assay. With this method it was possible to detect one microorganism in 150 ml fruit juice within 27 h.

INTRODUCTION

In the production of fruit juices considerable effort is put into microbiological quality control of the end products. Spoilage of the final products is most frequently caused by lactobacilli, yeasts and molds, which enter the fruit juice via the raw materials or during production.

Standard microbiological techniques to detect contamination of the end product involve incubation of the product in its original container for 48 - 72 h at the appropriate temperature. Thereafter, an aliquot is taken which is serially diluted and plated on a suitable solid growth medium and incubated again for 24 - 48 h. In general, microbiological quality control of the finished product takes between two and five days. It was the purpose of this study to develop a method which would allow detection of contaminating micro-organisms in fruit juices faster than with the currently used microbiological procedures. In this study, detection of contaminating bacteria in fruit juices is based on measurement of ATP from these micro-organisms. ATP is an energy rich intermediate in metabolism of organisms, the turnover of which is very fast (1, less than one second). The intracellular level of ATP is subject to very strict metabolic control and during cell death ATP is broken down through autolysis in a few minutes. Therefore, sterility testing of fruit juices can very well be performed with ATP measurement as only viable micro-organisms are detected with this technique (2, 3).

PRINCIPLE OF ATP MEASUREMENT

For measurement of ATP a purified luciferin-luciferase preparation from Photinus pyralis was used. In the reaction described below:

$$\text{LUCIFERIN+LUCIFERASE+ATP} \xrightarrow{Mg^{++}} \text{LUCIFERIN-LUCIFERASE-AMP+PYROPHOSPHATE}$$

$$\text{LUCIFERIN-LUCIFERASE-AMP} \xrightarrow{O_2} \text{DECARBOXYLUCIFERIN+LUCIFERASE+AMP+CO}_2\text{+LIGHT}$$

ATP is converted to AMP and light of 560 nm is generated. The amount of light produced is proportional to the ATP concentration of the sample.

MATERIALS AND METHODS

The luciferin-luciferase preparation used, Lumit^{R}PM was from LUMAC. F-NRS^R, L-NRB and SOMASE were also obtained from LUMAC. To measure the the light which is produced in the bioluminescent reaction, a BIOCOUNTER M2010 from LUMAC was used.
For inoculation of the fruit juice containers, an overnight culture of bakers' yeast was used. The number of viable cells in the inoculum was determined microscopically in a Bürker chamber. After inoculation, the containers were incubated at 28°C.

RESULTS AND DISCUSSION

When developing a bioluminescent method for fast detection of microbial contamination of fruit juices based on ATP specific bioluminescence, two major problems were encountered. Firstly, ATP is measured in an enzymatic reaction with an optimal pH of 7.75, while fruit juices are very acidic products (pH 3.5 - 4.5). Secondly the method is based on measurement of ATP from micro-organisms only, whereas the fruit juice also contains ATP from nonmicrobial sources (e.g. cell debris etc.) To overcome these problems, fruit juices were mixed with a buffered extracting reagent (F-NRS). This reagent will specifically extract ATP from nonbacterial sources and at the same time increase the final pH of the mixture to pH 6 - 8. Subsequently an ATP-ase enzyme (SOMASE) is added to the mixture of fruit juice and F-NRS in order to hydrolize all nonmicrobial ATP during a period of 45 minutes at room temperature. Thereafter, a 50 µl aliquot is taken from the mixture and to this 150 µl of L-NRB is added. This reagent will extract ATP quantitatively from microbial cells within 15 seconds. The ATP extracted from microbes is then reacted to a standardized preparation of luciferin-luciferase (Lumit PM) and light generated in the enzymatic reaction is measured in a photomultiplier instrument (BIOCOUNTER M2010) and is expressed as relative light units (RLU).

The sensitivity of the method described above was evaluated using containers of orange juice to which varying numbers of yeast cells were added. The results, as shown in fig. 1, clearly demonstrate that a linear relationship exists between the number of yeast cells present in the fruit juice and the amount of extractable yeast cell ATP - measured as the amount of light produced in the bioluminescent reaction. However, as can be seen from the same graph (fig. 1) the sensitivity of direct measurement, i.e. without preincubation of the juice, is not high enough to use direct measurement of microbial ATP for sterility testing of fruit juices.

Therefore, in another set of experiments, 6 containers with 150 ml. of orange juice were inoculated with 0, 1, 10, 50 and 100 yeast cells. These containers were incubated at 28°C and after various time intervals samples were withdrawn. The microbial ATP content was measured following sample preparation as described above. The results (fig. 2) show that after an incubation period of 24 - 30 h, microbial ATP had increased in all contaminated fruit juice containers. In practice, an incubation period of 24 - 30 h is sufficiently long for one microorganism to multiply and to reach cell numbers which will fall within the sensitivity range of the bioluminescent technique. Similar results have also been obtained with other fruit juices. Thus by using bioluminescent techniques for sterility testing of fruit juices, results are obtained in 30 h instead of 4 - 5 days. Additional experiments are currently being performed to evaluate the use of bioluminescent ATP measurement for sterility testing of other food stuffs.

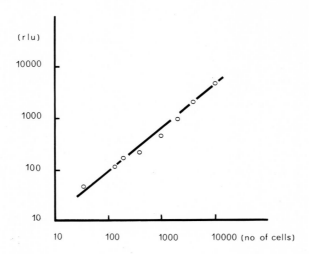

Figure 1: Sensitivity of direct measurement of viable yeast cells in fruit juice (no incubation).

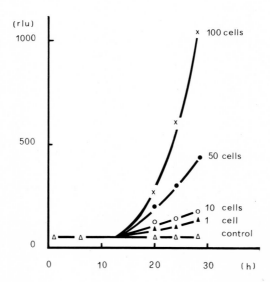

Figure 2: ATP increase in fruit juice containing 0 to 100 cells per 150 ml.

REFERENCES

1 A.G. Chapman and D.E. Atkinson, Advances in microbial Physiology 15 (1977), pp 253 - 306.
2 K. Nakao, H.E. Wade, T. Kamiyama et al, Nature, Lond. 194 (1962),877.
3 O. Holm-Hansen and C.R. Booth. Limnl Oceanog, 11 (1966), 510.

Amines, Food and Brain Function

P.RIEDERER, G.P. REYNOLDS, K. JELLINGER

Ludwig Boltzmann Institute of Clinical Neurobiology, Lainz-Hospital, A-1130 Vienna, Austria

There is little doubt that brain function is affected by our eating habits; our state of mind can vary according to what and how we eat. Compared to scientific progress in other fields, we have only a primitive understanding of many of these effects, although there have been several recent advances, two very different examples being the development of the concept of food allergy and the possible identification of migraine-inducing food constituents.

Since time inmemorial man has been aware of the neuroactive content of certain plants. The hallucinogenic properties of some mushrooms and cacti have been, and are still, used or abused in some cultures. The study of these substances, which include mescaline, psilocybin, harmine and related compounds have contributed to our understanding of brain (dys)function. While these substances are not usually found in the normal diet of western man, there are several vegetable sources of neuroactive compounds and their precursors. For example, the neurotransmitter dopamine is found in tropical fruits (1,2) and its precursor amino acid L-dopa is present in broad beans and in various organs (3-8). Fortunately, our bodies are equipped with protective mechanisms to prevent such compound reaching the central nervous system. The enzyme monoamine oxidase (MAO) is fairly ubiquitous within the human body and rapidly metabolises such biogenic amines to inactive products before they have a physiological effect (fig. 1). Secondly the so-called blood-brain barrier (BBB) regulates the access of these compounds and of many potential neurotoxic agents to the brain. This membrane barrier will, for example, permit lipophylic amines such as

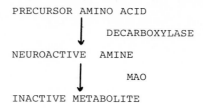

FIG. 1 Simplified general scheme of amine metabolism

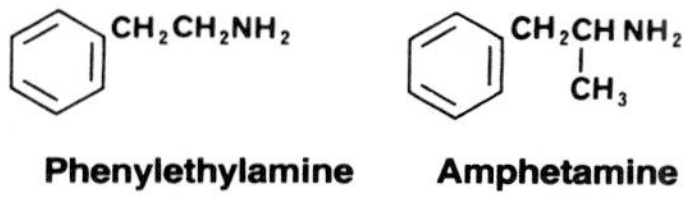

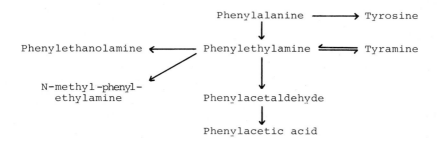

FIG. 2 Comparison of structures of phenylethylamine and amphetamine

```
                    Phenylalanine  ———→  Tyrosine
                          ↓
Phenylethanolamine ←—  Phenylethylamine  ⇌  Tyramine
                       ↓
  N-methyl-phenyl-
     ethylamine        Phenylacetaldehyde
                          ↓
                       Phenylacetic acid
```

FIG. 3 Possible routes of phenylethylamine metabolism

amphetamine and its endogenous analogue phenylethylamine (fig.2) to enter the brain while the more lipophobic amines, which include the neurotransmitters dopamine, noradrenaline and serotonin cannot pass through. However, there are active transport processes via the BBB which allow the amino acid precursors of all these compounds to reach the central nervous system (CNS). For example, L-dopa effectively replaces the dopamine deficit in Parkinson's disease, while dopamine administration has no effect (9). However, there can be situations where these protective mechanisms are inadequate. In phenylketonuria,

the inability of the liver to metabolize excess phenylalanine means
too much of this amino acid reaches the brain and inhibits the production of the catecholamine neurotransmitter, presumably by competition
with tyrosine, as well as leading to the formation of several toxic
metabolites like phenylethylamine and phenylacetic acid (fig. 3)(10).
Another pathological condition occurs in coma (11). Animal experiments
have shown that in a wide range of abnormal metabolic conditions the
BBB becomes more permeable to many compounds. For example the increase
in 5-HT penetration into the brain (table 1) may well be responsible
for the comatose state following severe hepatic dysfunction. Oedema
following brain infarction also leads to unwanted penetrations of the
BBB by neurotoxic species which in turn induces further oedema which,
if allowed to continue, is the major cause of death in such stroke
patients (12).

The inadequate function of MAO will also lead to neurophysiological
effects. The psychotic effects of harmine present in cacti is presumably
partly due to its inhibition of MAO resulting in a build-up of neuroactive amines (13), both transmitters and other amines, such as tyramine, acting on transmitter systems (fig.4). Therapeutic use of MAO

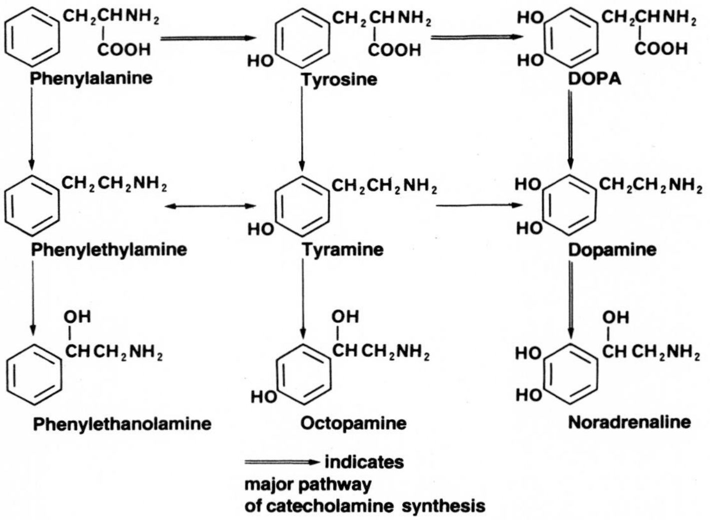

FIG 4   The relationship between catecholamine neurotransmitters
and trace amines

TABLE 1   THE INFLUENCE OF P.E. NUTRITION + L-VALINE ON THE BLOOD AMMONIA AND BRAIN INDOLES

Human brain area	Treatment	TRP µg/g	5-HT ng/g	5-HIAA ng/g	Blood valine µM/ml	Blood ammonia µg/100 ml; mean
Raphe + Form. reticularis	Hepatic coma without P.E. nutrition	80.0±22 (5)	2,660±220 (8)	18,145±9,189 (8)	0.24±0.056	384
	Hepatic coma + parenteral nutrition + L-valine	19.5±4,3(3)	251.0±120 (4)	1,585±573 (4)	2.17±0.348	203
		p < 0.01	p < 0.01	p < 0.01	p < 0.01	p < 0.01

Number of brains in parenthesis. Values are means ± standard error mean.
P.E.= parenteral.

inhibitors, for example in the treatment of depression (14,15), can thus bring about unwanted side effects. The most frequent type of these side effects is the so-called "cheese effect" (16). It represents a hypertensive reaction to certain foods, notably strong cheese. The exact mechanism is not fully understood, although the compound responsible for this effect is thought to be tyramine which, since its major removal pathway through MAO is inhibited, is thought to promote noradrenaline activity by releasing transmitter, from noradrenergic neurons. Thus patients receiving such MAO inhibitors need to eliminate such tyramine - (and presumably other amine -) containing foods from their diet.

These amines are also implicated in other neurophysiological disorders: migraine and headache (17). A substantial proportion of migraine sufferer's are sensitive to particular dietary components; red wine and chocolate are two examples. Again the patho-mechanisms involved are not fully understood, although the trace amines PEA and tyramine (fig. 4) and the putative neurotransmitter histamine are implicated.

Thus we have described some aspects of the action of particular dietary constituents on both normal brain function and neuropathological states.

REFERENCES

1 H. Stachelberger, E. Bancher, J. Washüttl, P. Riederer and A.Gold, Qual.Plant 27 (1977), 287
2 E.Bancher, J.Washüttl, H.Stachelberger, P.Riederer and A. Gold, Alimenta 14 (1975), 195
3 R.S.Andrews and J.B.Pridham, Nature 205 (1965), 1213
4 S.Udenfriend, W.Lovenberg and A.Sjoerdsma, Arch.Biochem.Biophys.85 (1959), 487
5 D.W.Bruce, Nature 188 (1960), 147
6 T.P.Waalkes, A.Sjoerdsma, C.P.Creveling, H.Weissbach and S.Udenfriend, Science 127 (1958),648
7 K.Mayer and G.Pause, Lebensm.-Wiss. und Technol. 5 (1972), 108
8 T.Nagasawa, H.Tagaki, K.Kawakami, T.Suziki and Y.Sahashi, Agr.Biol.Chem. 25 (1961), 441
9 W.Birkmayer and O.Hornykiewicz, Arch.Psychiat.Nervenkr. 203 (1962),560
10 G.P.Reynolds, Trends in Neurosciences 10 (1979), 1
11 P.Riederer, P.Kruzik, E.Kienzl, G.Kleinberger, K.Jellinger and

W.Wesemann, in: Transmitter Biochemistry of Human Brain Tissue (eds. P.Riederer, E.Usdin)(1981) J.Wiley, Chichester, pp 143
12 K.Jellinger and P.Riederer, in: Advances in Neurology (ed.: J.Cervos-Navarro), Vol 20 (1978) Raven Press New York, pp 535
13 A.Pletscher, K.F.Gey, and P.Zeller, in:Fortschritte der Arzneimittelforschung (ed.: E.Jucker) Band 2 (1961), Birkhäuser Verlag, Basel, pp 417
14 P.Riederer and W.Birkmayer, in: Enzymes and Neurotransmitters in Mental Disease (eds. E.Usdin, T.L.Sourkes, M.B.H.Youdim) (1980) Wiley J., Chichester, pp 261
15 H.Beckmann, Nervenarzt 52 (1981), 135
16 M.Sandler, V.Glover, A.Ashford and G.M.Stern, J.Neural Transm. 43 (1978), 209
17 M.Sandler, J.Neural Transm. Suppl. 14 (1978), 51

Determination of Subresidual Proteolytic Activities in Foods

P. RAUCH, L. FUKAL, J. KÁŠ

Department of Biochemistry and Microbiology, Institute of Chemical Technology, Suchbátarova 5, 166 28 Prague 6, Czechoslovakia

SUMMARY

The method using denatured ^{131}I-serum albumine as substrate was proved to be suitable for the measurement of very low proteolytic activities in various samples of food, as well as in other kinds of samples. For the determination of proteolytic activity of certain origine imunochemical methods were applied. It was demonstrated /on the example of beer samples/ that it is sometimes possible to prove the chosen enzyme by imunochemical methods even when its catalytical activity was lost and therefore the activity measurement became impossible.

INTRODUCTION

It is generally known, that the majority of processes taking place in raw material, semifinished products and even final products are closely related to different kinds of enzymatic reactions. Many enzymatic activities are very low, almost negligable though owing to their permanent and long lasting action can effect various properties of food.

Determination of low enzymatic activities is required in some other cases:
- it is necessary to determine the residual activities in food after different kinds of technological treatment

- to determine the remaining activities when technical enzyme preparations are used in processing
- to determine the contaminating activities in technical enzyme preparations /which can be desirable or undesirable/
- to determine the activities of enzymes desorbed from different kinds of their immobilized forms when used in technology
- to detect microbial contamination according to the increase of chosen enzymatic activities
- according to the activities of chosen enzymes it is possible to appreciate the technological conditions during processing

In addition the requirement to determine very low enzymatic activities is not only a problem of food technology, but of other branches of science and industry, as well, for example medicine and pharmacy. The necessity to determine low and very low enzymatic activities is not some special case, but the requirement of every day practice.

PROTEOLYTIC MEASUREMENT BY MEANS OF ^{125}I-SERUM ALBUMINE

Attention was paid to proteolytic activities since they occur most frequently in food material, besides being used very often in food technology in the form of technical preparation. For the measurement of low and very low proteolytic activity the method using ^{125}I-serum albumine as substrate was found the most suitable. The principle of this method is demonstrated on the scheme below.

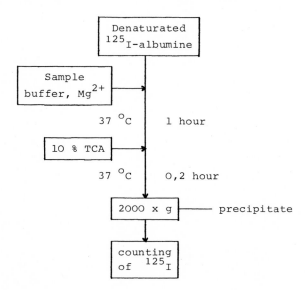

It is almost identical with methods using natural protein substrates, only the determination of protein splitting products is carried out by means of measurement of their radioactivity.

Following main advantages of this method arise from this experimental arrangement:
- possibility to use protein substrate, which is more suitable for the determination of proteolytic activity than any kind of artificial substrate
- while using the radioactive label it is possible to avoid unpleasant background of amino acids and peptides which are generally present in examined food material and cause the application of protein substrates for proteolytic activity determination difficult or even impossible
- radionuclide makes the method to be very sensitive

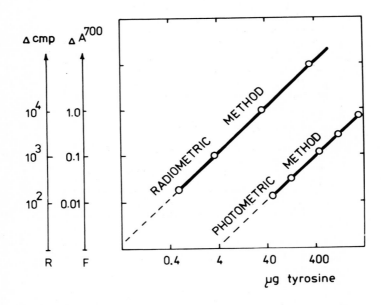

Fig.1 Comparison of two methods for proteolytic activity assay

Fig. 1 compares the sensitivity of the method when proteolytic splitting products are determined by Folin reagent and by measurement of radioactivity. Both values on the diagram are plotted in logarithimic scales. Concentration of protein splitting products are expressed in µg of tyrosine, enzymatic activity in absorbancy or in counts per min. It is evident, that the method using the labelled substrate is approximately a hundred times more sensitive. This sensitivity could be further increased by higher specific radioactivity and by chosing a substrate more susceptible to proteolytic breakdown.

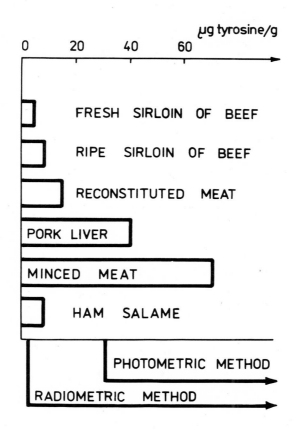

Fig.2 Proteolytic activities in meat products

Proteolytic activities found in different samples of meat and meat products are shown in Fig. 2. From the picture it is evident that the possibility to measure the proteolytic activity in meat and meat products by means of photometric method is very limited. On the contrary the method using a labelled substrate enables to measure these very low proteolytic activities without difficulty.

Proteolytic activities in tinned beef meat indicate that the content of older tins shows higher proteolytic activity. This is caused probably by reactivation of enzymes during storing. When the proteolytic activities of tinned beef produced in two different factories are compared it is possible to prove that the proteolytic activities in tinned beef from the factory with better processing conditions were always lower.

To be able to prove that the higher proteolytic activity in these tins is caused by enzymes of contaminating microflora we contaminated the content of laboratory prepared samples by psychrophilic microorganism Pseudomonas. Results obtained clearly prove that in contaminated tins /after sterilisation, of course/ higher proteolytic activity can be always found.

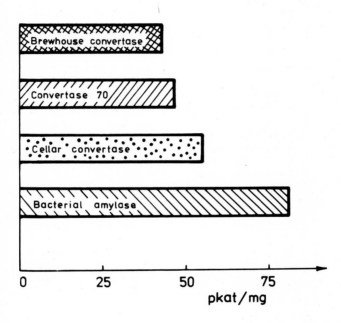

Fig.3 Residual proteolytic activity in different amylase preparations

Fig. 3 demonstrates the determination of proteolytic activities in various preparations of technical enzymes used in industrial processes of food technology. Fig. 4 shows one example of the aplicability of the method with labelled substrate outside of food industry. This method enables to prove that γ-globuline fraction produced by old technological procedure had low expiration time due to relatively high proteolytic activity present in the preparation. The γ-globuline fraction prepared by new separation technique has low proteolytic activity and, therefore, was more stable and expiration time could be prolonged.

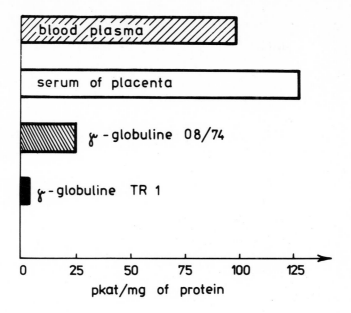

Fig.4 Proteolytic activity in blood plasma and γ-globuline preparations

ADVANTAGE OF IMMUNOCHEMICAL METHODS

The method with labelled substrate was proved to be suitable for proteolytic activity measurements of different kinds of foodstuffs, although the method cannot distinquish activities of different origin. Activities derived at least from four different sources can be found in food products: raw material, applied microorganisms, contaminated microflora, and added preparations.

This problem can be solved by immunochemical methods. The determination of papain in beer can serve as example. Preparations containing papain are used for beer stabilisation in the concentration of about 2 g/100 l. Owing to the content of papain in these preparations and variations of the content of papain in beer the sensibility of the method used should be at least 2 $\mu g/cm^3$.

The limits of a few methods for the determination of papain concentration in model samples are given (in $\mu g/cm^3$): Anson /400/, ^{125}I-HSA /0.1/, Ouchterlony or Mancini /20/, Ouchterlony or Mancini after protein concentration /2/, immunoprecipitation /5/, RIA /0.2/.

Anson's method is mentioned here only for comparison. The method with labelled albumin is very sensitive although not specific for papain determination. Double diffusion or radial diffusion methods enable the specific determination of papain with satisfactory sensitivity when samples are concentrated by amoniumsulphate precipitation or by some other suitable method. Direct immunoprecipitation is not suitable for practice owing to the high consumption of antiserum. Best results can be achieved with RIA-tests. They enable specific and very sensitive determination of papain.

Table I. Determination of papain concentration in beer

Sample	Method			
	LS		IA	RIA
	$\mu g/cm^3$	$nkat/cm^3$	$\mu g/cm^3$	$\mu g/cm^3$
A - 12 %	0.074	0.001	2.7	2.6
B - 12 %	0.079	0.001	3.2	2.8
C - 12 %	89	1.468	0.5	0.6
D - 12 %	63	1.040	0	0
E - 10 %	101	1.670	0	0

In Tab. I. are demonstrated the results of proteolytic activity determination in five samples of beer obtained by three methods /with labelled substrate, immunoassay and radioimmunoassay/. Only IA and RIA are reliable for the determination of papain because they enable the exact measuring of papain concentration. In the first three samples which contain papain good correspondence between IA and RIA was found, RIA can, however, be considered as a more precious method. The real proteolytic activity of all beer samples was measured under comparable conditions by means of labelled albumin. In samples A and B the proteolytic activities are very low. This finding indicates that papain is detectable by means of immunochemical methods even when it loses its proteolytic activity. On the contrary samples D and E show that in some cases the proteolytic activities can be quite high even when papain was not added to beer. These activities belong to yeast proteases. The previous results demonstrate that methods with labelled albumin, as well as similar methods using unspecific protein substrates, are not suitable for the detection and determination of papain in beer.

We would like to conclude that further development of suitable methods is desirable.

Applications of Isoelectric Focusing to the Analysis of Food Proteins

P.G. RIGHETTI[1] and ADRIANA BIANCHI BOSISIO[2]

[1] Department of Biochemistry, University of Milano, Via Celoria 2 and
[2] Clinical Chemistry Department, S. Carlo Borromeo Hospital, Via Pio II, Milano 20133, Italy

SUMMARY

After a brief historical survey, analytical aspects of isoelectric focusing (IEF) are reviewed, with particular emphasis on ultrathin polyacrylamide gel layer techniques, sample preparation before IEF, use of additives, artefacts and staining and destaining methodologies, including the most recent silver stain. The survey on food analysis includes screening of cheese products, pasta and macaroni, frozen fish fillets and canned crab meat, and determination of vegetable proteins in meats and sausages. Among the beverages, beer proteins and beer haze formation have been investigated by IEF. The review ends with a chapter on the effects of radiation on proteins, and on combined radiation and heat processes for food sterilization.

INTRODUCTION

On the genesis of isoelectric focusing

IEF, which has been the raging separation technique of the decade 1970-1980, as disc electrophoresis was in the decade 1960-1970, was born after an incredibly long gestation period. In 1954-56, Kolin (1-3) conceived the idea of "focusing ions in a continuous pH gradient", stabilized by a sucrose density gradient. He used the term "isoelectric spectrum" to denote a distribution pattern established by a sorting process. The

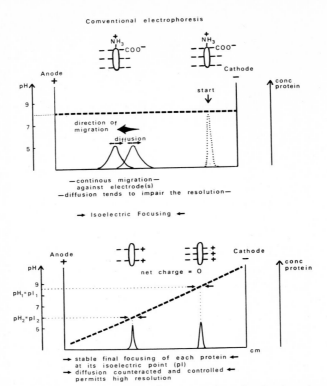

Fig. 1. Upper: schematic drawing of conventional electrophoresis of two proteins. The protein concentration at the start and after some time of migration toward the anode is indicated. Note the constant pH between anode and cathode during the experiment (heavy broken line). This means that the proteins have the same net charge all the time, as is illustrated in the symbols of the proteins in the upper part of the figure. Usually the pH of the buffer solution is selected so that the proteins acquire a negative net charge (pH>pI) which makes the proteins migrate toward the anode. The larger the difference (pH-pI) the higher is usually the negative net charge and the migration velocity. Lower: schematic drawing of IEF of two proteins when a stable final condition has been obtained. Irrespective of the position of the protein at the start, each protein migrates to and is focused at a pH=pI, where the net charge of the protein is zero. The diffusion is counteracted and controlled during the time the current is on. The pI is determined by measuring the pH where the protein has its maximum concentration (from Vesterberg[10]).

pH gradient was generated by placing the substance to be separated at the interface between an acidic and basic buffer, in a Tiselius-like apparatus, and allowing diffusion to proceed under an electric field. In these artificial pH gradient, Kolin was able to obtain "isoelectric line spectra" of dyes, proteins, cells, microorganisms and viruses on a time scale ranging from 40 sec up to a few min, a rapidity still unmatched in the field of electrophoresis. Concomitant with the pH gradient, a density gradient, an electric conductivity gradient and a vertical temperature gradient were acting upon the separation cell. The main drawback was the instability of these gradients, due to rapid migration of the buffering electrolytes during electrophoresis.

In the 1960's Svensson (alias Rilbe) (4-6) laid the foundations of IEF in its present form with a series of theoretical articles entitled: "isoelectric fractionation, analysis and characterization of ampholytes in natural pH gradients". He introduced the law of pH monotony, and studied the protolytic properties of carrier ampholytes, their conductivity, their buffering capacity, their titration curves and the resolving power of the technique in presence of suitable ampholytes, capable of generating stable pH gradients. These concepts were achieved in practice with the synthesis by Vesterberg (7) of carrier ampholytes with many of the properties described by Svensson.

In a recent historical article Kolin (8) has pointed out that, in a sense, IEF can be considered as a null method in which particles reach equilibrium in a focal zone of vanishing force. He has generalized the idea of IEF and shown that it may be considered as a member of a family of phenomena - isoperichoric (isos=equal and perichoron=environment) focusing effects - "in which particles migrating in a force field along a chemical concentration gradient are swept into sharp, stable stationary zones of vanishing force in which a physical parameter of the suspension medium matches that of the particle. Examples of such condensation phenomena are isopycnic-, isoconductivity-, isoedielectric-, isomagnetic-, isoparamagnetic- and isodiamagnetic- focusing". This article (8), as well as a similar historical article by Rilbe (9), on the "isoelectric" research group in Stockholm, make a fascinating reading.

The principle of isoelectric focusing, as compared to zone electrophoresis, is illustrated in Fig. 1 (ref. 10). In the latter, the protein zones travel on a support buffered at constant pH, therefore they maintain a constant surface charge and keep progressively diffusing during migration, with concomitant loss of resolution. In the former, a pH gradient is created and stabilized by the current on the support medium, so that proteins are progressively titrated to a zone of minimum charge (zero net charge), where they condense, or focus, in very sharp, stationary zones. An inherent advantage of IEF, as well as any other equilibrium technique, such as isopycnic centrifugation, is that the diffusion is counteracted and controlled, so that the band sharpness is maintained to the very end of the separation, by the combined action of several factors, such as voltage applied, slope of pH gradient, slope of the protein titration curve in the pI region, and medium viscosity. The focusing effect in IEF can be better appreciated in Fig. 2. If, to a prefocused polyacrylamide gel slab, the same protein is applied simultaneously in proximity of the anodic and cathodic regions, it is seen to converge and merge in the same isoelectric zone. This experiment is best done with a colored

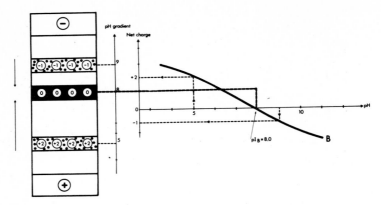

Fig. 2. Condensation or focusing effect in IEF. A protein with pI=8 is applied to the IEF column, in a preformed pH gradient, as an anionic (negatively charged, pH 9) as well as cationic (positively charged, pH 5) species. Both forms converge and merge in the same isoelectric zone (net charge zero) at pH 8. On the right, the protein titration curve is depicted (courtesy of LKB Produkter AB).

protein marker (e. g. hemoglobin) or with a colored amphoteric ion, such as doxorubicin (11).

Since its official introduction as a separation technique, around 1966, IEF has been the subject of several international meetings and a host of reviews, which cannot possibly be covered in the present article. For the reader willing to obtain a deeper insight in this topic, we suggest a laboratory manual, a monograph (12), which covers all the theoretical and practical aspects of IEF. After this book, we have published yearly a series of reviews covering the more recent developments of IEF (13-16). A very useful literature reference list, called Acta Ampholinae, is published by LKB Produkter AB. A volume covering the years 1960-1974 is supplemented yearly with references on both, IEF and isotachophoresis. As of this writing, more than 4,000 articles dealing with IEF have been listed in Acta Ampholinae.

Apparatus for analytical IEF

Since in the vast majority of cases food chemists deal with analytical applications, we will not review here preparative aspects of IEF. For versatility, ease of handling, staining, destaining and storage the slab technique in polyacrylamide gels (PAGIEF) is by far superior to any other methodology presently available. The gel may be held either vertically or horizontally between cooling plates. We prefer horizontal systems since the gel is subjected to less mechanical stress. Simple electrode arrangements are possible and there is more flexibility in methods for sample application. Separations may be conducted either along the short (10

Applications of Isoelectric Focusing to the Analysis of Food Proteins 251

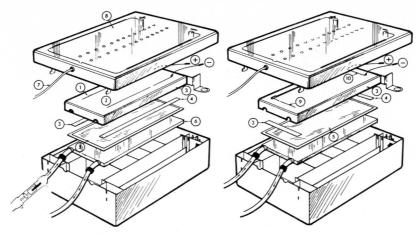

Multiphor for electrofocusing in thin layer polyacrylamide gel

Fig. 3. Drawing of the LKB Multiphor apparatus. Left: arrangement for IEF on the short side(10 cm). Right: arrangement for IEF on the long side (25 cm). (1): and (2): anodic and cathodic platinum wires, respectively, for runs in the short side; (3): platinum electrode holder and cover lid; (3') and (4): anodic and cathodic filter paper strips, respectively; (5): thin layer polyacrylamide gel cast on a 1 mm thick glass slab; (6): cooling block; (7): cable connection to power supply; (8): safety cover slip; (9) and (10): anodic and cathodic platinum wires, respectively, for runs on the long side (courtesy of LKB Produkter AB).

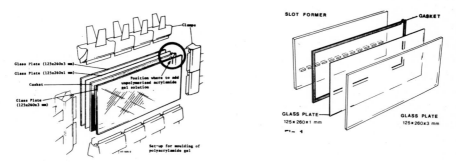

Fig. 4. Polyacrylamide moulding cell for the LKB Multiphor apparatus (left). Right: same cell, but with a cover lid for pocket formation (courtesy of LKB Produkter AB).

cm distance) or along the long axis (25 cm). Gels are usually cooled through their bottom surface and the electrolytes are soaked into filter paper strips, placed on the gel surface, on which a cover lid with platinum electrodes is seated. A typical apparatus design for thin layer is shown in Fig. 3. Similar apparatus are now available also from Desaga, Bio Rad and Pharmacia. In most systems, the gel is usually cast between two glass plates, separated by a rubber gasket (17). The plates are clamped together to form a water-tight chamber (Fig. 4, left). One of the two

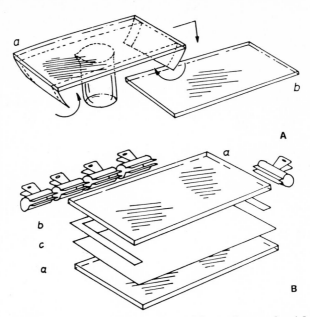

Fig. 5. (A): procedure for casting ultrathin polyacrylamide gels onto cellophane foils. A wet cellophane sheet, 2-3 cm wider than the 1 mm thin glass used for the gel cassette, is stretched and rolled flat on this glass (a), the excess film folded beneath and then (b) transferred onto the 3 mm thick glass for drying. (B): assembly of the gel cassette. (a): 3 mm thick glass; (b): U-shaped Parafilm gasket (1 layer = 120 μm); (c): 1 mm thin glass coated with cellophane film (from Görg et al.[18]).

plates may also be used as a template to imprint indentations into the gel for convenient sample application (Fig. 4, right). Originally, gels 2 mm thick were cast; subsequently, LKB introduced ready-made polyacrylamide gel plates (PAG-plates) which have a thickness of 1 mm. However, a real breakthrough has been made with the invention of ultrathin gels, as described by Görg et al. (18). Gels can be cast in three thickness ranges, 120, 240 or 360 μm, by using as gaskets 1, 2 or 3 layers, respectively, of Parafilm cut in a U-shape. Fig. 5 illustrates the polymerization process. Since such ultrathin gels are not resistant to tear, they are cast on cellophane foils, to which they adhere rather firmly. Alternatively, they can be covalently bound to a glass surface by using Silane A-174 (ref. 19), or can be supported onto partially hydrolyzed polyester film (20). The advantages of ultrathin gels are several fold: resolution is markedly improved; minimal temperature gradients are generated in the gel thickness; zymograms, staining, destaining and drying of the gels are accomplished in a fraction of the time needed for standard slabs. Recently, this method has been extended to two-dimensional techniques, both in single-pore gels and in porosity gradients, for the display of pI-MW protein maps (21). Ultrathin layer gels appear to be the technique of the future.

Sample preparation. Additives

Only in a few cases (e. g. frozen meats and fish) can the sample be analyzed after simple homogenization in appropriate buffer and removal of cell debris. In most cases (e. g. cooked meats and sausages, pasta, seed proteins) the samples can only be solubilized in presence of disaggregating agents, notably urea and detergents. These additives also prevent or lessen isoelectric precipitation during IEF analysis and are therefore incorporated in the IEF gel. In addition to stabilizing or solubilizing focused bands, some additives also have the advantage of reducing differences in osmolarity along the gradient caused by focused Ampholine. Non-ionic detergents have been successfully used in concentrations ranging from 0.1 to 5%. Many such detergents are now available under various trade names, e.g. Tween 80, Emasol, Brij 39, Triton X-100, Nonidet P-40 (NP-40). In addition to non-ionic, also zwitter-ionic surfactants, such as alkylbetaines: $R-\overset{+}{N}(CH_3)_2-CH_2-COO^-$ (R = C_{10} to C_{16}) or the sulphobetaines: $R-\overset{+}{N}(CH_3)_2-CH_2-CH_2-SO_3^-$ (R = C_{12}) can be effectively used in IEF (22).

Owing to their large pK differences ($\Delta pK = 11$) they are practically isoelectric between pH 3 and pH 10. Alper *et al.* (23) have used a similar approach by incorporating in the focusing system 0.2M taurine ($_3^+HN-CH_2-CH_2-SO_3^-$) which too is isoelectric over a wide pH range and serves to raise the medium osmotic pressure without contributing to its electrolyte concentration or to the conductivity of the liquid phase. Glycine, taurine or trimethylamino propane sulphonate have all been successfully used to maintain an iso-osmolar environment (300 mOms) for IEF of cells (24). Particularly attractive seems to be a combination of 9M urea and 2% NP-40 in the focusing medium (25-27). Urea dissociates aggregates and subunit assemblies and unfolds polypeptide chains, thereby reducing the influence of protein conformation on the focusing pattern and releasing possible non-covalently bound cofactors, while NP-40 helps disaggregating and solubilizing hydrophobic proteins. When using urea, in order to avoid possible carbamylation of proteins by cyanate, we suggest that the sample be solubilized in presence of 2% carrier ampholytes, which contain reactive amino groups and should scavenge cyanate. Moreover, a sample solubilized in urea should be analyzed within the same day, and never after storage, and should be applied to a prefocused gel at the anodic side, since no cyanate is produced, nor carbamylation occurs below pH 5 (ref. 28). Other disaggregating agents can be used in gel IEF: 50% dimethyl sulphoxide, 50% dimethyl formamide, 30-35% tetramethyl urea and up to 90% formamide (29).

Scientists have often been strongly disturbed by the impressive heterogeneity found by running any sample, no matter how purified, in gel IEF, and have often attributed this polydispersity to an artefact of the technique, due to binding of carrier ampholytes to the protein sample. We have recently dispelled this superstition (30): there is no such a thing as Ampholine-induced heterogeneity in IEF, the sample polydispersity being already inherent and present in the protein before the IEF analysis. It is true that we have reported a series of artefacts in IEF (31-33), but they regard only limiting structures, mostly polyanions, like heparin, and other sulphated or carboxylated polysaccharides, as well as polyglutamate, polyaspartate, polyphosphates, polysulphates and nucleic acids, such as t-RNA (34). The only protein known to give the same type of artefacts is the rat incisor phosphoprotein (RIP), but this is because RIP behaves in fact in solution as a polyanion, since it displays the same charge density as heparin.

Staining and destaining

Special procedures have to be used for staining proteins separated by IEF, since the carrier ampholytes form insoluble complexes with many protein stains. This problem can be overcome by first precipitating focused proteins in acid, e.g. 10% trichloroacetic acid (TCA), and eluting the acid-soluble ampholytes with extensive washing. After ampholyte removal, proteins can be detected by conventional methods. Today, several direct staining procedures have been developed to circumvent this laborious procedure, by solubilizing the dye-ampholyte complexes in alcoholic solvents and/or high temperatures. We will give here examples of the staining systems, which appear to be the most popular and sensitive: the two Coomassie Brilliant Blue (CBB) stains (R-250 and G-250) and the silver stain.

Coomassie Brilliant Blue R-250. Hayes and Wellner (35) fix the gel in 5% TCA-5% sulphosalicylic acid. THe excess acid is washed in H_2O and gels transferred to a solution of 0.066% CBB-R-250 in 0.2M Tris-HCl buffer, pH 7.7, for 3 to 4 hours. The background stain is removed in the same, diluted buffer (0.001M) for 24 to 48 hours. According to Spencer and King (36), gels are bathed in 0.01% CBB R-250 in 5% TCA, 5% sulphosalicylic acid and 20% methanol. It is less sensitive than other procedures, but background staining is very low. According to Vesterberg (37), a direct method, without Ampholine interference, consists of heating the gels at 60°C for 15 min in 0.1% dye in 28% methanol, 11% TCA and 3.5% sulphosalicylic acid. Destaining is also affected at 60°C in 28% ethanol-8% acetic acid. Beware, because the combination of alcohol and high temperatures can easily remove the dye from from the protein. A modification suggested by Söderholm et al. (38) consists in first leaching out Ampholine in 2% sulphosalicylic acid, 11% TCA and 27% methanol at 65°C. After 20 min, the gels are washed in destaining solution (9% acetic acid, 27% ethanol) and then stained in 0.1% dye dissolved in destaining solution. Recently, Vesterberg and Hansén (39) have described four additional variations, using various combinations of TCA, sulphosalicylic acid, perchloric acid (PCA) and urea in the fixing solution. It appears that urea increases the stain sensitivity by unfolding the protein chains. The detection limit appears to be around 0.2 μg protein per band. Righetti and Drysdale (40) have suggested to stain in an acid-alcohol solvent in the presence of 0.1% cupric sulphate to enhance the stain intensity by interaction of Cu^{++} with peptide bonds. Malik and Berrie (41) described a protein staining procedure with CBB R-250 prepared using sulphuric acid, KOH and TCA. This is a rapid staining procedure, but the gel background becomes significantly stained.

Coomassie Brilliant Blue G-250. Diezel et al. (42) have introduced the use of CBB G-250 (a dimethyl substitute of the R-250 dye) as a colloidal dispersion in 12.5% TCA. This causes a selective binding to the protein zones and prevents the dye's penetration into the gel network, thus avoiding background staining. Upon storage in 5% acetic acid, the proteins are intensified as the protein-bound dye becomes soluble, diffuses into the gel and binds again to the interior portions of the protein bands. Reisner et al. ((49) have found that 0.04% CBB G-250 in 3.5% PCA remains in a leuco form, but changes to the blue form when coupled to proteins. Ampholine does not precipitate, but the gel takes a faint orange to brownish background. By introducing an additional wash in 5% acetic acid after staining, Holbrook and Leaver (44) have claimed a 3 fold increase in sensitivity (but the background becomes heavier). Blakesley and Boezi (45) have described a recent modification which appears to be excellent for IEF in gels. To 400 ml of 0.2% dye in H_2O an equal volume of 2N H_2SO_4 is added. After standing for 3 hours, the precipitate is removed through a Whatmann 3MM paper. To the clear brown filtrate, 90 ml of 10N NaOH are added, producing a dark blue solution. To this, 120 ml of 100% TCA are added, and the greenish solution is ready for use. This dye can be used several times, provided the pH is maintained below 1.0 (in fact, for IEF gels, the used dye works much better than the fresh solution, probably because a trace impurity, which tends to precipitate neutral Ampholine, is eliminated). In 0.7 mm thick slabs, maximum color development is achieved after 5 hours (28) but the bands are already visible within 15 min. When placed in water, there is a marked color intensification. This dye works wonders for peptide analysis by IEF, allowing fixation and detection of chains as short as 12-15 amino acids (28).

Silver staining. This stain seems to achieve the ultimate sensitivity of any detection technique, since it approaches the detection limit of autoradiography (46). It is based on interaction of ammoniacal cupric-silver complexes with protein amino acids (possibly in the region of the peptide bond) and its sensitivity is said to be 100 times higher that the conventional CBB stain (e.g. detection of 0.38 vs. 38 ng/mm^2 of serum albumin). With some proteins, as little as 10^{-2} pmol/mm^2, or just a few molecules per cubic micrometer, could be detected. Visualization of molecules at such low concentration suggests a physical growth of the initially deposited silver on the protein zones, by an amplification mechanism similar to that occurring in developing a photographic emulsion.

FOOD, MEAT AND BEVERAGE ANALYSIS

At last we come to the applications of IEF to the analysis of foodstuff. The importance of analysis of packaged foods can never be overemphasized. Just as an example, in the case of meats, this product can be replaced or extended by other proteins, such as egg powder, dried milk powder, soya beans and field beans. In order to detect adulteration of the meat, or to ensure that the amounts of non meat proteins do not exceed recommended levels, fast and reliable analytical techniques have to be at hand. In other cases (e.g. frozen fish fillets), once the tell-tale features have been removed, it is taxonomically impossible to identify the species. Thus not unfrequently, in italian supermarkets, it was recently found that the gourmet sole fillets had been substituted with cheaper frozen fillets of *Pleuronectes platessa*. A minireview on the use of IEF for food analysis, in taxonomy and in agricolture has been written by Bishop (47).

Cheese

The clotting of milk in the production of cheese is brought about by the proteolytic enzyme rennin, which is found in the calf's fourth stomach. The large increase in cheese consumption and the simultaneous decrease in the amount of calf stomachs available for production have led to a world shortage of rennin. At present, the most common rennet substitute for milk-clotting in cheese production are bovine and pig pepsin and microbial enzymes isolated from fungi: acidic proteases from *Mucor pusillus*, *M. miehei* and *Endothia parasitica*. The cheese-maker needs to know the clotting activity of the rennet and its enzyme composition, since this plays an important role in the final quality of cheese, and needs to identify counterfeited rennets sold in the market as "calf rennet". The resolution and unequivocal characterization of all these types of rennets has been made possible by IEF in thin layer plates (48, 49): all these enzymes focus in the pH range 2.8 to 5.5 and give a highly characteristic spectrum of isozymes.

During cheese ripening, the casein micelle (which contains α and β chains in an approximate ratio of 3:2, plus small amounts of κ chains) is digested to smaller fragments by lytic enzymes from a host of microorganisms, such as micrococci, yeasts, molds and lactic acid bacteria. These casein digests can be fingerprinted by IEF followed by SDS electrophoresis, thus allowing the display of the pI and MW of each fragment. Righetti et al.(50) have used this approach for monitoring the digestion of casein by different strains of lactobacilli (*helveticus*, *lactis*, *bulgaricus*, *acidophilus* and *jugurti*). Most lactobacilli brake the casein micelle in the same way, yielding a series of small acidic fragments and two major,neutral chains of 14,000 daltons (pI 6.4) and 10,000 daltons (pI 7.1). A completely different picture is given by *L. jugurti*, which brakes down extensively all caseins, leaving only two core fragments undigested, a pI ≈9 band of about 4,000 daltons and a pI ≈3 band of ca. 5,500 daltons.

Pasta and macaroni

The analysis of pasta and macaroni is quite important for southern europeans. The main problem here is to detect soft wheat (*Triticum aestivum*) in durum wheat *(T. durum)* in semolina and macaroni products. The analysis if further complicated in presence of egg proteins or after heat denaturation of pasta. Resmini and DeBernardi (51-53) have proposed an analytical method based on the IEF, on a polyacrylamide gel slab, of the semolina extract at pH 4.5. After IEF, the focused bands can be visualized by simple immersion of the gel in 12% TCA: the protein zones appear as opalescent bands against a black background. 1% soft wheat could be detetcted at a semiquantitative level in durum wheat; 3% hexaploid in durum wheat was already amenable to exact determination. Soft wheat exhibits in the IEF pattern three typical bands, the two upper ones having pI's of 7.2 and 7.1, respectively. The presence of egg proteins does not seem to interfere with this determination, since these proteins (particularly ovalbumin) focus at much lower pH's (around pH 5). A similar analysis has been proposed by Windemann *et al.* (53) whe have also adopted two-dimensional separations (IEF followed by electrophoresis).

Fish

Conventional electrophoretic techniques generally lack the resolution and reproducibility needed for the reliable identification of fish species. Earlier data (54, 55) had shown the high potential of IEF for taxonomical studies of white fish. Sarcoplasmic proteins are extracted from fresh iced or frozen fish by simple homogenization in water or by mincing the flesh in water or simply by using the drip fluid from thawed frozen fish. In more recent studies (56, 57), by using a blind test, 16 unknown frozen fish fillets were identified with 93% accuracy by the IEF technique. It turned out that all wrong identifications were monkfish samples, perhaps due to the fact that, while all other samples were obtained as whole fish, monkfish is normally sold as monktail and different monkfishes from different areas could have been mixed. This IEF technique has now been adopted by the American Association of Official Analytical Chemists. Analysis of aqueous extracts of fish raw skeletal muscles has also been proposed by Kaiser *et al.* (58) who have adopted IEF in both, polyacrylamide gels and agarose matrices.

The problem is more complex if the fish meat has been cooked. This is the case with crabs which, in order to be frozen or canned for marketing, must be cooked and have the meat picked or rolled from the bodies, legs and claws. Once the species' distinguishable features (claws, carapace) have been removed, it is taxonomically impossible to identify the species. The problem has been solved by extracting the cooked meat in 5M urea and focusing in 5M urea gels. Species-specific patterns were thus produced in 8 species of crab: red, rock, jonah, blue, king, snow, european edible and dungeness (59). The species differences were more pronounced in the alkaline pH range (pH 7-9), but the banding pattern was altered by the degree of heat used in processing the crab meat. On the other hand, length of frozen storage did not change the banding patterns, which, among individuals of a single species, did not vary with size, sex or maturity.

Meats and sausages

There are two different aspects of meat analysis: one is the species identification of butchered meat, the other is the analysis of non-meat proteins in processed meats (e.g. pork, sausage, beefburgers). In the latter case, for instance, vegetable proteins (particularly soya beans) are now used in meats served in schools as well as in a large number of commercial and institutional catering establishments. Flaherty (60) and Llewellyn and Flaherty (61, 62) have devised a method for analysis of soybean proteins in meats. The meat products are heated, in a aqueous suspension, at 100°C for 15 min; this renders the meat proteins mostly insoluble, thereby enabling the selective extraction of the soya proteins in 6M urea-2% 2-mercaptoethanol. At 1% level in a fresh sausage, the soya could be detected but not quantified; at 5% the soya could be determined with a relative error of ∓20%. Frozen beefburgers containing soya proteins could be analyzed, but not canned ones: the high temperatures used in canning appear to render also the vegetable proteins insoluble in urea-mercaptoethanol. Meat analysis by IEF, for species identification, has also been proposed by Tinbergen and Olsman (63) and by Kaiser et al. (58).

Effects of radiation on proteins

One promising approach in food manipulation and storage is the possibility of prolonging their shelf life by irradiation with a radioactive source. Microorganisms can be effectively reduced or eliminated in a number of food products by a radiation treatment and microbiologically stable products can thus be obtained. Irradiation is used to extend storage life of high protein foods (meat, fish, egg) as well as of fruits and vegetables, to delay maturation or prevent fungal rotting. The radiation doses range from low levels (0.1-0.3 Mrad, radurization, which does not change the organolectic quality and extends the storage life by a factor of two to four) to high levels (3-5 Mrad, radappertization, which should produce microbiologically stable food by destroying most, or all, the microorganisms in food). However, even at the highest doses, autolytic enzymes remain still active so that, in irradiated meats, almost complete destruction of the sarcoplasmic proteins occurs during storage for 6 months at 37°C. Radola and Delincée (64) have studied this aspect of enzyme inactivation by IEF of horseradish peroxidase. This enzyme is present in many foods and residual activity is thought to be involved in loss of flavour in processed food upon storage. They have found that, even at the highest ionizing radiation doses, the enzyme aggregated, but still retained activity. The isozymes with the highest pI disappeared with increasing dose, but there was a parallel increase in the low pI species. A high synergistic effect was observed if irradiated enzymes were subsequently heat-inactivated: practically all activity was destroyed. Conversely, if the heat-treated sample is then irradiated, enzymatically active aggregates accumulate. This seems to be true also with sarcoplasmic proteins from irradiated beef and suggests a future strategy for a combined irradiation-heat processing of food.

Beer

IEF of beer proteins has enabled scientists to understand the mechanism of beer haze formation (the turbidity or precipitate that forms upon chilling). However, there seems to be a completely different mechanism in haze formation in english as compared to danish beers. According to Savage and Thompson (65-67), who have performed IEF of beer proteins and beer hazes, a group of acidic proteins (pI < 5.0, acidic fraction) are very important in haze formation. These pI < 5.0 proteins are the immediate precursors of beer haze, and are principally formed by the interaction of more basic proteins with simple polyphenol present in wort and beer. A completely different picture emerges from the Carlsberg group (68, 69). They too are able to resolve by IEF more than 30 discrete bands isoelectric between pH 3.5 and 10, with a much greater proportion of acidic bands (in the pH range 3.5 to 5.0). Subfractions of a class of acidic components (pI 4-5), with a molecular weight of ca. 44,000 daltons, were found to have an amino acid composition resembling barley albumins and globulins. These proteins (or large peptides) were found to have approximately half of the ε-amino groups of lysine residues blocked and to be quite rich in carbohydrates. It has been suggested that a large part of the beer proteins might be polypeptide chains cross-linked by carbohydrates, via a Maillard-type reaction. Practically no polyphenols could be detected in these acidic fractions. More work is clearly needed in this field to elucidate these aspects.

Conclusions

We hope that this brief survey of the field will give to the reader a feeling of the tremendous potential and the vast range of applications of the IEF technique. IEF has been already extensively used in analysis of proteins and enzymes from cereals, oil seeds and other plants and vegetables, and has been instrumental in genetic studies aimed at improving the nutritional value of these seeds. Surprisingly, not much has been done in the analysis of processed foodstuff, and we feel that there is still a lot of space in this area.

ACKNOWLEDGEMENTS

The literature survey has been completed in december-10-1980. Supported in part by grants from Consiglio Nazionale delle Ricerche (CNR) and Ministero della Pubblica Istruzione (MPI) to P.G.R. We are grateful to colleagues who have provided reprints of their work and the photographs reproduced herein.

REFERENCES

1. A. Kolin, J. Chem. Phys. 22 (1954) 1628-1629
2. A. Kolin, Proc. Natl. Acad. Sci. 41 (1955) 101-110
3. A. Kolin, in Methods of Biochemical Analysis (D. Glick, ed.) vol. 19, Wiley, Interscience, New York, pp. 259-288 (1971)
4. H. Svensson, Acta Chem. Scand. 15 (1961) 325-341
5. H. Svensson, Acta Chem. Scand. 16 (1962) 456-466
6. H. Svensson, Arch. Biochem. Biophys., Suppl. 1 (1962) 132-140
7. O. Vesterberg, Acta Chem. Scand., 23 (1969) 2653-2666
8. A. Kolin, in Electrofocusing and Isotachophoresis (B.J.Radola and D. Graesslin, eds.) de Gruyter, Berlin, pp. 3-33 (1977)
9. H. Rilbe, Sci. Tools 23 (1976) 18-22
10. O. Vesterberg, in VII Symposium on Chromatography and Electrophoresis, Publ. Press Acad. Europeennes, Brussel, pp. 81-98 (1973)
11. P.G. Righetti, M. Menozzi, E. Gianazza and L. Valentini, FEBS Letters 101 (1979) 51-55
12. P.G. Righetti and J.W. Drysdale, Isoelectric Focusing, North Holland /American Elsevier, Amsterdam and New York (1976)
13. P.G. Righetti, in Electrokinetic Separation Methods (P.G. Righetti, C.J. vanOss and J. Vanderhoff, eds.) Elsevier/North Holland Biomedical Press, Amsterdam and New York, pp. 389-441 (1979)
14. P.G. Righetti, E. Gianazza and A. Bianchi Bosisio, in Recent Developments in Chromatography and Electrophoresis (A. Frigerio and L. Renoz, eds.) Elsevier, Amsterdam, pp. 1-36 (1979)
15. P.G. Righetti, E. Gianazza and A. Bianchi Bosisio, in Recent Developments in Chromatography and Electrophoresis, vol. 10 (A. Frigerio and M. McCamish, eds.) Elsevier, Amsterdam, pp. 89-117 (1980)
16. P.G. Righetti, E. Gianazza and K. Ek, J. Chromatogr. 184 (1980) 415-456
17. H. Davies, in Isoelectric Focusing (J.C. Arbuthnott and J.A. Beeley, eds.) Butterworths, London, pp. 97-116 (1975)
18. A. Görg, W. Postel and R. Westermeier, Anal. Biochem. 89 (1978) 60-70
19. A. Bianchi Bosisio, C. Loeherlein, R.S. Snyder and P.G. Righetti, J. Chromatogr. 189 (1980) 317-330
20. B.J. Radola, Electrophoresis 1 (1980) 43-56
21. A. Görg, W. Postel, R. Westermeier, E. Gianazza and P.G. Righetti, J. Biochem. Biophys. Methods 3 (1980) 273-284
22. J.C. Allen and C. Humphries, in Isoelectric Focusing (J.C. Arbuthnott nad J.A. Beeley, eds.) Butterworths, London, pp.347-354 (1975)
23. C.A. Alper, M.J. Hobart and P.J. Lachmann, in Isoelectric Focusing (J.C. Arbuthnott and J.A. Beeley, eds.) Butterworths, London, pp. 306-312 (1975)
24. J. McGuire, T. Miller, R. Tips, R.S. Snyder and P.G. Righetti, J. Chromatogr. 194 (1980) 323-333
25. P. O'Farrell, J. Biol. Chem. 250 (1975) 4007-4021
26. K. Zechel, Anal. Biochem. 83 (1977) 240-251
27. P.G. Righetti, E. Gianazza, A.M. Gianni, P. Comi, B. Giglioni, S. S.tolenghi, C. Secchi and L. Rossi Bernardi, J. Biochem. Biophys. Methods 1 (1979) 45-55

28 P.G. Righetti and F. Chillemi, J. Chromatogr. 157 (1978) 243-251
29 P.G. Righetti, E. Gianazza, O. Brenna and E. Galante, J. Chromatogr. 137 (1977) 171-181
30 E. Gianazza and P.G. Righetti, in Electrophoresis '79 (B.J. Radola, ed.) de Gruyter, Berlin, pp.129-140 (1980)
31 P.G. Righetti and E.Gianazza, Biochim. Biophys. Acta 532 (1978) 137-146
32 E. Gianazza and P.G. Righetti, Biochim. Biophys. Acta 540 (1978) 357-364
33 P.G. Righetti, R. Brown and A.L. Stone, Biochim. Biophys. Acta 542 (1978) 232-243
34 E. Galante, T. Caravaggio and P.G. Righetti, Biochim. Biophys. Acta 442 (1976) 309-315
35 M.B. Hayes and D. Wellner, J. Biol. Chem. 244 (1969) 6636-6644
36 E.M.Spencer and P.T. King, J. Biol. Chem. 246 (1971) 201-208
37 O. Vesterberg, Biochim. Biophys. Acta 257 (1972) 11-19
38 J. Söderholm, P. Allestam and T. Wadström, FEBS Letters 24 (1972) 89-92
39 O. Vesterberg and L. Hansen, in Electrofocusing and Isotachophoresis (B.J. Radola and D. Graesslin, eds.) de Gruyter, Berlin, pp. 123-133 (1977)
40 P.G. Righetti and J.W. Drysdale, J. Chromatogr. 98 (1974) 271-321
41 N. Malik and A. Berrie, Anal. Biochem. 49 (1972) 173-176
42 W. Diezel, G. Kopperschläger and E.Hofmann,Anal. Biochem. 48 (1972) 617-620
43 A.H. Reisner, P. Nemes and C. Bucholtz, Anal. Biochem. 64 (1975) 509-516
44 I.B. Holbrook and A.G. Leaver, Anal. Biochem. 75 (1976) 634-636
45 R.W. Blakesley and J.A. Boezi, Anal. Biochem. 82 (1977) 580-582
46 R.C. Switzer III, C.R. Merril and S. Shifrin, Anal. Biochem. 98 (1979) 231-237
47 R. Bishop, Sci. Tools 26 (1979) 2-8
48 P.J. de Koning and J. Th. M. Draaisma, Neth. Milk Dairy J. 27 (1973) 368-378
49 P.G. Righetti, B.M. Molinari and G. Molinari, J. Dairy Res. 44 (1977) 69-72
50 P.G. Righetti, P. Muneroni, R. Todesco and S. Carini, Electrophoresi 1 (1980) 37-42
51 P. Resmini and G. DeBernardi, Tecnica Molitoria 27 (1976) 97-109
52 P. Resmini and G. DeBernardi, Tecnica Molitoria 28 (1977) 139-143
53 H. Windemann, U. Müller and E. Baumgartner, Z. Lebensm. Unters. Forsch. 153 (1973) 17-22
54 B.M. Djupsund, LKB Application Note No. 243, Oct. 1976
55 R.C. Lundstrom and S.A. Roderick, Sci. Tools 26 (1979) 38-43
56 R.C. Lundstrom, J. Assoc. Off. Anal. Chem. 62 (1979) 624-629
57 R.C. Lundstrom, J. Assoc. Off. Anal. Chem. 63 (1980) 69-73
58 K.P. Kaiser, G. Matheis, C. Kmita-Dürrmann and H.D. Belitz, Z. Lebensm. Unters. Forsch. 170 (1980) 334-342

59 J. Krzynowek and K. Wiggin, .J. Assoc. Off. Anal. Chem. 62 (1979) 630-636
60 B. Flaherty, Chem. Ind. (1975) 495-497
61 J.W. Llewellyn and B. Flaherty, J. Food Technol. 11 (1976) 555-563
62 J.W. Llewellyn, Proceedings of the Analytical Division of the Chemical Society (1977) 75-76
63 B.J. Tinbergen and W.J. Olsman, Die Fleischwittschaft 56 (1976) 1495-1498
64 B.J. Radola and H. Delincée, Ann. Technol. Agric. 21 (1972) 473-486
65 D.J. Savage and C.C. Thompson, J. Inst. Brew. 77 (1971) 371-375
66 D.J. Savage and C.C. Thompson, J. Inst. Brew. 78 (1972) 472-476
67 D.J. Savage, C.C. Thompson and S.J. Anderson, in Isoelectric Focusing (J.P. Arbuthnott and J.A. Beeley, eds.) Butterworths, London, pp. 329-337 (1975)
68 J. Hejgaard and S.B. Sørensen, Compt. Rend. Trav. Lab. Carlsberg 40 (1975) 187-203
69 S.B. Sørensen and M. Ottesen, Carlsberg Res. Commun. 43 (1978) 133-144

Ultrathin-Layer Isoelectric Focusing, Electrophoresis and Protein Mapping of Must and Wine Proteins

A. GÖRG, W. POSTEL, R. WESTERMEIER and G. GÜNTHER

Lehrstuhl für Allgemeine Lebensmitteltechnologie der Technischen Universität München, D - 8o5o Freising - Weihenstephan

SUMMARY

An ultrathin-layer horizontal polyacrylamide gel system for isoelectric focusing, pore gradient electrophoresis and 2D-techniques is described and demonstrated by separating must and wine proteins. The influence of technological treatments on must and wine proteins, glycoproteins and enzymes are demonstrated and discussed.

The proteins of must and wine, which amount to only about 1o - 15 % of the total nitrogen substances, are in spite of their quantitatively small portion important for taxonomy as well as for cellarage economy. Protein-instable wines tend to develop undesired after-turbidities during their storage. As there is not necessarily a linear positive correlation between the probability of turbidity and the protein content, the differentation of wine proteins due to their electrophoretic mobilities, isoelectric points, and molecular weights as well as a distinction in their enzymes and glycoproteins are not only of scientific, but also of practical interest.

Several authors (1-6) have already demonstrated that the so-called wine protein is not a homogeneous substance, as it was supposed at first. It can, however, be separated into a distinct number of fractions, due to the applied method of separation.

An important methodical advance in the electrophoretical characterization of proteins has been made possible by the new technique of ultrathin-layer polyacrylamide gels polymerized on foils. The advantages of ultrathin-layer gels, theoretically described by Righetti and Drysdale in 1976 (7), could be put into practice and confirmed by us for the first time, as we succeeded in reducing the conventional gel thickness to one tenth. Preparation, applications and advantages of ultrathin-layer gels polymerized on foils were described by us in 1978 for the first time (8). Isoelectric focusing in ultrathin-layer gels did not only provide substantially improved separation results, but also considerably shorter separation, staining, destaining and drying periods. The handling of the gels is substantially facilitated, because they stick to a foil support; the reduced consumption of reagents causes cost cutting. This was reason enough for us to investigate whether ultrathin-layer polyacrylamide gels can be used universally for the rest of the electrophoretic methods. The following separation techniques with ultrathin-layer polyacrylamide gels have been described by us in various publications (8-13), and are to be explained now for the characterization of must and wine proteins:

Ultrathin-layer isoelectric focusing
Ultrathin-layer electrophoresis
Ultrathin-layer SDS electrophoresis
Ultrathin-layer gradient gel electrophoresis
Ultrathin-layer SDS gradient gel electrophoresis
Ultrathin-layer 2D-electrophoresis
Ultrathin-layer high resolution 2D-electrophoresis

With the help of these methods a differentation of must and wine proteins due to their isoelectric points, electrophoretical mobilities and molecular weights is possible. Specific staining techniques additionally differentiate among the proteins between enzymes and glycoproteins. The high number of analytical data is easily obtained with the methods shown here, as we succeeded in a high standardization and simplification of these methods.

All separation techniques are performed horizontally, also the 2D-techniques, so that only a single separation chamber is needed. The ultrathin-layer gels polymerized on foils are always prepared according to the same principle with only small modifications for isoelectric focusing, electrophoresis (with or without a gel gradient) as well as for 2D-techniques.

Preparation of Ultrathin-Layer PAA-Gels:
The polymerization of the PAA-gels is performed in a glass chamber (LKB's system, also called "clamps technique"), whereby merely one of the two glass plates was previously covered with a foil support, and instead of a conventional rubber gasket parafilm is used. The gel thickness can be varied by the number of the layers of parafilm (1 layer of parafilm ≙ 120 μm, etc.).

For the preparation of ultrathin-layer PAA-gels for isoelectric focusing, electrophoresis (with or without stacking gel, with or without gel gradient) and 2D-techniques see Figure 1A-F (8-13).

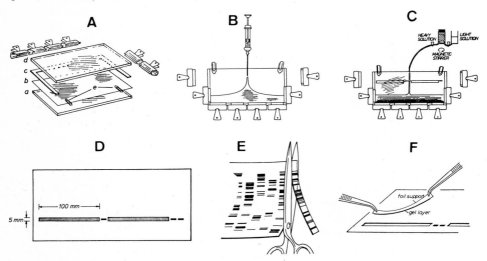

Fig. 1: Preparation of Ultrathin-Layer PAA-Gels.
(A) Polymerization chamber a) glass plate b) foil-support c) U-shaped gasket (Parafilm) d) glass-plate or template. e) spacers (B) Filling of the polymerization chamber. The polymerization solution is poured into the chamber to about half of the height. When the spacers are removed and the remaining two clamps installed, the liquid level rises uniformly to the upper edge. (C) Pouring of an ultrathin-layer exponential gradient gel. (D) Template for 2D-Techniques (E) Procedure for 2D-Electrophoresis: Cutting-off the focusing gel strip and (F) Transfer of the tear-proof focusing strip to the electrophoresis gel. For further details see Ref. (8-13).

Fig. 1C shows the casting of an ultrathin-layer gradient gel as we use it for the one-dimensional SDS gradient gel electrophoresis or for the high-resolution two-dimensional electrophoresis according to O'Farrell (14). The analogous chamber system is clearly evident, as we originally described for ultrathin-layer isoelectric focusing (8). As a new approach, we use a gradient mixer for small quantities instead of a syringe for filling-in the acrylamide solution. The casting of ultrathin-layer gradient gels (Fig. 1C) was developed by us in cooperation with Righetti and Gianazza in 1979 (13). Thus, the clamps technique is usable universally for all ultrathin-layer polyacrylamide gels for focusing as well as for electrophoresis (with or without a stacking gel, with or without a gel gradient), which is not possible with other recently modified casting techniques, as for instance the so-called "flip-flap technique" of Radola (15). The clamps technique is universal and thus reduces the instrumental as well as the methodical effort to a minimum.

Results

Fig. 2 shows the ultrathin-layer isoelectric focusing of must and wine proteins of the variety Rieslaner. The two arrows point to the only banding differences between must and wine. The must has two protein bands with the isoelectric points at pH 4.o; in contrast, the wine fermented from this must, shows two protein bands in the isoelectric-point region of pH 3.8. It is clearly evident, that must and wine proteins do not substantially differ, contrary to other reported opinions. The major part of the protein fractions of this variety is situated in the pH region between 3.8 and 7.o.

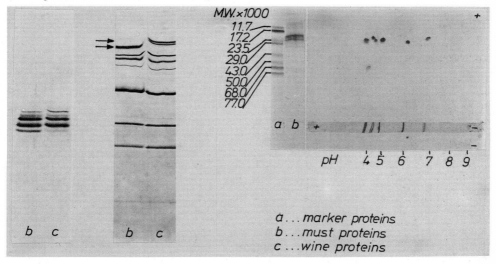

Fig. 2. Ultrathin-layer isoelectric focusing of must and wine proteins of the variety Rieslaner.
Fig. 3. Horizontal ultrathin-layer electrophoresis of Rieslaner must.
Fig. 4. Horizontal high-resolution 2D-electrophoresis (urea-IEF x SDS gradient gel electrophoresis) of Rieslaner must.

The protein, glycoprotein and the protein/glycoprotein double-staining of the pherograms shows that the major part of the must and wine proteins are glycoproteins (not illustrated here).

Isoelectric focusing of pasteurized or Bentonit-refined musts and wines shows that at first the neutral and later the weakly acidic proteins are denaturated or adsorbed. The acidic proteins, especially glycoproteins, are the most stable ones (no figure).

A distinct dependence of the peroxidase isoenzymes of the must and the wine on technological treatments was also demonstrated (no figure).

Figure 3 shows the ultrathin-layer electrophoresis of Rieslaner must. Whoever knows pherograms of disc electrophoretically separated must and wine proteins will certainly agree **that in this pherogram the bands are** separated very sharply, which is caused by the ultrathin-layer gel.

The horizontal SDS gradient gel electrophoresis (no figure) for molecular weight determination demonstrates that must and wine proteins are for the major part of low molecular size, mostly about 2o.ooo daltons. For the Rieslaner must additional fractions with 1o.ooo and 68.ooo daltons can be shown. The ultrathin-layer SDS gradient gel electrophoresis permit molecular weight determinations with only a small amount of sample, which is of interest especially for must **and** wine proteins.

Figure 4 shows the horizontal high resolution 2D-electrophoresis (urea IEF x SDS gradient gel electrophoresis) of Rieslaner must. It provides the unique possibility to characterize small protein amounts simultaneously in one pherogram according to two independent physico-chemical criteria, the molecular weights and the isoelectric points, without preparative steps. The pherogram resembles a coordinate system: the x-axis shows the isoelectric points, the y-axis the molecular weights. Every spot of the 2D-separation can be appointed in this coordinate system. The spots demonstrate clearly what could already be observed in the one-dimensional separations by isoelectric focusing and SDS gradient gel electrophoresis. The Rieslaner must proteins show above all a heterogeneity of their isoelectric points, whereas the molecular weights are mainly about 2o.ooo daltons, with exceptions of 68.ooo and 1o.ooo daltons.
This should only be a small part of the high number of possible applications of the above-described methods. The methodical and instrumental effort could be reduced to a minimum, so there may be no more obstacles for broad range applications of horizontal ultrathin-layer methods.

References

1. W. Diemair, J. Koch and E. Sajak, Z. Lebensm. Unters.-Forsch. 116 (1962), 7
2. R.H. Moretti and H.W. Berg, American J. Enol. Viticult. 16 (1965), 69
3. F. C. Bayly and H.W. Berg, American J. Enol. Viticult. 18 (1967), 18
4. L. Jacob, Dtsch. Weinbau-Jahrbuch 1969
5. R. Ebermann, J. Barna and F. Prillinger, Mitt. Klosterneuburg 22 (1972), 414
6. J. Barna, Mitt. Rebe, Wein, Obstbau 24 (1975), 413
7. P.G. Righetti and J.W. Drysdale, Isoelectric Focusing. North Holland, Amsterdam, Oxford and American Elsevier, New York 1976
8. A. Görg, W. Postel and R. Westermeier, Anal. Biochem. 89 (1978), 60
9. A. Görg, W. Postel and R. Westermeier, Z. Lebensm. Unters.-Forsch. 168 (1979), 25
10. A. Görg, W. Postel and R. Westermeier, GIT Labor Medizin 1 (1979), 32
11. A. Görg, W. Postel and R. Westermeier in: Electrophoresis'79, Walter de Gruyter, Berlin-New York 1980
12. A. Görg, W. Postel and R. Westermeier, Science Tools 27 (1980), 14
13. A. Görg, W. Postel, R. Westermeier, E. Gianazza and P.G. Righetti, J. Biochem. Biophys. Methods 3 (1980), 273
14. P.H. o'Farrell, J. Biol. Chem. 250 (1975), 4007
15. **B.J. Radola** in: Electrophoresis'79, Walter de Gruyter, Berlin-New York 1980

Enzymes in Food Analysis

Christiane MERCIER

Institut National de la Recherche Agronomique, Centre de Recherches Agro-Alimentaires, Laboratoire de Biochimie des Aliments, Chemin de la Géraudière, 44072 Nantes Cédex France.

SUMMARY

The use of enzymes in food analysis has been reviewed. A classification of the methods has been suggested over six different groups. Methods using the enzyme-substrate specificity have been largely applied to carbohydrates (mono-, di-, oligo- and polysaccharides), alcohols and acetaldehyde, organic acids, amino acids, organic bases and triglycerides. Methods using enzymes as metalloproteins, based on their activation or inhibition, have been developed to determine trace quantities of metals, pesticides and insecticides. Enzymes have been recently used as a tool for structural analysis of macromolecules, such as triglycerides, glycogen and starch. They also can be involved not directly but as specific desintegrating agents in order to minimize the modification of the other constituents of food, such as in dietary fiber determination. The nutritive value of food may be estimated by *in vitro* tests, based on the enzyme susceptibility of proteins and starch, for instance. The recent technique of enzyme immobilization will lead in the future, to the development of fast and routine methods in food analysis, using enzymes electrodes and fibre entrapped enzymes.

INTRODUCTION

Food analysis involves the determination of a component in raw and processed materials. Therefore, a satisfactory analytical technique must be specific, reproducible, fast with an appropriate level of sensitivity. In general, chemical determinations are rarely specific in complex media and are time-consuming since previous extraction and purification are usually required to obtain good reproducibility.

Due to their specificity, enzymes became rapidly the ideal tool for analytical purposes and reviews are available on their use in food analysis (BERGMEYER, 1970, 1978; GUILBAULT, 1970, 1976; BOEHRINGER, 1971; WISEMAN, 1978; DRAPRON, 1979).

Although the use of enzymes for analysis does not necessitate the understanding of the complex kinetics of the enzyme, two basic properties explain the increase of their use : their specificity and their ability to act fastly under certain conditions of pH and temperature which do not alter the other constituents of the material, except their substrate.

Contrary to chemical catalyst, enzyme leads, as a biocatalyst, to only one type of reaction, acting on one exclusive substrate. For example, β-D-glucose oxydoreductase oxydises only β-D-glucose. β-galactoside - galactohydrolase hydrolyses β-galactosides and not β-glucosides which differ only by the hydroxyl group in position 4. In a racemic mixture, enzyme are able to distinguish between optical isomers as the L-lactico dehydrogenase, which recognizes only the L-lactic acid. However, such a specificity is not always as strict. Certain enzymes are able to catalyse the transformation of similar substrates. The oxydation of ethanol and others alcohols into aldehyde is obtained by the same alcohol dehydrogenase. Enzymes, acting on macromolecules, may recognize one type of linkage among others of the same nature, but with a different steric hindrance. For instance, chymotrypsin is able to hydrolyse the peptidic linkage of proteins exclusively located towards the terminal carbon of phenylalanine and tyrosine, amino acids which possess a similar steric hindrance; the same phenomenon is observed with trypsin which hydrolyses specifically linkages located towards the right of arginine and lysine.

The efficiency of enzyme hydrolysis has also to be taken into account for analysis. Whereas the acid hydrolysis of peptidic linkage, at 38°C, needs 2000 hours, the pepsin hydrolysis, at the same temperature, will be obtained within 2 hours. At 20°C, one molecule of lipoxygenase oxydises, in 1 second, 300 linoleic acid molecules, whereas, the oxydation with Cu^{++} needs more than 100 000 seconds (DRAPRON, 1974).

For food analysis, it is usually the substrate of one enzyme that is to be determined. The appropriate enzyme will select this substrate from a complex mixture and convert it to the product at a characteristic rate which depends on pH and temperature and on the concentrations of the enzyme and substrate, in absence of activators and inhibitors.

In order to obtain the define conditions to develop an enzymic determination the different parameters precited have to be precised.

Since the method has to be applied to food, which is generally complex media, the enzyme must be free of enzymic contaminants susceptible to hydrolyse others constituents of the material. In some cases, enzymic preparation can be used without any purification, provided that, under the experimental conditions, no action on others substrates could be detected. Partly purified enzymes are sometimes needed : for instance, amyloglucosidase, for starch determination, has to be contaminated with traces of α-amylase in order to be able to transform completely the α-glucan into glucose. But in most of the enzymic methods described, enzyme has to be highly purified and even crystallized, although a crystalline enzyme is not necessarily pure. In any case, therefore, the use of a new batch of enzyme, purified or not, for an enzymic determination, must be verified in terms of purity, eventual contaminants and specific activity.

The substrate to be measured, must be soluble in the digest in order to form the enzyme substrate-complex, to liberate the end-reaction product and to recover the total enzyme involved in the procedure. With macromolecules such as starch, β-glucans, proteins, a pretreatment is generally necessary to disperse and/or to solubilize the substrate either by heat treatments (swelling, autoclave) or by aqueous organic solvent (alcohol, DMSO ...), treatments which in the same time, inactivate endogenous enzymes.

The pH of the dispersed food must be known before any enzymic determination as well as the presence of various salts or other soluble compounds susceptible to play the role of inhibitor of the enzyme. Since some enzymes are very sensitive to ions, a previous extraction of substrate can be needed and bi-distilled water is often used for the digest.

The temperature for which the method has been developed can be either the maximum temperature of the reaction or a defined temperature very often around 30°C. The final conditions of hydrolysis must correspond to the irreversibility of the reaction, and the eventual inhibition by concentrations of substrate and end-product have to be studied. The total action of the enzyme, after the adequate time has elapsed for the defined pH and temperature conditions, can be measured either as the extent of loss of the substrate or as the appearance of a measured quantity of the end-product, whatever is more convenient. Alternatively, another enzyme can be used to convert one of the products to another material, which may be coupled with the reduction of the co-factor, allowing to use spectrophotometric or colorimetric measurement.

The purpose of this paper is not to describe all the enzymic methods developed in the literature for the determination of the various constituents of foods since recent and good reviews are available. But, an effort has been made to classify the methods based on the use of enzymes 1) on specific substrate; 2) for their activation and inhibition; 3) as a tool for structural analysis; 4) as specific desintegrating agents; 5) as *in vitro* digestibility test; 6) as electrodes. Each chapter has been emphasized with examples.

1. METHODS USING THE ENZYME-SUBSTRATE SPECIFICITY

1.1. <u>CARBOHYDRATES</u>. Methods have been developed for the determination of carbohydrates, such as mono-, di-, oligo- and polysaccharides in the presence of each other by using highly specific oxidase, dehydrogenase and hydrolase alone or in a coupled system. Tables 1 and 2 list the enzymes used to measure glucose, fructose, galactose, sucrose, maltose, lactose, melibiose, raffinose, maltodextrins, glucose syrups, starch and glycogen, the type of reaction involved and the possible measurements which depend upon the specificity of the enzyme. The comparison of monosaccharides

TABLE 1

ENZYMIC DETERMINATION OF MONOSACCHARIDES IN FOOD

Monosaccharide	Enzyme	Type of reaction	Measurement	Sensitivity
Glucose (8, 33, 35)[*]	Glucose - oxidase (GOD) [**]E C 1.1.3.4 Peroxidase (POD) E C 1.11.1.7	Glucose + H_2O $\xrightarrow{GOD}_{O_2}$ gluconic Ac + H_2O_2 H_2O_2 + DH_2 $\xrightarrow{POD}$ $2H_2O$ + oxidized-chromogen	O_2 consumption gluconic acid titrimetry $\left\{\begin{array}{l}\text{oxidized chromogen}\\\text{orthodianisidine 540 nm}\\\text{ABTS 560 nm}\end{array}\right.$	5-10 µg
Glucose (8, 12)	Hexokinase (HK) E C 2.7.1.1 Glucose-6-phosphate dehydrogenase (G-6-PDH) E C 1.1.1.49	Glucose + ATP $\xrightarrow{HK}$ Glucose-6-phosphate+ADP G-6-P + $NADP^+$ $\xrightarrow{G-6\ PDH}$ 6-phospho-gluconolactone + NADPH	$NADP^+$ - NADPH (UV)	5 µg
Fructose (8, 12)	Hexokinase (HK) E C 2.7.1.1 Glucophosphate isomerase (GPI) E C 5.3.1.9 Glucose-6-phosphate dehydrogenase (G-6-PDH) E C 1.1.1.49	Fructose + ATP $\xrightarrow{HK}$ Fructose-6-phosphate+ADP F-6-P $\xrightarrow{GPI}$ G-6-P G-6-P + $NADP^+$ $\xrightarrow{G-6-PDH}$ 6-phospho-gluconolactone + NADPH	$NADP^+$ - NADPH (UV)	5 µg
Galactose (56)	Galactose - oxidase (GAO) E C 1.1.3.9 Peroxidase (POD) E C 1.11.1.7	Galactose + H_2O $\xrightarrow{GAO}$ galactohexodialdose + H_2O_2 H_2O_2 + DH_2 $\xrightarrow{POD}$ H_2O + oxidized chromogen	$\left\{\begin{array}{l}\text{oxidized chromogen}\\\text{orthodianisidine 540 nm}\\\text{ABTS 560 nm}\end{array}\right.$	5-8 µg
Galactose (8, 12)	Galactose dehydrogenase (GaDH) E C 1.1.1.48	Galactose + NAD^+ $\xrightarrow{GaDH}$ galactonolactone + NADH	NAD^+ - NADH (UV)	3 µg

[*] numbers in brackets correspond to references cited

[**] Enzyme classification (20)

TABLE 2

ENZYMIC DETERMINATION OF OLIGO- AND POLYSACCHARIDES IN FOOD

Oligo- and polysaccharide	Enzyme	Type of reaction	Measurement
Sucrose [*] (8, 12, 13) α-D-glucopyranosyl 1,2 β-D-fructofuranose	β-D-fructofuranosidase or invertase [**] E C 3.2.1.26	sucrose $\xrightarrow{\text{invertase}}$ glucose + fructose	idem to glucose or fructose
Maltose (8) α-D-glucopyranosyl 1,4 β-D-glucopyranose	α-D-glucosidase or maltase E C 3.2.1.20	maltose $\xrightarrow{\text{maltase}}$ glucose + glucose	idem to glucose
Lactose (8, 12, 15) β-galactopyranosyl 1,4 α-D-glucopyranose	β-D-galactosidase or lactase E C 3.2.1.23	lactose $\xrightarrow{\text{lactase}}$ glucose + galactose	idem to glucose or galactose
Melibiose (8) α-galactopyranosyl 1,6 α-D-glucopyranose	α-D-galactosidase or melibiase E C 3.2.1.22	melibiose $\xrightarrow{\text{melibiase}}$ galactose + glucose	idem to glucose or galactose
Raffinose (8, 60) α-galactopyranosyl 1,6 α-D-glucopyranosyl 1,2 β-D-fructofuranose	α-D-galactosidase or melibiase E C 3.2.1.22	raffinose $\xrightarrow{\text{melibiase}}$ galactose + sucrose	idem to galactose
Starch (8, 65) Glycogen Glucose syrups Maltodextrins	Exo-1,4-α-D-glucosidase or amyloglucosidase, glucoamylase and γ-amylase E C 3.2.1.3	$\left\{\begin{array}{l}\text{starch} \\ \text{maltodextrins}\end{array}\right\} \xrightarrow{\text{amyloglucosidase}}$ glucose	idem to glucose

[*] numbers in brackets correspond to references cited
[**] Enzyme classification (20)

and sucrose determinations, their advantages and limits and their sensitivity are not discussed in detail since they have been well documented recently by WISEMAN (1978).

The development of a high specific starch determination (THIVEND et al., 1972) in complex media is due to the isolation and purification of the enzyme, amyloglucosidase (E C 3.2.1.3). Such enzyme catalyzes the hydrolysis of α-D-1,4, α-D-1,6 and α-D-1,3 linkages liberating D-glucose which is determined specifically with glucose-oxidase.

A transglucosidase action of amyloglucosidase observed with *Aspergillus niger* and with *Rhizopus delemar* preparation could interfer by forming isomaltose from the liberating D-glucose which can not be determined by glucose-oxidase and therefore reduce the starch amount. But traces of α-amylase in the enzyme are needed to reach complete degradation. The rate of hydrolysis depends on the linkage, being faster for α-D-1,4 than for α-D-1,6 and α-D-1,3 glucosidic linkages, and increases with the degree of polymerisation ($\overline{DP}$) of the substrate. Moreover, in the case of the α-D-1,3 linkage, the rate is independent of the size of the molecule. Under these conditions, the presence of glucoamylase-degradable α-D-1,3 glucans (such as dextrans) does not constitute an important error. However, glycogen, maltodextrins, glucose syrups and maltose can not be differentiate by the enzyme, whereas the presence of polysaccharides formed from residues other than D-glucose nor from D-glucose residues bound by others linkages do not interfer. The method is directly applicable in complex media, including starchy food and feed and digestive contains with a reproducibility of $\pm$ 1 %. In presence of α-1,4 and α-1,6 soluble maltodextrins (such as baby-foods), aqueous-ethanol extraction is necessary before analysis. The action of amyloglucosidase is complete only on dispersed starch which is obtained by swelling followed by pressure heating in case of high-amylose content starch (amylomaize, wrinkled pea) or by DMSO solubilization, amyloglucosidase being totally active in the range up to 40 % aqueous DMSO (v/v). Acidic dispersion, as used in chemical method (EVERS method) has been suggested but the specificity is reduced.

ANDERSON et al. (1978) have recently described a direct and specific method for the determination of 1,3 - 1,4 β-glucans in barley and other cereals. In the procedure, amylase-free bacterial 1,3 - 1,4 β-glucan hydrolase is used to depolymerize the 1,3 - 1,4 β-glucan in autoclaved and ethanol extracted flour prepared from whole grain. The liberated oligo-glucosides are extracted with 80 % (v/v) ethanol and, following acid hydrolysis, are measured by the glucose-oxidase method. The method can be used to measure total β-glucan and β-glucan water-soluble and water-insoluble fractions of cereal grains.

1.2 <u>ALCOHOLS AND ACETALDEHYDE</u>. Methods to determine ethanol, glycerol, sorbitol, acetaldehyde in beverages (fruit juice, wine, beer, japanese sake) use specific dehydrogenase and kinase (Table 3). They are mainly available from BOEHRINGER (1971) and have been discussed in detail by WISEMAN (1978). Determination of cholesterol using cholesterol oxidase have been reported by WORTBERG (1975) including its use in egg pasta.

TABLE 3

ENZYMIC DETERMINATION OF ALCOHOLS AND ACETALDEHYDE IN FOOD

Alcohol Acetaldehyde	Enzyme	Type of reaction	Measurement	Application
Sorbitol (8)*	Sorbitol dehydrogenase (SDH) ** E C 1.1.1.14	D-Sorbitol + NAD^+ $\xrightarrow{SDH}$ fructose + NADH	idem to fructose	wine
Ethanol (8)	Alcohol dehydrogenase (ADH) E C 1.1.1.1	Ethanol + NAD^+ $\xrightarrow{ADH}$ acetaldehyde + NADH	NAD^+ - NADH (UV)	fruit juice wine beer meat
Glycerol (8)	Glycero kinase (GK) E C 2.7.1.30 Pyruvate kinase (PK) E C 2.7.1.40 Lactase dehydrogenase (LDH) E C 1.1.1.27	Glycerol + ATP $\xrightarrow{GK}$ glycerol-1-phosphate + ADP ADP + phosphoenol pyruvate $\xrightarrow{PK}$ ATP + pyruvate Pyruvate + NADH $\xrightarrow{LDH}$ lactate + NAD^+	NADH - NAD^+ (UV)	wine
Cholesterol (72)	Cholesterol oxidase (CO) E C 1.1.3.6	Cholesterol + O_2 $\xrightarrow{CO}$ 4 - cholesten - 3 - one +(?)	O_2 consumption	egg pasta
Acetaldehyde (36)	Aldehyde dehydrogenase (AD) E C 1.2.1.3	Aldehyde + NAD^+ + H_2O $\xrightarrow{AD}$ acetic acid + NADH	NAD^+ - NADH (UV)	wine

* numbers in brackets correspond to references cited
** Enzyme classification (20)

1.3 ORGANIC ACIDS. Enzymic assays of several organic acids have been achieved for wine, beer, milk, molasses, vegetables, meat and cheese. Dehydrogenase-based methods, measuring absorption due to NADH at 340 nm, have been reported for pyruvate, D-lactate, L-lactate, L-malate, D-gluconate, acetate, citrate, isocitrate, succinate (Table 4). Carbonic acid (CO_2) has been determined using carbonic anhydrase in carbonated wines and soft drinks (WISEMAN, 1978). REINEFELD and BLIESENER (1977) have developed a method to measure formic acid in molasses and MARCHESINI *et al*. (1974) used the ascorbic acid oxidase to determine ascorbic acid (vitamin C) and dehydro-ascorbic acid in spinach. Gluconolactone, as additive in meat products, is determined by a method based on phosphorylation to gluconic acid-6-phosphate by ATP in the presence of gluconate kinase. The gluconic acid-6-phosphate is then oxidized to ribulose-5-phosphate by NADP in the presence of 6-phosphogluconate dehydrogenase and the NADPH formed during the reaction is determined by U V spectrophotometry (INKLAAR, 1977).

1.4 AMINO ACIDS AND ORGANIC BASES. Several natural amino acids derived from food can be determined by enzyme methods, only whether bacterial L-amino acid decarboxylases are available (Table 5). An accurate and precise automated method has been developed, (WALL and GEHRKE, 1977; ROY, 1979), to determine lysine in grain hydrolysates. L-lysine decarboxylase specifically catalyzes removal of the carboxyl group from L-lysine, producing an amine and CO_2. The CO_2 produced is selectively dialysed into a stream of carbonate with phenolphtalein by a dialysis block containing a CO_2 gas dialysis membrane. BOEHRINGER (1971) uses L-glutamate dehydrogenase to determine L-glutamate in beer and wort, coupled to conversion of NAD^+ to NADH. From L-aspartate, aspartate aminotransferase produces oxaloacetate which is measured by conversion to malate with production of NAD^+ from NADH, under the action of malate dehydrogenase. Few enzymic determinations of amines found in some cheeses and meat have been developed. Creatine is measured, in meat, by a three-enzyme steps, using successively creatine kinase, pyruvate kinase and lactate dehydrogenase leading to a conversion of NADH into NAD^+ (Table 5). The trimethylamine determination involves the coupling of trimethylamine dehydrogenase with phenazine methosulphate (PMS), followed by the coupling to the reduction of a dye, from which the colour is read at 600 nm.

Urea and ammonia salts in meat products can be determined with the ammonia electrode, in the range of 0.1 to 50 µg/ml by the use of the enzyme urease.

1.5 TRIGLYCERIDES. Lipid analysis is followed by the liberation of glycerol from triglycerides, carried out either using lipases (NAZIR *et al*, 1976) or by saponification with ethanolic KOH as it has been developed by KOHLER (1978), for a routine fat determination in soft cheese. Then the resultant glycerol is enzymatically determined with a BOEHRINGER test kit containing glycerol kinase, pyruvate kinase and lactate dehydrogenase (Table 3).

1.6 OTHER COMPONENTS. Other components in food may be determined by the direct action of enzymes on their substrates, such as nitrate, inorganic pyrophosphate, ochratoxin using carboxypeptidase and the enzymic determination of hydrogen peroxides (WISEMAN, 1978). Traces of nitrate in water (50 p.p.b. to 7.5 p.p.m.) are reduced to nitrite, under the action of the NADH-dependent nitrate reductase (EC 1.6.6.1).

TABLE 4

ENZYMIC DETERMINATION OF ORGANIC ACIDS IN FOOD

Acid	Enzyme	Type of reaction	Measurement	Application
Formic (53)[*]	Formate dehydrogenase (FDH) [**] E C 1.2.1.2	Formate + NAD^+ $\xrightarrow{FDH}$ CO_2 + NADH	NAD^+ − NADH (UV)	molasses
Acetic (8)	Acetate kinase (AK) E C 2.7.2.1	Acetate + ATP $\xrightarrow{AK}$ $CH_3COOPO_3H_2$ + ADP $CH_3COOPO_3H_2$ + H_2NOH $\xrightarrow{FeCl_3}$ $CH_3CONHOH$ + PO_4H_3	540 − 550 nm	fruit juice wine beer sour dough
Acetic (29,44) Pyruvic (8,31,64)	Acetate kinase (AK) E C 2.7.2.1 Pyruvate kinase (PK) E C 2.7.1.40 Lactate dehydrogenase (LDH) E C 1.1.1.27	Acetate + ATP $\xrightarrow{AK}$ acetyl phosphate + ADP ADP + phosphoenolpyruvate $\xrightarrow{PK}$ ATP + pyruvate Pyruvate + NADH $\xrightarrow{LDH}$ lactate + NAD^+	NADH − NAD^+ (UV)	wine wort beer
Lactic (8, 52)	L-lactate dehydrogenase (L-LDH) E C 1.1.1.27 D-lactate dehydrogenase (D-LDH) E C 1.1.1.28	L-lactate + NAD^+ $\xrightarrow{L-LDH}$ pyruvate + NADH D-lactate + NAD^+ $\xrightarrow{D-LDH}$ pyruvate + NADH	NAD^+ − NADH (UV)	milk sour dough
Malic (8, 15, 30)	Malate dehydrogenase (MDH) E C 1.1.1.40	L-malate + $NADP^+$ $\xrightarrow{MDH}$ pyruvate + NADPH	$NADP^+$ − NADPH (UV)	musts vines
Citric (8, 50)	Citrate (pro-3S) lyase (CL) E C 4.1.3.6 Malate dehydrogenase (MDH) E C 1.1.1.40	Citrate $\xrightarrow{CL}$ oxaloacetate + acetate Oxaloacetate + NADPH $\xrightarrow{MDH}$ malate + $NADP^+$	NADPH − $NADP^+$ (UV)	citrus fruit milk
Isocitric (7)	Isocitrate dehydrogenase (IDH) E C 1.1.1.42	Threo − De − Isocitrate + $NADP^+$ $\xrightarrow{IDH}$ 2 oxoglutarate + CO_2 + NADPH	$NADP^+$ − NADPH (UV)	citrus fruit
Succinic (37, 67)	Succinyl CoA synthetase (SCS) E C 6.2.1.5 Pyruvate kinase (PK) E C 2.7.1.40	Succinate + CoA + ITP $\xrightarrow{SCS}$ succinyl − CoA + IDP IDP + phosphoenolpyruvate $\xrightarrow{PK}$ ITP + pyruvate	idem as lactate	wine
Gluconic (8, 34)	Gluconokinase (GK) E C 2.7.1.12 6-phosphogluconate dehydrogenase (6-PGDH) E C 1.1.1.44	D-gluconate + ATP $\xrightarrow{GK}$ 6-phospho-D-gluconate + ADP 6-phospho-D-gluconate + $NADP^+$ $\xrightarrow{6-PGDH}$ D-ribulose 5-phosphate + CO_2 + NADPH	$NADP^+$ − NADPH (UV)	wine fruit juice meat cheese
Ascorbic (Vitamin C) (45)	Ascorbate oxidase (AO) or ascorbase E C 1.10.3.3	2 L-Ascorbate + O_2 $\xrightarrow{AO}$ 2-dehydroascorbate + $2H_2O$	O_2 consumption (Clark type electrode)	spinach

[*] numbers in brackets correspond to references cited
[**] Enzyme classification (20)

TABLE 5

ENZYMIC DETERMINATION OF AMINO ACIDS AND ORGANIC BASES IN FOOD

Amino acid organic base	Enzyme	Type of reaction	Measurement	Application
Aspartic [8]*	Aspartate amino transferase or glutamic oxaloacetic transaminase (GOT) ** E C 2.6.1.1	L-aspartate + α-oxoglutarate $\xrightarrow{GOT}$ oxalacetate + L-glutamate		
	Malate dehydrogenase (MDH) E C 1.1.1.40	Oxalacetate + NADPH + CO_2 $\xrightarrow{MDH}$ malate + $NADP^+$	NADPH-$NADP^+$ (UV)	
Glutamic [8]	Glutamate dehydrogenase (GLDH) E C 1.4.1.2	L-glutamate + H_2O + NAD^+ $\xrightarrow{GLDH}$ 2-oxoglutarate + NH_3 + NADH	NAD^+-NADH (UV)	
Lysine [57, 69]	Lysine decarboxylase (LD) E C 4.1.1.18	L-lysine $\xrightarrow{LD}$ cadaverine + CO_2	CO_2 with glass electrode	grain meat
Creatine [8]	Creatine kinase (CK) E C 2.7.3.2	Creatine + ATP $\xrightarrow{CK}$ Phosphocreatine + ADP		
	Pyruvate kinase (PK) E C 2.7.1.40	ADP + phosphoenol pyruvate $\xrightarrow{PK}$ pyruvate + ATP		meat
	Lactate dehydrogenase (LDH) E C 1.1.1.27	Pyruvate + NADH $\xrightarrow{LDH}$ L-lactate + NAD^+	NADH-NAD^+ (UV)	
Trimethylamine [42]	Trimethylamine dehydrogenase (TD) E C 1.5.99.7	$(CH_3)_3NH + H_2O$ + PMS $\xrightarrow{TD}$ $(CH_3)_2NH_2^+$ + HCHO +$PMSH_2$ $PMSH_2$ + DCPIP $\longrightarrow$ PMS + $DCPIPH_2$ (colour less) coloured	PMS=phenazine methosulphate DCPIP=2,6-dichlorophenolindophenol 600 nm	meat

* numbers in brackets correspond to references cited
** Enzyme classification (20)

The rate of disappearance of NADH is followed by a decrease of fluorescence intensity (KIANG et al, 1978).

By the catalytic action of the inorganic pyrophosphatase, inorganic pyrophosphate is split to orthophosphate which, in acid solution, produces a coloured complex with ammonium molybdate and ammonium vanadate measured between 360 and 420 nm (INKLAAR, 1977). Using phosphorylase a, phosphate reacts with glycogen to form glucose-1-phosphate, which is converted to glucose-6-phosphate by the action of phosphoglucomutase. Glucose-6-phosphate is determined, using NAD^+ and glucose-6-phosphate dehydrogenase at 340 nm. A fluorescent assay of NADH can measure inorganic phosphate in the range from 2×10^{-11} to 2×10^{-10} M (SCHULZ et al, 1967).

The cyanogenic glucosides present in parenchymal tissue, peel and leaves from Cassava are hydrolyzed by incubation with exogenous linamarase, to free cyanide, which is estimated spectrophotometrically (COOKE, 1978). The acid extraction solution inactivates endogenous linamarase, and assay of aliquots without enzyme treatment gives the free (non-glycosidic) cyanide contents of the extract. The detection limit is less than 0.01 mg per 100 g fresh material and 40 - 50 samples per a day can be handled easily.

2. METHODS USING THE ACTIVATION OR INHIBITION OF ENZYMES

Enzymes which are metalloproteins, can be activated or inhibited with trace quantities of metals. GUILBAULT (1970) has listed 57 enzymes and 31 inhibitory substances including heavy metals, which can be used in the determination of those substances. Among them, a useful enzyme for the determination of Mg^{2+} (10 p.p.b.) is luciferase. The green chemiluminescence produced by the conversion of luciferin to oxyluciferin in the presence of oxygen allows an extremely sensitive measurement.

The inhibition of invertase can be used for the accurate determination of silver $(2-10.10^{-7} M)$ and mercury $(2-10.10^{-8} M)$ in the presence of other metals. Cyanide and sulphide, by decreasing the inhibition, and thiourea, increasing it, can be measured with invertase.

The enzyme glyceraldehyde-3-phosphate dehydrogenase is used to perform an oxidative arsenolysis of D-glyceraldehyde-3-phosphate. The rate of the reaction, of first order in arsenic, is measured by fluorescence and has been applied for the determination of arsenic in water (GOODE and MATTHEWS, 1978).

Some pesticides and insecticides have been reported to inhibit enzymes. As little as 100 p.p.b. of aldrin, chlorodane, DDT and heptachlor can be detected, in the presence of other pesticides, by inhibition of yeast hexokinase, with a precision and accuracy of $\pm$ 2 % (SADER and GUILBAULT, 1971) The assay for hexokinase is developed in Table 1.

Organic phosphorus insecticides inhibit the action of cholinesterase, the activity being determined from the rate of splitting acetylthiocholine into acetate and thiocholine. Thiocholine reacts with dithiobisnitrobenzoic acid to form thionitrobenzoic acid, whose colour intensity is measured between 400 and 420 nm (BOEHRINGER, 1971). Folazon and its metabolites have been measured in apples using this effect

(FATEEVA and POPOVA, 1978).

Detection of tetrachlorvinphos and diazinon in cabbage, brussels sprouts and carrots extracts have been measured by inhibition of the acylcholine acylhydrolase (WEDŽISZ, 1979).

Bovine liver carboxylesterase inhibition by organochlorine insecticides have been used to assay lindane at 500 ng, where other insecticides needed at least 10 µg for significant inhibition. Carbamates inhibit the pig liver esterase which is followed by spectrophotometric detection of the hydrolysis of indophenyl acetate.

3. METHODS USING ENZYMES AS A TOOL FOR STRUCTURAL ANALYSIS

The distribution of fatty acids in a triglyceride is of interest for the physical, chemical and biological behaviour of food, in particular on their metabolism pathway in animal tissues. Stereospecific analysis determines how the saturated and/or unsaturated fatty acids are distributed over the 3 different positions of the glycerol. The demonstration that pancreatic lipase hydrolyses specifically the fatty acids esterified with the primary hydroxyl groups of glycerol, serves as the basis for such a method. Moreover, by acylation of partial glycerides and hydrolysis of the resulting triglyceride with pancreatic lipase, it is possible to determine the structure of mono- and diglycerides (BROCKENHOFF, 1971). A lipase from *Geotrichum candidum* (KROLL and FRANKZE, 1974) have been reported to degrade natural triglycerides and is also used for structural studies.

In the area of carbohydrates, the knowledge of the distribution of the α-1,6 bonds in polymers of α-1,4 linked anhydroglucose residues such as starch, amylopectin, maltodextrins, glycogen, is of importance for the technological behaviour and the nutritive value of food.

A two-dimensional paper chromatographic technique has been developed by FRENCH *et al*. (1966), which is particularly useful to rapidly survey enzyme action on individual members of homologous series. LEE *et al*. (1968) and MERCIER and WHELAN (1970) have then quantified this technique, by using successively the action of debranching (α-1,6) enzyme, pullulanase (E C 3.2.1.41) and/or isoamylase (E C 3.2.1.68), of saccharifying β-amylase (E C 3.2.1.2) which splits the α-1,4 linkage from the non-reducing end liberating β-maltose and of the α-1,4, α-1,6 amyloglucosidase transforming the α-glucan into glucose. The method allows to number the α-1,6 linkages in the polysaccharide. The gel permeation chromatography of the debranched-pullulanase and degraded-β-amylase-debranched materials permits to determine the number and the length of the constitutive chains of the macromolecule, from what schematic models have been suggested for glycogen (MERCIER and WHELAN, 1970; GUNJA-SMITH *et al*, 1970) amylopectin and starch (MERCIER, 1973; ROBIN *et al*, 1974, 1975).

4. METHODS USING ENZYMES AS SPECIFIC DESINTEGRATING AGENTS

Enzymes are used, in recent techniques, not directly but as a tool to remove some constituents of foods in order to concentrate into the substrate which is then quantified by gravimetric or chemical methods. For instance, the determination of "dietary fibre", including cellulose, hemicelluloses and lignin of various plants is actually carried out after the complete removal of starch under the action of α-amylase (Takadiastase, pancreatic) and amyloglucosidase and of protein by pepsin or papain (ASP, 1978; SOUTHGATE, 1979). However, attention must be paid during the enzymatic predigest. The dietary fiber content can be underestimated either by the lost of water-soluble fraction during dialysis of the hydrolysis products (glucose, amino acids and peptides) and/or by the use of enzymes contaminated by glucanases and hemicellulases susceptible to degrade structural polysaccharides.

In some cases, enzymes can be used to desintegrate cell-walls such as endopolygalacturonase for vegetables tissues (ZETELAKI-HORVATH and URBANYI, 1978). The release of chlorogenic acid in leaf protein has also been improved by a proteolytic digestion (LAHIRY and SATTERLEE, 1975) and the determination of allyl mustard oil is carried out in presence of myrosinase (yellow mustard seed) to liberate allyl isothiocyanate from the glycoside sinigrin (HILS, 1979).

5. METHODS USING ENZYMES AS *IN VITRO* DIGESTIBILITY TEST

Food analysis involves not only the determination of their composition but also their nutritive value. Raw materials undergo during storage and processings some modifications which may affect their digestibility. Tests have been developed for starch and proteins based on the comparative action of enzyme on the raw and processed food and under conditions which simulate the *in vivo* digestibility.

The "*in vitro* digestibility" of starch is followed as described by TOLLIER and GUILBOT (1971), under the action of *Bacillus subtilis* α-amylase in function of time at 37°C and pH 6.9, by a colorimetric measurement of the liberated oligosaccharides. The residual starch can also be quantified after centrifugation by weight. Kinetics parameters such as initial and final rate of the reaction, allows to distinguish between damaged and intact starch granules (MERCIER, 1977).

Based on this principle, several methods have been developed for the enzymic determination of damaged starch in flour, cereals and cereal products. Methods using the action of β-amylase followed by the reducing power of the liberated maltose (GREER and STEWART, 1959) or bacterial α-amylase (DONELSON and YAMASAKI, 1962; AUDIDIER *et al*, 1966) and colorimetric determination of the formed oligosaccharides have been compared on flour wheats. Each of these methods leads to different damaged starch content due to the different action of enzymes. However, the value compared to the unprocessed material is of interest and allows to quantify the effect of processings.

The digestive acceptability of a protein can be determined by measuring the intensity and rate of proteolysis with proteolytic enzymes. Different methods exist using pepsin, trypsin, pancreatin or papain (SATTERLEE *et al*, 1977). These enzymes, as endopeptidases, split the protein to a different number of peptides but do not give complete hydrolysis. Therefore the soluble fraction can be measured either as nitrogen content or using microbiological determination of some amino acids (FORD, 1965). CAMUS *et al*. (1972, 1973) have developed *in vitro* pepsin and trypsin proteolysis tests for food and have shown a correlation with the intestinal absorption of proteins by man. The pepsin proteolysis is followed, at pH 1.9, 37°C, under magnetic stirring, by the liberation of perchlorosoluble nitrogen with a crystalline pepsin. The trypsin proteolysis, at pH 7.0, 37°C, under magnetic stirring, is measured at constant pH, by titrimetry of the liberated protons during 5 minutes, with a lyophilized trypsin. Using those methods, it has been shown that the digestibility of proteins decreases with the degree of denaturation which also causes changes in the nutritive value of proteins.

6. METHODS USING ENZYMES ELECTRODES

The use of immobilized enzymes in analysis is increasing due to the improvement of the pH and temperature stability of the enzyme and the possibility to recover the biocatalyst. But the main use of immobilization has been introduced under the enzyme electrode by UPDIKE and HICKS (1967), and since then there have been many applications of this technique.

Recently, BAKER and SOMERS (1978) have up-to-dated the different enzymes electrodes and enzymes based sensors. The earliest enzymes electrodes, with an electrochemical sensor, a semi-permeable membrane and an intermediate layer containing the enzyme, required diffusion of the substrate through the membrane, slowing the response time. Later enzyme electrodes used enzymes immobilized on easily accessible surfaces of inorganic supports. The most widely used solute electrode is the glass pH electrode, which is limited in enzyme electrode because most enzyme reactions are not linear over a broad pH range, and for accurate results the reaction media must have a low buffering capacity. Variation of the glass composition can be used to produce ion electrodes, such as the ammonium, cyanide and phosphate specific electrodes.

The constant potential polarographic measurements (amperometry), in which the current is proportional to a species reduced or oxidised at a fixed potential has a wide application for enzyme electrodes, such as the CLARK pO_2 electrode. The pCO_2 electrode has also some application to enzyme electrodes.

Electrodes based on the measurement of oxygen consumption or hydrogen peroxide production have been attractive for the development of enzymes electrodes such as glucosidases and amino acid oxidases. Electrodes based on peroxide determination amperometrically have been described for the determination of alcohol, uric acid, xanthine, galactose and ascorbic acid. Urease and amino acid decarboxylase based enzymes electrodes have been also tried and the use of a pCO_2 electrode for the specific analysis of L-lysine in cereal grains has just been described by SKOGBERG and RICHARDSON (1979).

Fibre-entrapped enzymes (MARCONI *et al*, 1974) have also attracted considerable interest in connection with food analysis and control.

REFERENCES

1. M.A. Anderson, J.A. Cook and B.A. Stone, J. Inst. Brew. $\underline{84}$ (1978), 233
2. N.G. Asp, J. Plant Foods $\underline{3}$ (1978), 21
3. Y. Audidier, J.F. de la Guérivière, Y. Seince and K. Benoualid, Ind. Alim. Agric. $\underline{83}$ (1966), 1597
4. A.S. Barker and P.J. Somers, in "Topics in enzyme and fermentation biotechnology" Vol 2, pp 120-151, Ed. A. Wiseman, Pub. Horwood/Wiley (1978)
5. H.U. Bergmeyer, in "Methods of enzymatic analysis". Verlag Chemie, Weinheim, New-York (1970)
6. H.U. Bergmeyer and K. Gawehn, Principles of enzymatic analysis. Verlag Chemie, Weinheim, New-York (1978)
7. B. Bergner-Lang, Deutsche Lebensmittel Rundschau, $\underline{73}$ (1977), 211
8. Boehringer Mannheim GmBh, Food analysis manual (1971)
9. H. Brockerhoff, Lipids $\underline{6}$ (1971), 942
10. M.C. Camus, J.C. Laporte and C. Sautier, Ann. Biol. anim. Bioch. Biophys. $\underline{13}$ (1973), 193
11. M.C. Camus and C. Sautier, Ann. Biol. anim. Bioch. Biophys. $\underline{12}$ (1972), 281
12. A. Cantafora, D.A. Villalobos and R. Rodini, Riv. Soc. Ital. Sci. Aliment. $\underline{7}$ (1978), 131
13. J. Cerning-Beroard, Cereal Chem. $\underline{52}$ (1975), 431
14. R.D. Cooke, J. Sci. Food Agric. $\underline{29}$ (1978), 345
15. F. Corradini and R. Pellegrini, Vin d'Italie $\underline{20}$ (1978), 225
16. E. Davies, E. Bourke and J. Costello, Analyst $\underline{100}$ (1975), 758
17. J.R. Donelson and W.T. Yamazaki, Cereal Chem. $\underline{39}$ (1962), 460
18. R. Drapron, l'Actualité chimique $\underline{4}$ (1974), 16
19. R. Drapron, Cahiers de Nut. Diet. $\underline{14}$ (1979), 99
20. Enzyme nomenclature 1978. IUB, Pub. Academic Press, New-York, London (1979)
21. O.F. Fateeva and V.A. Popova, Khimiya V Sel' Skom Khozyaistve $\underline{16}$ (1978), 64
22. J.E. Ford, Brit. J. Nutr. $\underline{19}$ (1965), 277
23. D. French, A.O. Pulley, M. Abdullah and J.C. Linden, J. Chromatog. $\underline{24}$ (1966), 271
24. S.R. Goode and R.J. Matthews, Anal. Chem. $\underline{50}$ (1978), 1608
25. E.N. Greer and B.A. Stewart, J. Sci. Food Agric. $\underline{10}$ (1959), 248
26. G.G. Guilbault, Enzymatic methods of analysis, Pergamon Press, Oxford (1970)
27. G.G. Guilbault, Handbook of enzymatic methods of analysis. Marcel Dekker New-York and Basel (1976)
28. Z. Gunja-Smith, J.J. Marshall, C. Mercier, E.E. Smith and W.J. Whelan, Febs letters $\underline{12}$ (1970), 101
29. M. Hara, A. Totsuka and N. Takenaka, J. Soc. Brew. (Japan) $\underline{73}$ (1978), 741
30. M. Hara, A. Totsuka, Y. Takahashi and H. Iki, J. Soc. Brew. (Japan) $\underline{73}$ (1978), 744
31. W. Heeschen, SIK Rapport $\underline{435}$ (1978), 17
32. A.K.A. Hils, Lebensmittel Runds. $\underline{75}$ (1979), 123
33. A.S.O. Huggett and D.A. Nixon, Biochem. J. $\underline{66}$ (1957), 12
34. P.A. Inklaar, Voedings middelen technologie $\underline{10}$ (1977), 9

35 M. Jemmali and R. Rodriguez-Kabana, Anal. Biochem. 37 (1970), 253
36 A. Joyeux and S. Lafon-Lafourcade, Ann. Fals. Exp. Chim. 72 (1979), 321
37 A. Joyeux and S. Lafon-Lafourcade, Ann. Fals. Exp. Chim. 72 (1979), 317
38 C.H. Kiang, S.S. Kuan and G.G. Guilbault, Anal. Chem. 50 (1978), 1323
39 P. Kohler, Mitteil. Geb. Lebensmittel Unters- Hyg. 69 (1978), 85
40 J. Kroll and C. Frankze, Fette Seifen Anstrmittel 76 (1974), 385
41 N.L. Lahiry and L.D. Satterlee, J. Food Sci. 40 (1975), 1326
42 P.J. Large and H. McDougall, Anal. Biochem. 64 (1975), 304
43 E.Y.C. Lee, C. Mercier and W.J. Whelan, Arch. Biochem. Biophys. 125 (1968), 1028
44 L.P. McCloskey, J. Agric. Food Chem. 24 (1976), 523
45 A. Marchesini, F. Monturori, D. Muffato and D. Maestri, J. Food Sci. 39 (1974), 568
46 W. Marconi, F. Bartoli, S. Gulinelli and F. Morisi, Process Biochem. 9 (1974), 22
47 C. Mercier, Die Stärke 25 (1973), 78
48 C. Mercier, Bull. Anc. El. ENSMIC 278 (1977), 3
49 C. Mercier and W.J. Whelan, Eur. J. Biochem. 16 (1970), 579
50 I.D. Mutzelburg, Aust. J. Dairy Techn. 34 (1979), 82
51 D.J. Nazir, B.J. Moorecroft and M.A. Mishkel, Am. J. Clin. Nut. 29 (1976), 331
52 E. Rabe, Getreide, Mehl und Brot 31 (1977), 230
53 E. Reinefeld and K.M. Bliesener, Zucker 30 (1977), 650
54 J.P. Robin, C. Mercier, R. Charbonnière and A. Guilbot, Cereal Chem. 51 (1974), 389
55 J.P. Robin, C. Mercier, F. Duprat, R. Charbonnière and A. Guilbot, Die Stärke 27 (1975), 36
56 H. Roth, S. Segal and D. Bertoli, Anal. Chem. 10 (1965), 33
57 R.B. Roy, J. Food Sci. 44 (1979), 480
58 M.H. Sader and G.G. Guilbault, J. Agric. Food Chem. 19 (1971), 357
59 L.D. Satterlee, J.G. Kendrick and G.A. Miller, Nut. Rep. Int. 16 (1977), 187
60 H. Schiweck and L. Buesching, Zucker 28 (1975), 242
61 D.W. Schulz, J.V. Pasonneau and O.H. Lowry, Anal. Biochem. 19 (1967), 300
62 D. Skogberg and T. Richardson, Cereal Chem. 56 (1979), 147
63 D.A.T. Southgate, in "Sugar : Science and Technology" ed. G.G. Birch and K.J. Parker. Appl. Sci. Publ. Ltd London (1979)
64 G. Suhren, W. Heeschen and A. Tolle, XX Internat. Dairy Congress (1978), 205
65 P. Thivend, C. Mercier and A. Guilbot, in "Methods in Carbohydrate Chem.", Vol VI (1972), 100
66 M.Th. Tollier and A. Guilbot, Ann. Zootech. 20 (1971), 633
67 A. Totsuka, M. Nakane and T. Hara, J. Soc. Brew. (Japan) 73 (1978), 811
68 S.J. Updike and G.P. Hicks, Nature 214 (1967), 986
69 L.L.Sr. Wall and C.W. Gehrke, J. AOAC, 57 (1974), 1098
70 A. Wedzisz, Bromot. Chem. Toksykol 12 (1979), 135
71 A. Wiseman, in "Development in Food Analysis Techniques" ed. King (1978), 179
72 B. Wortberg, Z. Lebensmittel Unters- Forsch. 157 (1975), 333
73 K. Zetelaki-Horwath and G. Urbanyi, Acta Aliment. 7 (1978), 68

A version of this lecture has also been published in Ernährung 5 (1981) by agreement of the publishers.

Fast Method for Enzymatic Determination of Starch Using an Oxygen Probe

J.C. CUBER

Laboratoire de Physiologie de la Nutrition - I.N.R.A. - C.N.R.Z.
78350 JOUY-en-JOSAS , France.

SUMMARY

A rapid method was developed for the estimation of the amount of starch in various products. The method was suitable for samples of flour in the range 10-100mg. The procedure -which is given in detail- involves first solubilizing the starch by an aqueous solution of dimethylsulfoxyde (90 per cent) during one hour at 120°C and then subjecting an aliquot in the presence of both amyloglucosidase and glucose oxidase reagents. The α-D-glucose was converted by mutarotase in β-D-glucose which was oxidized by glucose oxidase. The oxygen consumption was recorded by an oxygen probe. The starch determination was shown to be reproducible. In addition, this technique was very fast (less than 10 min. per sample) and necessitated few manipulations. Results obtained with this one-step assay were compared with those provided by two enzymic methods.

INTRODUCTION

Usual methods for enzymatic determination of starch (1,2,3) involve a two-step and time-consuming procedure. The solubilized starch is submitted to an enzymic hydrolysis with a glucoamylase during 2(1) or 24hours (3). The glucose released is then oxidized by glucose oxidase. The hydrogen peroxide produced during this reaction is detected by the mean of a chromogenic system (peroxidase and one or two dyes). However, the action of peroxidase doesn't allow a full specificity. There are indeed many redox substances which compete with the chromogen for H_2O_2 (4). Therefore, a new procedure was developed to avoid such interferences and to permit a rapid one-step determination.

PROCEDURE

1) _Principle_ : hydrolysis of starch, defined as all the α linked glucose polymers, is performed by glucoamylase. The β-D-glucose liberated is oxidized by glucose oxidase as it is released and α-D-glucose is converted by mutarotase in β-D-glucose which is oxidized by the glucose oxidase. The oxygen consumption during this step is recorded by an oxygen probe (Clark electrode).
Theoritically, one mole of oxygen is consumed during the oxidation of one mole of β-D-glucose. It's realizable in practice only if decomposition of the peroxide to oxygen and water is prevented. The decomposition is catalyzed by the enzyme catalase which is normally present as a contaminant in glucose oxidase. In the presence of ammonium molybdate, peroxide is rapidly reduced by iodide (5). Any remaining catalase-catalyzed decomposition of peroxyde leads to formation of acetaldehyde instead of release of oxygen in the presence of ethanol (6). The incorporation of iodide molybdate and ethanol in the buffer thus gives maximum sensitivity since oxygen depletion rather than peroxide formation is measured no secondary chromogenic reaction must be used and all of the many problems inherent in the determination of peroxide are avoided.

2) Reagents and apparatus

a) **Buffer** : 0.1M acetate buffer - pH 5.0. It contained for 2L : sodium chloride : 2.92g; glacial acetic acid (pH adjusted to 5.0 with NaOH after this addition): 5.68mL; ammonium molybdate (10g/L): 0.5mL; ethanolic iodine (10g/L): 2mL; ethanol 95%: 200mL; mercuric iodide: 4.0mg; formaldehyde 30%: 7.5mL; ethanol: 4 drops (antifoaming reagent). Mercure iodide and formaldehyde are preservatives. The buffer was equilibrated with atmospheric oxygen at least one night at 40°C in a water bath before use and kept at this temperature all day long.

b) **Enzymes** : 150mg of glucoamylase (Merck) were dissolved in 2ml of distilled water. 3ml of isopropanol were added at room temperature. The precipitate was collected, reprecipitated once and dissolved in 1ml of the buffer. 10mg of glucose oxidase type II (Sigma) were added in order to destroy any remaining trace of glucose. This solution was kept at least one night at room temperature. Complete removal of glucose was verified by assay with the oxygen electrode. 150mg of glucose oxidase type II (Sigma) were dissolved in 1ml of the buffer. Mutarotase has been purchased from Boehringer and used without previous preparation.

c) **Standard-D-glucose solution** : 4g of pure D-glucose per liter. It may be prepared by dissolving in water the appropriate weighed amount of glucose with benzoic acid to provide a concentration of 2g/l of that preservative. Freshly prepared solution was permitted to age at room temperature for at least twelve hours to ensure mutarotation equilibrium.

d) **Apparatus** : the oxygen consumption was measured by a polarographic oxygen analyzer with oxygen electrode (used in the model of the K-IC oxygraph, Gilson Medical Electronics, Inc. Middleton, Wisconsin) equipped with a controlled temperature (40 ± 0.1°C) water bath.

3) **Method**

a) **Sample preparation and solubilization** : The starchy compounds were ground to obtain a particle size smaller than 0.5mm. The sample (100mg) was mixed with 25ml of 90% dimethylsulfoxyde (DMSO). The solubilization was achieved during one hour at 120°C with an apparatus built in our laboratory and which permitted the treatment of 100 samples at the same time.

b) **Conditions of the standard assay procedure** : In the reaction cell, were placed equilibrated buffer (1.4ml), 20µl of glucoamylase solution and 20µl of glucose oxidase. After setting the meter reading of the oxygen analyzer at a proper position, while continuous stirring with a stirring bar covered with Teflon, we added 10 µl of the sample solution or standard glucose solution. A few minutes later, 20µl of mutarotase solution were added to ensure completion of the reaction. The results were read against a standard glucose curve.

c) **Calculations** :

$$\% \text{ of starch} = 0.9 \times \frac{DMG}{DMS} \times \frac{A1}{A2} \times 100$$

in which DMG is the percentage dry weight of the glucose powder, DMS is the percentage dry weight of the sample, A1 is the total amount of oxygen consumed per 10µl of the solubilized sample (read in millimeters on the tracing), A2 is the total amount of oxygen consumed per 10µl of a 4g/l glucose solution (read in millimeters on the tracing).

d) **Comparison with other techniques** : This new enzymic-polarographic assay was compared with the THIVEND et al. technique (1) and the method of CUBER and LAPLACE (7) modified after HOLZ (8).

RESULTS AND DISCUSSION

1) Calibration curve

Typical tracing illustrating the way in which enzymes were utilized to measure D-glucose and its anomers is shown in Fig.1. Glucoamylase (arrow 1) and glucose oxidase (arrow 2) are added into the buffer solution. After equilibration, a standard glucose solution is injected in the cell. There is an immediate utilization of oxygen (A) in the buffer until approximately 2 min. after the injection the reaction is sharply decreased (B). After mutarotase addition, the oxygen consumption is increased (C) and the completion of the reaction is obtained in less than 5 min. There is a linear relationship between the increasing concentrations of glucose standards and the amount of oxygen consumed at the completion of the reaction.

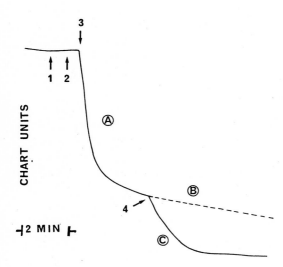

Figure 1

Representative curve showing the method utilized for calibration of the apparatus with standard-D-glucose solutions. The curve is derived by adding 20µl of glucoamylase (arrow 1) 20µl of glucose oxidase (arrow 2) to 1.4ml of 0.1M acetate buffer (pH 5.0). 10µl of the sample (glucose standard solution) are added at the point indicated by the arrow 4. After the rapid utilization of oxygen (A), a slower step of oxygen consumption takes place (B). After mutarotase addition the oxygen consumption is increased (C).

2) Measurement of starch in various products

Addition of starch sample after glucoamylase and glucose oxidase injection in the cell provokes an immediate utilization of oxygen (Fig.2). 93 and 97 p.cent of the starch amount added in the cell are oxidized in 2 and 4 min. respectively. After this step, a slower one takes place. Mutarotase addition (arrow 4) increases and completes oxygen consumption. We determined the starch content of a variety of products by the enzymic polarographic method. Typical results are shown in Table 1; individual results showed a reproduceability of ± 1.7 except for maize (4%).

A comparison with other assay techniques was also carried out, and the results obtained by two enzymic methods are also included in this table.

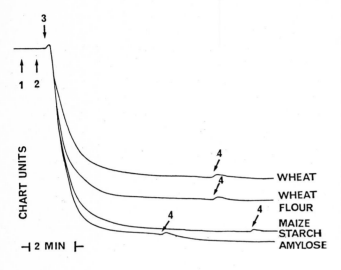

Figure 2

Oxygen level after starchy compounds additions. Reaction mixture contains 1.4ml of 0.1M acetate buffer. 20µl of glucoamylase (arrow 1) and 20µl of glucose oxidase (arrow 2) are added in the cell. After equilibrium is reached under a continuous stirring, 10µl of the sample (solubilized in DMSO as indicated in "methods") are injected in the cell (arrow 3). 20µl of mutarotase (arrow 4) are added for completion of oxygen consumption.

Table 2 : Typical results for the starch content (% on a dry matter basis) of some products obtained by the enzymic-polarographic method, and a comparison with other assay procedures.

Product	ASSAY METHOD		
	E.P.(1)	E.C.T.(2)	E.C.A.(3)
Amylose (Merck)	96.55	88.8	96.8
Waxy maize starch	99.25	99.3	99.3 *
Maize strach	97.1	97.0	95.4
Wheat starch	96.8	96.6	95.3
Potatoe starch	94.1	93.2	93.8
Manioc starch	97.6	96.6	93.0
Wheat flour	82.5	80.0	81.6
Manioc flour	83.0	81.6	74.6
Wheat	67.5	65.3	64.7
Barley	58.5	57.1	57.4
Maize	68.3	69.3	58.8
Wheat bran	20.1	19.8	18.9

(1) Enzymic-polarographic method described here - 6 replicates
(2) Enzymic-colorimetric method of THIVEND et al. (1) - 6 replicates
(3) Automated enzymic-colorimetric method of CUBER and LAPLACE (7) modified after HOLZ (8) - 5 replicates
 * Waxy maize starch is a standard in this method

It can be seen that the enzymic method of THIVEND et al. (1) is not applicable to purified amylose because recovery is not complete. Otherwise the agreement between this method and the enzymic-polarographic method is good, notwithstanding the time-consuming nature of the assay of THIVEND et al. (1). In contrast, falsely low values are obtained by the method of CUBER and LAPLACE (7) modified after HOLZ (8) for starch content of manioc starch, manioc flour and maize. No explanation for this discrepancy has been found in the case of manioc products. For maize, it seems that some damaged granules stay attached to other material (membranes...) after the step of starch dispersion. These particles are separated from the solubilized starch during centrifugation. This hypothesis is supported by the fact that complete recovery of starch is obtained in the case of the enzymic-polarographic method which uses the same way of starch dispersion but doesn't need centrifugation after starch solubilization. Starch measurement with the enzymic-polarographic method before and after centrifugation confirms this point of view. We would suggest that the enzymic-polarographic assay developed here can be used routinely to estimate accurately the amount of starch in any starchy compound.

REFERENCES

1. P. Thivend, C. Mercier and A. Guilbot, Methods in Carbohydrate Chemistry, vol. VI,(1972) ed. R.L. Whistler, Academic Press New York and London, 100.
2. W. Banks, C.T. Greenwood and D.D. Muir, Die Stärke, 22,(1970), 105.
3. J.C. MacRae and D.G. Armstrong, J. Sci. Fd. Agric., 19, (1968), 578.
4. M. Hjelm and C.H. de Verdier, Scand. J. Clin. Lab. Invest., 15 (1963) 415.
5. H.V. Malmstadt and H.L. Pardue, Analyt. Chem., 33, (1961), 1040.
6. D. Keilin and E.F. Hartree, Biochem. J., 42, (1948), 230.
7. J.C. Cuber and J.P. Laplace, Ann. Zootech., 28, (1979), 173.
8. F. Holz, Landw. Forsch. Suppl., 33-11 (1977), 228.

A Heat Resistant Trypsin Inhibitor in Beans (Phaseolus vulgaris)

U.M. LANFER-MARQUEZ[1], K. RUBACH[2], W. BALTES[2]

[1] Depto. de Alimentos e Nutricao Experimental da Faculdade de Ciencias Farmaceuticas - USP, Caixa Postal 30.786, 05508 Sao Paulo, S.P., Brazil

[2] Institut für Lebensmittelchemie der Technischen Universität Berlin, D-1000, Berlin 12

SUMMARY

Bean proteins are known as carriers of trypsin inhibitors. Although it has been assumed until now that those inhibitors are destroyed by heating, we found a heat resistant type in the albumin fraction with trypsin-inhibitor activity. Because of its high content of sulphur-containing amino acids, the albumin fraction which includes about 30% of the proteins is very interesting regarding the alimentary, physiological aspect. It contains about 75% of the entire trypsin-inhibitor activity of the bean. The destruction of the trypsin-inhibitor by heating the fraction in the autoclave succeeded only in basic medium while the activity in the acidic medium remained nearly constant. We are reporting on the isolation and behaviour of this trypsin-inhibitor.

INTRODUCTION

Because of their high protein content, beans are one of the most important protein sources in Latin America and belong to the six most cultivated legumes of the world. Some bean constituents such as trypsin-inhibitors, phytohemagglutinins and antiamylases, however limit the biological value of legume proteins. [1] [2] [3] [4] [5]. The various anti-physiological effects of these factors can be somewhat reduced by heat treatment. Elias et al. (1979) [6] reported on a heat resistant protease (enzyme-) inhibitor from the seed coat of various colored beans. Not a protein, this inhibitor has a phenolic nature linked with the tannins of the seed coat. With respect to good human nutrition, information about conditions for destroying these toxic compounds without minimizing the biological value of food components (e.g. by overheating) is important. Therefore, we aim to show the conditions for complete inactivation of the inhibitors and to characterize their constitution and reactivity.

EXPERIMENTAL

Material
Carioca bean (Phaseolus vulgaris, L.) seeds obtained from the Agronomy School of Lavras, M.G., Brazil. Beans milled to a fine powder were used.

Isolation
Albumins were prepared from the bean flour as demonstrated by the diagram outlined in Figure 1. A 10% suspension of the flour in distilled water was mixed for 1 h at room temperature, and the extract filtered through a cheese cloth and centrifuged at 1,000 rpm for 30 min. The supernatant was dialysed at $4^\circ C$ for 5 days against distilled water and lyophilized. The albumin fraction was then resuspended in water, adjusted to pH=3.0 with 0.1 N HCl and autoclaved at $121^\circ C$/30 min., followed by centrifugation at 4,000 rpm/30 min. The residue was discarded and the trypsin-inhibitor activity of the supernatant was measured.

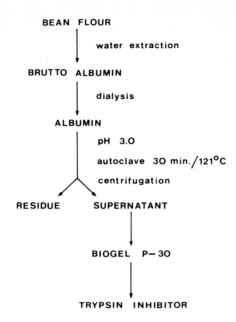

Figure 1: Schematic Diagram for Fractionation of Trypsin-Inhibitor

Determination of Trypsin-Inhibitor (TI) Activity
TI activity was determined by the method of Kakade et al. (1969) [7]. Bovine trypsin (2 x crystallized, salt free, lyophilized, Serva) was used to hydrolize the substrate benzoil-DL-arginine-p-nitroanilide (BAPA).

Chromatography on Biogel P-30
Chromatography on Biogel P-30 was carried out in a column (2.5 x 100 cm) equilibrated with a 50 mM phosphate buffer pH=7.6 and eluated at a flow rate of 1 ml/min. The eluate was continuously monitored at 254 nm by a Uvicord II monitor, LKB, Sweden. TI activity of each of the 7 ml fractions was measured, and all active fractions were pooled for further experiments.

Amino Acid Analysis
Amino acid analysis was performed with a Biotronic LC 6000 E amino acid analyzer according to Moore's & Stein's method (1963) [8]. The samples were hydrolyzed in 6 N HCl under N_2 at 110°C for 22 h in sealed tubes. According to Moore's method (1962) [9], protein oxidized by performic acid was used for the determination of half-cystine as cysteic acid.

Isoelectric focusing
Isoelectric focusing was performed in acrylamide gels containing 2% (w/v) ampholines, LKB, Sweden, (pH 3.5-9.5 or pH 4.0-6.0), at $4°C$. The proteins were stained with Serva Blue G. With a 110 ml LKB-column, preparative-column-isoelectric-focusing experiments were carried out in a sucrose gradient at $4°C$. The column was drained at a flow rate of 2 ml/min., and the eluate collected in 1 ml fractions. The pH and the TI activity of each fraction was measured.

RESULTS AND DISCUSSION

Experiments for the Inactivation of the TI
As our experiments showed, the TI destruction in boiling water with pH values between 2-10 did not exceed 20%. Therefore we tested TI stability after heating at $121°C$ in an autoclave. Hereby we found a graduate destruction depending on pH and on the time of heating. Our results are shown in table I.

| | \multicolumn{6}{c}{minutes of heating} |
pH	0	10	20	40	60	80
3.0	200	164	141	118	118	118
5.0	220	120	104	79	37	26
7.0	229	14	9	4	0	0
9.0	300	4	0	0	0	0

values in µg TI/mg protein.

Table I: Effects of pH and Time upon the Thermal Inactivation of Bean-Albumin Trypsin-Inhibitor in Water (T=$121°C$).

As can be seen, 70% of TI activity remained after treatment for 40 min. at pH=3.0 . Nonetheless, TI decomposition was accelerated with increased alkalinity, whereby the TI was destroyed nearly completely at pH=9.0 after 10 min. of heating. At room temperature, however, TI activity by ascending pH-values increased by about 50%. Antunes & Sgarbieri (1980) [10] show that a heating time of 90 min. at $97°C$ is sufficient to destroy the TI activity of the whole bean completely. On the other side, Pusztai

[11] described a TI in Phaseolus vulgaris in 1968 which maintained its activity after boiling for 2 h in a water bath. Moreover, Gatfield (1980) [12] described a TI in soaking water of beans which maintained up to 56.5% of its activity after 60 min. heating at 100°C.

Isolation

The isolation of the TI-system was carried out as shown in Figure 1. Heating the albumin fraction at pH=3.0 caused a 80% precipitation of the proteins, while 85% of the active TI remained in solution. This step of cleaning seemed to be favorable because it caused an elimination of a large part of interferring proteins. Further cleaning was achieved by column chromatography on Biogel P-30 with 0.05 M phosphate buffer at pH 7.6 (figure 2). TI activity was found exclusively in one peak (fractions 25-38) representing a compound of a molecular weight of 13,000 Daltons whereby marker proteins were used as comparison. Despite the very low absorption at 254 nm, these fractions showed a very high specific activity. The fractions were collected and lyophilized for further experiments.

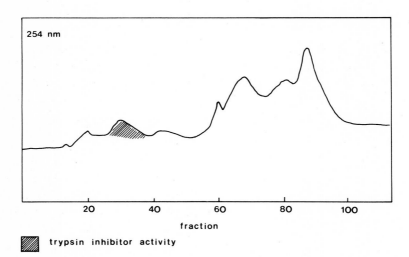

Figure 2: Chromatography of Heated Albumin Fraction on a Biogel P-30 Column.

Chemical and Physical Properties

By the amino acid analysis of the TI fraction, all physiological amino acids were found having a high content of aspartic acid, glutamic acid, serine, threonine and an unexpected high content of cysteine, which practically represented nearly 100% of the entire cysteine content of the whole bean. The distribution of the S-containing amino acids in the various protein fractions can be seen in table II. The high content of cysteine and probably of the S-S bonds in the TI of beans responsible for a stabilisation of the molecule had been cited at several occasions (11) (13) (14).

fraction	g/100g protein	
	MET	1/2 CYS
GLOBULIN G_1	0.3	0
GLOBULIN G_2	0	trace
GLUTELIN	0.7	0
ALBUMIN	0.8	3.2
Trypsin Inhibitor	0.2	3.2

Table II: Distribution of Methionine and Cysteine in Bean Protein and in the Trypsin-Inhibitor

The isoelectric thin-layer focusing of the albumin fraction showed a number of bands, stained with Serva Blue, between isoelectric points 3.5-6.0. After preparative column focusing the TI activity of the isolated TI fraction was proven by enzyme-inhibitor tests.
The results are represented in Figure 3.

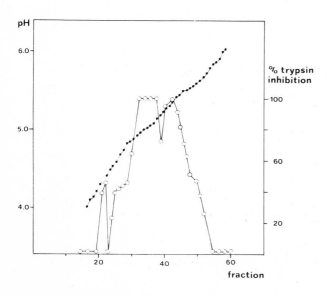

Figure 3:
Isoelectric Focusing of Trypsin Inhibitor on an Ampholine Column.
o-o Trypsin Inhibitor Activity; ●-● pH Gradient

As is demonstrated, the components with high TI-activity possess isoelectric points between 4-6. Therefore we assume that this is more than one heat resistant component with a TI-activity in the bean. However, we were not able to detect these components after a second cleaning by isoelectric thin-layer focusing by protein-dying methods. Simultaneously we found out that the TI became very labile after isoelectric focusing and lost its activity after a few days in aqueous solution. Dialysis of the focused active fraction through a Spectrapor^R Membrane (exclusion volume 3,500 Daltons) also caused a rapid decrease of activity, although a small part of the TI activity was found outside. Apparently, purification of the TI causes reduction of molecular weight and affects its stability.

According to these results, we assume that the TI is a substance of phenolic nature which causes an unspecific inhibition of proteases and which is able to form aggregates with proteins under certain conditions by altering the functional properties of the single components. Supporting this hypothesis Elias & col. (1979) [6] reported on two inhibitors: a real, thermolabile inhibitor and a heat resistant factor probably being tannin from the seed coat. Lately few articles have been published about legume tannins and other polyphenolic substances assumed to act as thermoresistant TI [15] [16]. Nevertheless, it is assumed that these tannins can have negative effects on the biological value of food by forming indigestible complexes with proteins [17]. We think that these complexes are also responsible for growth inhibition and an increased N-elimination which result from a bean diet [18].

Acknowledgments

The authors thank Mr. Dipl.Ing. F. Köhler, Institute of Food Technology (Cereal Technology of the Technical University Berlin) for carrying out the amino acid analysis, and the Alexander von Humboldt-Stiftung, Bonn, for financial support.

REFERENCES

1) J.J. Rackis, In "Soybeans: Chemistry and Technology", Ed. Smith A.K. and Circle, S.J. Vol 1. Avi Publ. Co., Westport, Conn.,1978
2) J.J. Rackis, J. Am. Oil Chemist's Soc. 51 (1974), 161A
3) R. Bressani, L.G. Elias, Food and Nutr. Bull. UNU, 1(4) (1979), 23
4) M.L. Kakade, J. agr. Food Chem. 22 (1974), 550
5) M.L. Kakade, R.J. Evans, J. Nutr. 90 (1966), 191

6) L.G. Elias, D.G. de Fernandez, R. Bressani, J. Food Sci. __44__ (1979), 524
7) M.L. Kakade, N.R. Simons, I.E. Liener, Cereal Chem. __8__ (1969), 518
8) S. Moore, W. Stein, Meth. Enzymol. __6__ (1963), 819
9) S. Moore, J. biol. Chem. __238__ (1963), 235
10) P.L. Antunes, V.C. Sgarbieri, J. agr. Food Chem. __28__ (1980), 935
11) A. Pusztai, Eur. J. Biochem. __5__ (1968), 252
12) I.L. Gatfield, Lebensm.-Wiss. u. -Technol. __13__ (1980), 46
13) Y. Birk, Proteinase Inhibitors. Proc. Int. Res. Conf. 2nd (Bayer Symp. V) Grosse Leder 1973, p 355, Springer Verlag Berlin and New York, 1974
14) H. Gerstenberg, H.D. Belitz, J.K.P. Weder, Z. Lebensm. Unters. Forsch. __171__ (1980), 28
15) M.C. Mondragon, D.I. Gonzalez, Arch. latinoamer. Nutr. __28__ (1978), 41
16) B.O. de Lumen, L.A. Salamat, J. agric. Food Chem. __28__ (1980), 533
17) D.W. Griffiths, G. Moseley, J. Sci. Food. Agric. __31__ (1980), 255
18) U.M. Lanfer Marquez, F.M. Lajolo, J. agric. Food Chem. 1981 (in press)

Post Mortem Changes of Enzymatic Activities and Morphology in Meat

V. VÁŇA, P. RAUCH, J. KÁŠ

Department of Biochemistry and Microbiology, Institute of Chemical Technology, Suchbátarova 5, 166 28 Prague 6, Czechoslovakia

SUMMARY

The changes of activity of marker enzymes were followed under post mortem conditions and compared with morphological changes of cell structures.

INTRODUCTION

Post mortem processes taking place in animal tissues used as food /e.g. liver and meat/ are very interesting for basic research, as well as, for food technology. In spite of the complexity of this problem it is important to improve our knowledge in all ways One important field of investigation is the comparison of morphological changes with some biochemical characteristics related to the investigated morphological structure. This kind of study helps us better to understand the mutual relationship between morphological and biochemical changes during the post mortem period. Morphological changes at the cellular and subcellular level are directly caused by the action of enzymes and on the contrary the released enzymes promote further morphological changes. Our research group was being engaged during several past years in this kind of research.

POST MORTEM PROCESSES

Different animal tissues /guinea pig liver, rat liver, beef liver, beef sirloin, pork joint and carp muscle/ were investigated by means of biochemical/ measurement of enzymatic activities, neasurement of pH/ and microscopical methods/ microscopy in polarized light, fluorescence microscopy, electron microscopy/.

The main results of our study may be summarized as follows:
- the studied interval of post mortem processes is divided into four periods with following characteristics:

Period	Storing at 4°C /days/	at 37°C /hours/	Typical change
I. rigor mortis	0-1	0-1	decrease of ATP synthesis, decrease of pH, very slow proteolysis, shrinkage of mitochondria, decrease of the activity of some enzymes
II. "ripening" in meat technology	2-7	1-6	irreversible swelling of mitochondria, disappearing of ATP, increase of activity of mitochondrial enzymes, minimal pH /after 5 days/, disruption of lysosomes, proteolysis
III. "ripe meat"	8-12	6-8	increase of pH, inactivation of mitochondrial enzymes, decrease of proteolysis, /neutral proteases/
IV.	12-	8-	uncontrolled processes, deep decay and bacterial contamination

In this paper we should like to demonstrate the correlative biochemical and microscopical study performed on skeletal muscle of beef /beef sirloin-musculus psoas maior/. We investigated samples of meat taken off just after slaughter and transport into laboratory and samples stored in the refrigerator at the temperature about $0-4^{o}C$/. Cell fractions were prepared by routine methods in sucrose gradient. Enzyme activities were determined in meat juice obtained by pressing meat between stainless steel plates and in meat homogenate. During storage we examined changes of pH in situ by means of inserted electrode. Nicotinamide nucleotide-linked dehydrogenase activities were assayed spectrophotometrically at 340 nm/malatedehydrogenase-MDH and glutamate dehydrogenase-GDH/. Cytochrome oxidase activity was determined using oxygen electrode by method of Wharton and Griffith. Activity of acid phosphatase was assayed spectrophotometrically. For morphological studies we fixed material both for light microscopy and electron microscopy in glutaraldehyde and used routine methods of dehydration, embedding and staining. When one studies muscle tissue post mortem mitochondrial and lysosomal enzymes are the most suitable for the correla-

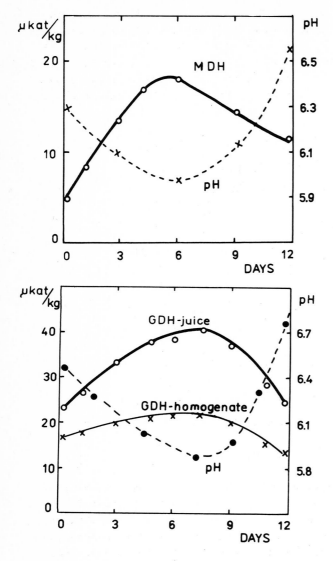

Fig. 1. Activity of malate dehydrogenase during meat storing at 4°C

Fig. 2. Activity of glutamate dehydrogenase during meat storing at 4°C

tive study of biochemical and morphological changes. These enzymes play distinct role in the metabolism of muscle tissue, routine methods for the assay of enzyme activities are known and morphological changes of the corresponding organelles are well defined. In mitochondria soluble and structurally bond enzymes show different behavior. We studied activities of two soluble enzymes /malate dehydrogenase and glutamate dehydrogenase/ and one structurally bond enzyme /cytochrome oxidase/. It is better to assay activities of soluble enzymes in meat juice than in meat homogenate. The homogenation of the tissue always leads to the damage of part of mitochondrial population and therefore results are not reproducible. This was confirmed by our results. In meat juice we found maximum activity of MDH and GDH when mitochondria are disrupted and mitochondrial matrix is released into tissue /i.e. after 5th day of storage, Fig. 1 and 2/. The maximum activity for the structurally bond cytochrome oxidase is reached somewhat later /about 8 th day of storage, Fig. 3/

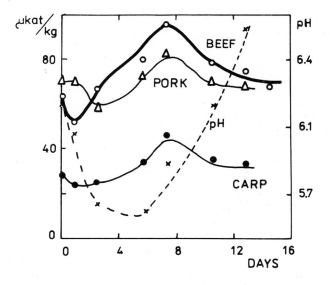

Fig. 3. Activity of cytochromoxidase during meat storing at $4^{\circ}C$

If we compare these results with the morphological changes observed on electronograms /Fig. 4/ we see that the maximum activity after 5th day of storage corresponds with the irreversible swelling of mitochondria and disruption of their outer membranes in situ. /Similar changes can be observed in isolated mitochondrial fraction/. Later-about 8th day - - mitochondria are much more damaged, cristae are fragmented into vesiculi and therefore maximum activity for bond enzyme was found. If we summarize the results and compare them with pH changes we see that the maximum activity of soluble enzymes corresponds with minimal value of pH. We can assume that low pH value activates some hydrolases which act during destruction of mitochondrial membranes. The stability and behavior of lysosomes under post mortem conditions is best characterized by the changes of activity of acid phosphatase /Fig. 5/. The activity was assayed using p-nitrophenylphosphate /β-glycerophosphate was not

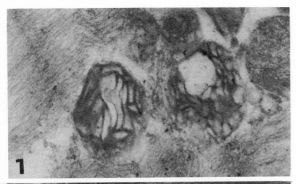
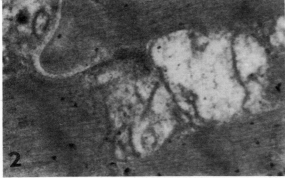
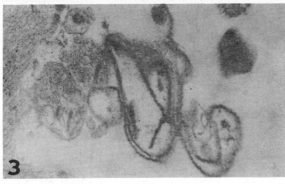

Fig. 4. Ultrastructure of mitochondria
of beef sirloin
1-fresh sample
2-5 days of storing at $4^{\circ}C$
3-9 days of storing at $4^{\circ}C$
Electron microscopy, magnification
approx. 50000 x .

used because of high phosphate concentration in meat/. This substrate is cleaved also by microsomal glucose-6-phosphatase, therefore we see two peaks on the curve, first of which belongs to acid phosphatase and corresponds with the disruption of lysosomes /3rd to 4th day of storing/ Fig. 6

The described changes of studied organelles during two weeks of storing are accompainied by the changes of basic structural and functional units of skeletal muscle /striated/. By means of light microscopy we observed changes of crossstriation of muscle fiber,in electron microscopy changes in arrangement of contractile protein in myofibril, especially in isotropic zone with so-called Z-line.

The results presented here show close relationship of biochemical and morphological changes of animal tissue during the post mortem period.

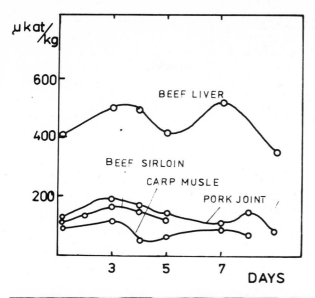

Fig. 5. Activity of acid phosphatase during meat storing at $4°C$

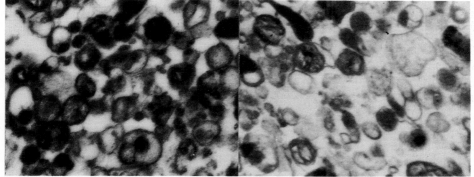

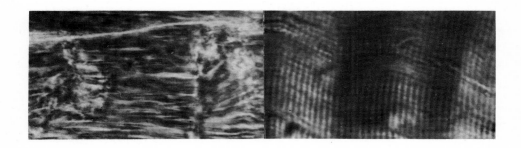

Rapid Chromatographic Analysis of Enzymes and Other Proteins

O. MIKEŠ

Institute of Organic Chemistry and Biochemistry, Czechoslovak Academy of Sciences, 166 10 Prague 6, Czechoslovakia

SUMMARY

The problems of classic column chromatography of proteins are treated briefly and the requirements necessary for modern packings to be used for high (or medium) pressure liquid chromatography (HPLC, MPLC) of proteins are described. Various commercial chromatographic supports are characterized by examples of chromatography of proteins (enzymes). The instrumentation used for this type of chromatography is outlined and the principle of post-column enzyme detectors is explained. The outlooks of HPLC (MPLC) of proteins in food chemistry are discussed.

INTRODUCTION

This lecture on rapid column separation of proteins has already appeared in the printed version (1) as a condensed review with 112 citations. The aim of this shorter oral version is to illustrate the topic by using several selected data and figures.

The history of classic liquid chromatography is a relatively long one, since the chromatography of biopolymers - especially of proteins - has been problematic for many years. Inorganic supports irreversibly sorb proteins; column packings with aromatic matrices are too hydrophobic and denature proteins. The only exception is the methacrylate ion-exchanger Amberlite IRC 50 and similar resins.

Because these low-capacity packings are microporous, the functional groups on outer surface of the particles are active only. In the middle of the fifties Peterson and Sober (2, 3) introduced hydrophilic ion-exchange cellulose derivatives and Porath, Flodin and Lindner (4, 5) the macroporous derivatives of crosslinked polydextran. A rapid development of liquid chromatography of proteins followed afterwards. Recently crosslinked derivatives of agarose were introduced (6). Thousands of papers have been published describing successful separations of biopolymer mixtures on these materials. - However, they are not found suitable for rapid protein analysis.

The general disadvantages of column packings based on polysaccharide matrices are the following. - First: They are too soft and do not allow the use of higher pressures. They cannot be used for HPCL and the separation is lengthy. - Second: They are digested by some enzymes from cultivation media. This is the reason why these packings are attacked also by microbes. - Third: Some of them - for instance the ion-exchange derivatives of crosslinked polydextran - change their volume with ionic strength and the bed shrinks during regeneration. Repeated use of columns once packed is impossible. - Fourth: Mixed aqueous-organic solvents cannot be used.

We have witnessed during the past few years the rapid penetration of modern HPLC-methods into many branches of chemistry. No wonder that suitable packings have also been sought for rapid separation and isolation of enzymes and other proteins.

COLUMN PACKINGS FOR HPLC (MPLC) OF PROTEINS AND THEIR USE

The requirements necessary for modern packings to act as supports in pressure chromatography of biopolymers are the following. The material must be macroporous in order to permit the penetration of biopolymers, and hydrophilic in order not to denature them. The material must be rigid and resisting to high pressures in columns. Fine and size-homogeneous spherical beads are the best. The particles must be chemically and biochemically resistant and must not change their volume. The supports should be repeatedly usable.

Various packings have been developed for this purpose, but not all of them meet all of the requirements mentioned. The first paper on the HPLC of proteins on inactivated silica gel was published by Shechter (7) in the seventies. In our Laboratory (8, 9) we prepared and tested for MPLC of biopolymers the ion-exchange derivatives of Spheron, a

TABLE I. CONTROLLED PORE GLASS

(Manufactured by Corning, USA, distributed by Pierce)

Designation	Operating range (10^3)	Pore diameter (Å)	Pore volume (cm^3/g)	Surface area (m^2/g)
CPG-40	1-8	40	0.1	190
CPG-100	1-30	100	0.4	170
CPG-250	2.5-125	250	1.0	130
CPG-550	11-350	550	1.0	70
CPG-2500	200-1500	2500	1.5	10

Glycophase G^{TM}: $-\overset{|}{\underset{|}{Si}}CH_2CH_2CH_2OCH_2\overset{OH}{\underset{|}{CH}}-\overset{OH}{\underset{|}{CH_2}}$

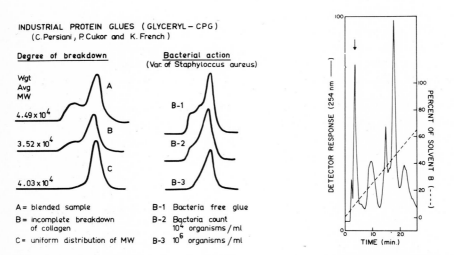

Fig.1. Size exclusion chromatography of industrial process glues (23).

Fig.2. Rapid separation of trypsin from contaminants (14).

fully synthetic hydrophylic polymer (10), prepared in 1973 by Čoupek and coworkers (11) in Prague. Suitable derivatives of porous glass and microparticulate silicas were developed by Regnier et al. (12-16). Various hydrophilic and hydrophobic or ion-exchange modifications of silicas have been reported since by other authors (17-21). Fundamental information on these materials and examples of their use in the HPLC of proteins will be given below.

TABLE II. GLYCOPHASE ION EXCHANGERS

Anion exchangers

DEAE-Glycophase
0.1 meq/g

$-\underset{|}{\mathrm{Si}}(CH_2)_3 OCH_2 \underset{OH}{\mathrm{CH}} - \underset{OCH_2CH_2N(Et)_2}{\mathrm{CH}_2}$

QAE-Glycophase
0.05 meq/g

$-\underset{|}{\mathrm{Si}}(CH_2)_3 OCH_2 \underset{OH}{\mathrm{CH}} - \underset{OCH_2CH_2\overset{+}{N}(Et)_3 Cl^-}{\mathrm{CH}_2}$

Cation exchangers

CM-Glycophase
0.1 meq/g

$-\underset{|}{\mathrm{Si}}(CH_2)_3 OCH_2 \underset{OH}{\mathrm{CH}} - \underset{OCH_2COOH}{\mathrm{CH}_2}$

SP-Glycophase
0.05 meq/g

$-\underset{|}{\mathrm{Si}}(CH_2)_3 OCH_2 \underset{OH}{\mathrm{CH}} - \underset{OCH_2CH_2CH_2SO_3H}{\mathrm{CH}_2}$

TABLE III. SPHERON GELS

(Manufactured by Lachema, distributed through Chemapol, Ltd., Prague)

Name	Exclusion limit (M_r) (water-dextran)		Surface area (m^2/g)	Particle size (swollen state)
	(a)	(b)		
Spheron 40	20000-60000		50-150	<25, 25-40, 40-63, 63-100
Spheron 100	70000-250000	100000	50-100	<25, 25-40, 40-63, 63-100
Spheron 300	260000-700000	501000	50-150	<25, 25-40, 40-63, 63-100, 100-200, 200-600
Spheron 1000	800000-5000000	1000000	50-150	<25, 25-40, 40-63, 63-100, 100-200, 200-600
Spheron 10000	10000000[c]		220-270	<25, 25-40, 40-63, 63-100

[a] Data from the producer. [b] Data determined in our study (24). [c] The exclusion limit was determined by means of phage sd, a host of Esch. coli.

Various derivatives of controlled pore glass (Tab. I) have been used for the chromatography of biopolymers. Unmodified macroporous glass (22) has a large inner surface and irreversibly adsorbs proteins. When provided with inner and outer Glycophase coating, i.e.

with glycerol bound via propyl silan to the glass surface, it receives hydrophilic properties. It can thus be used for exclusion chromatography of proteins. Moreover, one hydroxyl group can be ionogenically modified. LiChrosorb DIOL and SynChropak GPC are the derivatives of silica of the same composition of the hydrophilic layer. - The practical application of ionogenically unsubstituted glyceryl derivatives of Controlled Pore Glass (CPG) is illustrated (Fig. 1) by the size exclusion chromatography of industrial process glues (23). The left curves represent the possibilities of checking the breakdown-degree of collagen in the plant process. The right-hand curves illustrate the chromatographic control of the influence of bacterial contamination which also causes the partial degradation of the protein.

Table II presents a survey of ion-exchange derivatives of Glycophase on Controlled Pore Glass. These weakly and strongly basic anion-exchangers and weakly and strongly acidic cation-exchangers have been used for the chromatography of biopolymers. However, they have relatively low nominal ion-exchange capacities. - As an example of rapid separation of proteins on these ion-exchangers a chromatogram from Regnier's laboratory (14) is shown (Fig. 2). Commercial trypsin contains several inactive components which were separated on the DEAE-derivative in 30 minutes from the active enzyme indicated here by the arrow.

Materials found very suitable in our Laboratory are the ion-exchange derivatives of Spheron (24-27). This fully synthetic macroreticular polymer is prepared by copolymerization of glycol methacrylate and bis-glycol-methacrylate as a crosslinking agent. The densily crosslinked and microporous xerogel is very rigid and mechanically stable. The microstructure of the copolymer is characterized by repeating units of trimethyl acetic acid ester, which is extremely chemically stable. The copolymer contains many free hydroxyl groups of glycol and is therefore hydrophilic in principle. In a special process of suspension copolymerization this material is first separated from the solution in the form of very small submicroscopical drops - the so-called microspheres. The latter aggregate into larger drops - the beads - and continue to polymerize. The beads - the so-called macrospheres - are therefore composed of very small microspheres. The cavities between microspheres are macropores allowing the penetration of proteins. The macroporosity can be controlled over a wide range. Such a macrostructure of aerogel type has large inner surface area about 100 m^2/g. Many hydroxyl groups can be ionogenically modified.

TABLE IV. SPHERON ION EXCHANGERS

Matrix: Spheron 300

(Specific pore size 0.60 cm^3/g, diameter of most frequent pores 250 $\overset{o}{A}$)

Cation exchangers

CM-Spheron 300	carboxymethyl derivative	2.0 meq/g
P-Spheron 300	phosphate derivative	4.0 meq/g
S-Spheron 300	sulfate derivative	1.5 meq/g

Anion exchangers

DEAE-Spheron 300	diethylaminoethyl derivative	2.0 meq/g
QAE-Spheron 300	quaternary derivative	2.0 meq/g

Matrix: Spheron 1000

(Specific pore size 1.69 cm^3/g, diameter of most frequent pores 370 $\overset{o}{A}$)

Cation exchangers

Spheron C 1000	carboxyl derivative	2.0 meq/g
Spheron Phosphate 1000		3.5 meq/g
Spheron S 1000	sulfopropyl derivative	1.5 meq/g

Anion exchangers

Spheron DEAE 1000	diethylaminoethyl derivative	1.5 meq/g
Spheron TEAE 1000	triethylammonium derivative	1.5 meq/g

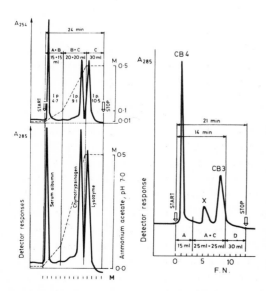

Fig.3. MPLC of proteins and their fragments on CM-Spheron 300 (9).

Table III summarizes various Spheron gels which are produced in Czechoslovakia now. They broadly differ in exclusion limits. We have selected two of them, with exclusion limits half a million and one million M_r, for experiments with ionogenic substitution. Unsubstituted Spheron can also be used directly for hydrophobic chromatography of proteins (28, 29). - Table IV shows a list of Spheron ion-exchangers prepared so far. The top part of the Table shows ion-exchangers prepared on a laboratory scale only. The bottom part summarizes ion--exchangers whose commercial production is being prepared in Lachema, Brno. They have an exclusion limit of 1 million M_r.

MICROPARTICULATE POLYETHYLENIMINE ANION EXCHANGER (A.J. Alpert and F.E. Regnier)

WEAK BUFFER:
10 mM KH_2PO_4, 10 mM glucose, 5 mM $MgCl_2$, 0.5 mM dithioerythritol (pH 7.2)

STRONG BUFFER:
0.6 M KCl in weak buffer

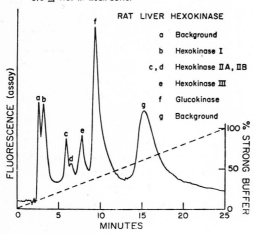

Fig. 4. HPLC of hexokinase isoenzymes on PEI 6-LiChrospher Si 500 (15).

Weakly, medium and strongly acidic cation-exchangers and weakly and strongly basic anion-exchangers have been prepared.

These rigid ion-exchangers can be regenerated by 2 molar hydrochloric acid and 2 molar sodium hydroxide. The bed volume does not depend on ionic strength. Mixed aqueous and organic solvent-solutions can be used for chromatography and washing. The ion-exchangers are not atacked by any enzyme known so far. - Fig. 3 shows examples of rapid separation of proteins and their fragments on carboxymethyl-Spheron (9). The column dimensions were 20 by 0.8 cm. The left-hand chromatogram proves that the separation of proteins follows their isoelectric points, so that it is a true ion-exchange chromatography. The right-hand diagram represents rapid separation of two serum albumin fragments containing 31 and 185 amino acids.

TABLE V. MICROPARTICULATE ANION EXCHANGERS

(A.J. Alpert and F.E. Regnier (15))

Crosslinked polyethylenimine
...NH.$(CH_2)_2$.NH.$(CH_2)_2$.NH.$(CH_2)_2$.NH...

Supports

LiChrosorb
LiChrospher
Partisil
Chromosorb LC-6
Controlled-pore Glass (CPG)
Porous titania
Spherisorb alumina

Crosslinkers

1,3-Dibromopropane
Epoxy resins
2-Methyl-2-nitro-1,3-propanediol
Dithiobis(succinimidylpropionate)
Cyanuric chloride
Dimethyl adipimidate dihydrochloride

The chemical principles of microparticulate anion-exchangers developed in American laboratories (15) are explained in Table V. A very thin layer of polyethylenimine is formed on the surface of supports by evaporation and the dry imine is subsequently crosslinked by some of the agents given in the Table. It is interesting that the crosslinked layer is not covalently bound to the cationic inorganic matrix. It behaves like a slightly basic anion-exchanger. SynChropak AX with oxirane crosslinkers, a high capacity support produced by the American firm SynChrom, is of this type. - An example (15) of HPLC of hexokinase isoenzymes on a microparticulate polyethylenimine anion-exchanger of the general structure just described is given in   Figure 4. A simple linear gradient was used for the elution and fluorescence monitoring for post-column detection of enzymes.

Of the Japanese TSK gels (30) the so-called G 2 to 4 thousands SW Gels are advantageous for the separation of proteins by size exclusion chromatography (Tab. VI). They were tested not only in buffers, but also in the solutions of urea, sodium dodecyl sulfate and guanidine hydrochloride. The exact chemical composition of these gels is not yet known. - An example of their application is given in Fig. 5. Commercial enzyme preparations of catalase, alcohol dehydrogenase, and peroxidase were chromatographed on TSK gels (21). Isocratic elution and a phosphate buffer were used. The dimensions of the stainless-steel column were 60 by 0.7 cm, particle size 10 micrones, necessary pressure 110 atmospheres.

TABLE VI. TSK-GELS

(Produced by Toya Soda, Tokyo, Japan, for M_r-distribution measurements: PW-Gels for oligomers and polymers, SW-Gels for proteins)

SW-Gels

Microparticulate silica, bonded with hydrophilic compounds.
(Y. Kato, K. Komia, H. Sasaki and T. Hashimoto (30))

		(In buffers)	(In 6M G-HCl)
G 2000 SW	M_r	$0.5 - 10 \times 10^4$	$0.1 - 2.5 \times 10^4$
G 3000 SW		$10 - 50 \times 10^4$	$0.2 - 7 \times 10^4$
G 4000 SW		$20 - 70 \times 10^4$	$0.3 - 40 \times 10^4$

SWG = preparative columns for SEC

Supports with chemically bonded phases are used for reversed-phase chromatography of proteins (31, 32). This method is based on hydrophobic interactions between hydrocarbon chains and hydrophobic domains of the protein molecules chromatographed. The mobile phase is polar. The gradual elution of the individual compounds of the mixture can be achieved by lowering the polarity of the mobile phase by the addition of alcohol or acetonitrile. - An example (31) is shown in Figure 6. Tryptophan, a hydrophobic amino acid, used as internal standard, and a mixture of six simple proteins were perfectly separated in 50 minutes on octadecyl phase bound to Hypersil. A short 10 by 0.5 cm column was used. The dotted line represents the gradient of acetonitrile. An injection artifact is seen at the start of the chromatogram.

INSTRUMENTATION FOR HPLC (MPLC) OF PROTEINS AND POST-COLUMN ENZYME DETECTORS

For medium-pressure liquid chromatography - which is often sufficient when particles larger than 20 microns are used - a simple "home-made" protein analyzer can be used (26). Thick-walled glass columns and parts of amino acid or sugar analyzers can be used to advantage for such a purpose. - A schematic representation of such an versatile apparatus for biopolymer analysis, which was built in the author's Laboratory in Prague, is shown in Figure 7. Using Spheron beads 20 to 40 microns in diameter and 20 cm columns, the necessary pressure does not exceed 10 atmospheres. - The majority of authors who studied the HPLC of proteins used various commercial liquid chromatographs or their parts developed for the separation of low molecular weight substances which they combined or modified for their own purposes. Recently the Waters Company introduced the so-called Protein Separation System PSS which seems to be effective for many purposes. None of the instruments developed so far can provide for automatic regeneration of ion-exchange columns by solutions of extreme pH-values.

I am going to explain now the principle of post-column enzyme detectors according to Chang, Gooding and Regnier (13). It was developed for the detection of isoenzymes present in human plasma. Isoenzymes have the same specificity but differ in their retention. The chromatographic patterns of isoenzymes can be used for diagnostic purposes in medicine. - Plasma contains various proteins in large quantities. An efficient chromatographic isolation of small amounts of particular isoenzymes from other proteins in one run is not possible in spite of the fact that the individual isoenzymes are separated from one another. Their peaks are overlapped by other proteins present. Therefore a detector was developed, which can distinguish the isoenzymes only and "overlooks" the other proteins.

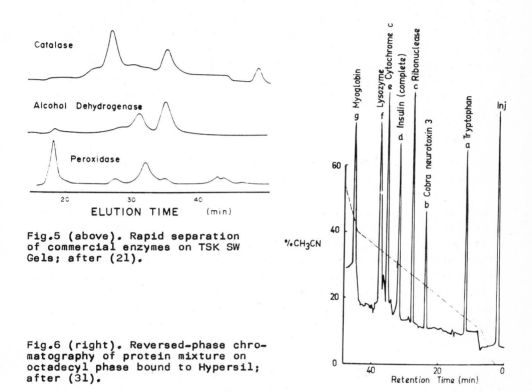

Fig.5 (above). Rapid separation of commercial enzymes on TSK SW Gels; after (21).

Fig.6 (right). Reversed-phase chromatography of protein mixture on octadecyl phase bound to Hypersil; after (31).

The monitoring system (Fig. 8) consists of two columns (13). The analytical column is packed by chromatography supports. The proteins are separated in this column. The effluent is mixed with a specific substrate. The mixture passes through the second thermostated column, filled with hydrophilized microparticular non-porous compact beads only. In this column the components are mixed and the enzyme reaction proceeds. A substrate is selected which permits measurements of the reaction products independently of the presence of proteins. It is then possible to detect only the isoenzymes separated. The system has been developed further by other authors. The principle will be illustrated by lactate dehydrogenase isoenzymes as an example. These enzymes oxidize lactic acid to pyruvic acid and nicotine adenine dinucleo-

VERSATILE EQUIPMENT FOR MPLC

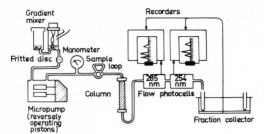

Fig.7. Versatile apparatus for MPLC biopolymer analysis

tide is converted to its reduced form. The latter can simply be detected by ultraviolet absorbance at 340 nm or by fluorescence measurement at 457 nm:

$$CH_3CH(OH)COOH + NAD^+ \xrightarrow{LD} CH_3.CO.COOH + NADH + H^+ \quad (A_{340}, E_{457})$$

Figure 9, taken from the paper by Schlabach and coworkers (33), shows the pattern of lactate dehydrogenase isoenzymes in normal human serum when chromatographed on a column of microparticulate anion-exchanger. All the five isoenzymes known are seen with the exception of the peak of serum albumin. This protein is present in extremely high quantities in the serum and interferes a little. Let us compare this profile with the following one (Fig. 10). This is the lactate dehydrogenase profile after myocardial infarction (33). A great difference can be seen at a first sight. This change of the profile appears in serum very quickly, already one minute after the infarction. Also various internal injuries after accidents can be detected quickly by isoenzyme analysis of the serum before operation of the patient.

These detection methods seem to have no relation to food-chemistry applications. But I think that these examples may stimulate efforts to develop similar assay for biochemical industry.

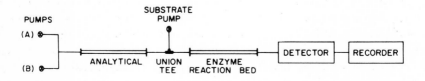

Fig.8. Principle of post-column on-line enzyme detector (13).

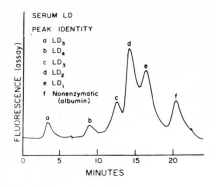

Fig.9 (below). Pattern of lactate dehydrogenase isoenzymes in normal human serum (33)

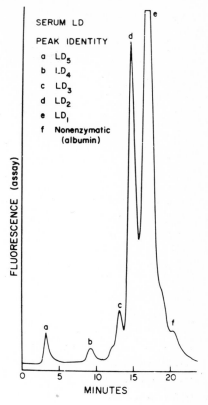

Fig.10 (right). Lactate dehydrogenase isoenzymes after myocardial infarction (33)

OUTLOOKS OF HPLC (MPLC) OF ENZYMES AND OTHER PROTEINS IN FOOD INDUSTRY ANALYSIS

The fermentation processes of all kinds and the technological procedures for enzyme preparations require rapid and precise analytical diagnostic methods of a similar type. Not only isoenzymes but also all the multiple forms of enzymes are technically important. We meet with them often in technical preparations. The separation of pectolytic enzymes by MPLC on Spheron ion-exchangers, developed in Prague in cooperation with Rexová from Bratislava (34), can serve as an example (Fig. 11). A technical pectolytic enzyme, Pectinex Ultra, was chromatographed on a 20 by 0.8 cm column of Spheron sulfate 1000, particle size 20 - 40 microns, using a system of ionic strength gradients. The necessary pressure varied around 5 atmospheres. L designates pectin-lyase, N endopolygalacturonanase and S pectin-esterase.

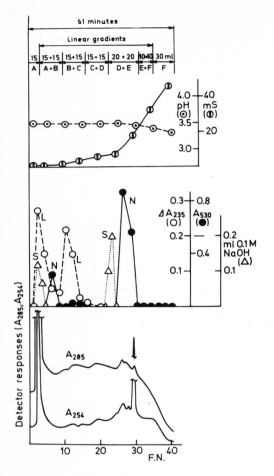

Fig.11. Rapid MPLC of a technical pectolytic enzyme Pectinex Ultra on Spheron S 1000 (34)

The polyplicity of enzyme forms cannot always be explained only by the existence of isoenzymes coded for genetically. In some cases these forms arise - without a loss of activity - by additional modification of the enzyme during the process of its production, for instance by limited proteolysis or, in the case of glycoproteins, by modifications of the oligosaccharide side chains. Therefore the detection of the multiple forms of enzymes opens the way to sensitive chromatographic diagnostics in all branches of biochemical industry. - Simple rapid chromatography of technical enzymes and other proteins, disregarding the multiple forms, is essential for food industry. The rapid chromatography of technical protease or glucose oxidase on CM- or DEAE-Spheron (Figs 1 or 4 in ref. 35) may serve as an example.

Let me show now how the process of enzymic degradation of a macromolecular substrate can be checked by a series of six very rapid analyses using SEC (size exclusion chromatography). This example is taken from the work of Chang and coworkers (13). - Deoxyribonucleic acid was digested by deoxyribonuclease (Fig. 12). Chromatogram $\underline{a}$ represents the original substrate, i.e. the intact nucleic acid, $\underline{b}$ is the sample after 2 minutes of digestion, $\underline{c}$ after half an hour, $\underline{d}$ after one hour, $\underline{e}$ after 3 hours and $\underline{f}$ after 4 hours, where already no high molecular weight substrate is present. A 30 by 0.4 cm column was packed with LiChrospher Si-1000, particle size 10 microns, coated with Glycophase G. The pressure was 45 atmospheres. The example illustrates the possibility of a rapid HPLC check of a fermentation kinetics. - Another example describing the slow decomposition of beta-lactoglobulin by alpha-chymotrypsin can be found in the paper by Fukano et al. (21).

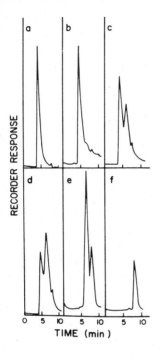

Fig.12. Series of six rapid SEC analyses of deoxyribonucleic acid digests by deoxyribonuclease (13)

CONCLUSIONS

At the end of my lecture let me summarize the advantages of HPLC or MPLC of proteins.-

A) These procedures enable a rapid and a better characterization of raw materials and products in food technology.

B) They pave the way to biochemical diagnostics of technology processes and ripening as well as storage processes.

C) These rapid methods permit us not only to follow but also to influence running macromolecular fermentations of all kinds, i.e. to act during the process and not only after it has been completed and cannot be affected any more. -

It is therefore very probable that these modern chromatography methods will be still more and widely applied in food technology analyses in the near future.

ACKNOWLEDGEMENT

The lecturer is most grateful to the authors cited and to the publishers of Journals listed below for their kind permission to use some of the Figures and data from their papers.

J. Chromatogr. Science (Preston Publications, Inc., Niles, Il.)
Analyt. Chem. (Amer. Chemical Society, Washington D.C.)
J. Chromatogr. (Elsevier, Amsterdam)
Int. J. Peptide Protein Res. (Munskgaard, Copenhagen)
Clinical Chemistry (Amer. Assoc. of Clinical Chemists, New York)

REFERENCES

1. O. Mikeš, Ernährung/Nutrition 5 (1981), 88
2. H.A. Sober and E.A. Peterson, J. Amer. Chem. Soc. 76 (1954), 1711
3. E.A. Peterson and H.A. Sober, J. Amer. Chem. Soc. 78 (1956), 751
4. J. Porath and P. Flodin, Nature (London) 183 (1959), 1657
5. J. Porath and E.B. Lindner, Nature (London) 191 (1961) 69
6. J. Porath, T. Låås and J.Ch. Janson, J. Chromatogr. 103 (1975) 49
7. I. Shechter, Analyt. Biochem. 58 (1974), 30
8. O. Mikeš, P. Štrop, J. Zbrožek and J. Čoupek, J. Chromatogr. 119 (1976), 339
9. O. Mikeš, Int. J. Peptide Protein Res. 14 (1979), 393
10. J. Janák, J. Čoupek, M. Krejčí, O. Mikeš and J. Turková, p. 189 in Z. Deyl, K. Macek and J. Janák (Editors): Liquid Column Chromatography, Elsevier, Amsterdam 1975
11. J. Čoupek, M. Křiváková and S. Pokorný, J. Polymer Sci., Polymer. Symp. 42 (1973), 185
12. F.E. Regnier and R. Noel, J. Chromatogr. Sci. 14 (1976), 316
13. S.H. Chang, K.M. Goodin and F.E. Regnier, J. Chromatogr. 125 (1976), 103 and 321
14. S.H. Chang, R. Noel and F.E. Regnier, Analyt. Chem. 48 (1976), 1839
15. A.J. Alpert and F.E. Regnier, J. Chromatogr. 185 (1979), 375
16. F.E. Regnier and K.M. Gooding, Anal. Biochem. 103 (1980), 1
17. K. Unger, J.J. Schick-Kalb and K.F. Krebs, J. Chromatogr. 83, (1973), 5
18. P. Roumeliotis and K.K. Unger, J. Chromatogr. 149 (1978), 211; 185 (1979), 445
19. H. Engelhardt and D. Mathes, J. Chromatogr. 142 (1977), 311
20. T. Hashimoto, H. Sasaki, M. Aiura and Y. Kato, J. Chromatogr. 160 (1978), 301
21. K. Fukano, K. Komiya, H. Sasaki and T. Hashimoto, J. Chromatogr. 166 (1978), 47
22. W. Haller, Nature (London) 206 (1965), 693

23 C. Persiani, P. Cukor and K. French, J. Chromatogr. Science 14 (1976), 417
24 O. Mikeš, P. Štrop and J. Čoupek, J. Chromatogr. 153 (1978), 23
25 O. Mikeš, P. Štrop, J. Zbrožek and J. Čoupek, J. Chromatogr. 180 (1979), 17
26 O. Mikeš, P. Štrop, M. Smrž and J. Čoupek, J. Chromatogr. 192 (1980), 159
27 O. Mikeš (Editor): Laboratory Handbook of Chromatographic and Allied Methods. Elis Horwood, Ltd. (Halsted Press, J. Wiley & Sons), Chichester 1979. Pp. 260-1, 346-7, 403-5
28 P. Štrop, F. Mikeš and Z. Chytilová, J. Chromatogr. 156 (1978), 239
29 P. Štrop and D. Čechová, J. Chromatogr. 207 (1981), 55
30 Y. Kato, K. Komiya, H. Sasaki and T. Hashimoto, J. Chromatogr. 190 (1980), 297; 193 (1980), 29 and 458
31 M.J. O'Hare and E.C. Nice, J. Chromatogr. 171 (1979), 209
32 M.T.W. Hearn and W.S. Hancock, J. Chromatogr. Sci. 12 (1979), 243
33 T.D. Schlabach, A.J. Alpert and F.E. Regnier, Clinical Chemistry 24 (1978), 1351
34 O. Mikeš, J. Sedláčková, Ľ. Rexová-Benková and J. Omelková, J. Chromatogr. 207 (1981), 99
35 O. Mikeš, P. Štrop and J. Sedláčková, J. Chromatogr. 148 (1978), 237

A version of this lecture has also been published in Ernährung 5 (1981) by agreement of the publishers.

New Chlorine Containing Organic Compounds in Protein Hydrolysates

J. DAVÍDEK, J. VELÍŠEK, V. KUBELKA[+] and G. JANÍČEK

Department of Food Chemistry and Analysis, [+]Department of Mass Spectrometry, Institute of Chemical Technology, CS - 166 28 Prague, Czechoslovakia

SUMMARY

Crude liquid neutralized protein hydrolysate, its filtrate and filter cake (melanoidins) were extracted with diethyl ether, the extract was separated by column chromatography on silica gel and the obtained individual fractions were analyzed by thin - layer chromatography, gas chromatography and by recording their mass, infrared and nuclear magnetic resonance spectra. It was found that the first two analyzed samples contained various proportions of glycerol chlorohydrins esters with higher fatty acids. These esters represent a new class of endogenous food contaminants.

INTRODUCTION

Chemical hydrolysates of proteins have become important commodities in many countries all over the world for the improvement of the flavour of various foods. Currently, vegetable raw materials, e.g. wheat and maize gl utens, soybean meal and flour and occasionally some other raw materials have been employed in the process of hydrolysis.

All of these raw materials contain residual lipids which are mainly present in the form of triacylglycerols and phospholipids. Their amount represents about 1 to 3 % of the raw material weight.

Some time ago, the neutral fractions of several types of commercial and laboratory - made protein hydrolysates were analyzed in our laboratory and three not previously reported chlorine - containing alcohols were identified (3-chloro-1-propanol, 1,3-dichloro-2-propanol, 2,3-dichloro-1-propanol). As it was shown, all the three alcohols together with some new compounds, e.g. 3-chloro-1,2-propanediol, were formed in the reaction of glycerol with hydrochloric acid under the conditions used for the manufacture of protein hydrolysates (1). They were also identified as the hydrolytic products of triacylglycerols (triacetin, tributyrin, tripalmitin, tristearin, triolein) and phospholipids (soya lecithin) with hydrochloric acid (2, 3).

Besides these alcohols and diol their corresponding esters with fatty acids were also found. Their structure can be expressed as

$$\begin{array}{ccc} CH_2 - CH - CH_2 \\ | & | & | \\ Cl & R_2 & R_1 \end{array}$$

where R_1 is acyl or chlorine atom and R_2 acyl, chlorine atom or hydroxyl group. The main reaction products were diesters of 3-chloro-1,2-propanediol, 1-monoesters of 3-chloro-1,2-propanediol and esters of 1,3-dichloro-2-propanol.

Dichlorohydrins as well as monochlorohydrins of glycerol which were found in protein hydrolysates are only intermediate products of the hydrolysis of lipids with hydrochloric acid, their esters with fatty acids might also form during the production of protein hydrolysates analogically with the model experiments previously described.

EXPERIMENTAL

Liquid neutralized protein hydrolysate (1.5 kg), its filtrate (3 kg) and the corresponding filtr cake - melanoidins (0.5 kg) were, therefore, repeatedly extracted with diethyl ether and then with solutions of sodium bicarbonate and sodium hydroxide. The residual neutral fraction dissolved in chloroform was subjected to column chromatography on silica gel (500 x 10 mm) and eluted with petroleum ether - diethyl ether mixtures. The individual fractions were then analyzed by thin - layer chromatography, gas chromatography and by recording their mass, infrared and nuclear magnetic resonance spectra (4).

RESULTS AND DISCUSSION

A thin - layer chromatographic separation of diethyl ether extracts of the analyzed samples is presented on Fig. 1.

Fig. 1 Thin - layer ch romatogram of samples from various stages of protein hydrolysate production

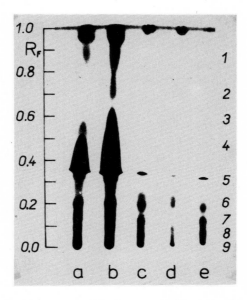

a = neutralized protein hydrolysate, b = filter cake, c = filtrate, d = filtrate without phenols, e = phenols from filtrate, 1 = esters of 1,3-dichloro-2-propanol, 2 = diesters of 3-chloro-1,2-propanediol, 3 = triacylglycerols, 4 = free fatty acids, 5 = 1,3-dichloro-2-propanol, 6 = 1-monoesters of 3-chloro-1,2-propanediol, 7 = 1,3-diacylglycerols, 8 = 1,2-diacylglycerols, 9 = monoacylglycerols and 3-chloro-1,2-propanediol

It was found that only the samples a and b contained glycerol chlorohydrins esters with higher fatty acids. These esters were detected neither in the filtrate of the neutralized protein hydrolysate nor in the final product - commercial protein hydrolysate.

Tab. 1 presents the amounts of glycerol chlorohydrins esters in the neutralized protein hydrolysate and in filter cake as well as the amounts of triacylglycerols and free fatty acids. It is evident that the hydrolysis of lipids was not completed for the original triacylglycerols were still present in the analyzed samples. In filter cake the contents of all of the analyzed compounds were higher (approximately 10 times) than those in the neutralized protein hydrolysate. Filter cake represents about one tenth of the weight of the neutralized hydrolysate and, therefore, it is possible that glycerol chlorohydrins esters remain on the filter together with melanoidins and other solids as they are non - polar compounds, practically undissolved in the hydrolysate where they form a solid layer on its surface on cooling.

Tab. 1 Concentrations in mg/kg of some constituents of the analyzed samples

Compound	Neutralized hydrolysate	Melanoidin
Esters of 1,3-dichloro-2-propanol	8	65
Diesters of 3-chloro-1,2-propanediol	4	35
1-Monoesters of 3-chloro-1,2-propanediol	35	205
Triacylglycerols	40	315
Free fatty acids	1290	14000

The qualitative composition of fatty acids bound in glycerol chlorohydrins esters did not differ from that of lipids occuring in the raw materials used for the production of protein hydrolysates. The main acids were linoleic and oleic acids, palmitic, stearic and linolenic acids were present in lower quantities, some other acids, i.e. myristic, palmitooleic, arachidic, gadoleic and behenic acids were present in trace amounts not exceeding 1 %.

It will be necessary to apreciate the newly identified esters of glycerol chlorohydrins with higher fatty acids from a hygienic - toxicologic point of view, owing to the fact that these substances are chlorine - containing organic compounds. They represent a new class of endogenous food contaminants.

REFERENCES

1 J. Velíšek, J. Davídek, J. Hajšlová, V. Kubelka and G. Janíček, Z. Lebensm. Unters. - Forsch. 167 (1978), 241
2 J. Velíšek, J. Davídek, V. Kubelka, J. Bartošová, A. Tučková, J. Hajšlová and G. Janíček, Lebensm. - Wiss. u. - Technol. 12 (1979), 234
3 J. Davídek, J. Velíšek, V. Kubelka, G. Janíček and Z. Šimicová, Z. Lebensm. Unters. - Forsch. 171 (1980), 14
4 J. Velíšek, J. Davídek, V. Kubelka, J. Janíček, Z. Svobodová and Z. Šimicová, J. Agr. Food Chem. 28 (1980), 1142

Method for the Qualitative and Quantitative Analysis of Gelling and Thickening Agents

U. PECHANEK, W. PFANNHAUSER, H. WOIDICH
Forschungsinstitut der Ernährungswirtschaft
1190 Wien, Blaasstraße 29

SUMMARY

This article presents an introduction to an electrophoretic method for qualitative and quantitative determination of gelling and thickening agents using cellulose acetate membranes. The identification is based on their migration behaviour, staining abilities and shapes of their electrophoretic zones. Quantification is done with an TL scanner. The principle of isolation out of food is also described.

INTRODUCTION

The macromolecular polysaccharides originating from algae, plant seeds and exudates are used as gelling agents and thickeners in precooked food, icecreams, instant drinks and puddings showing multifunctional qualities.
In Austria, type and amount of these gelling agents and thickeners in food are determined by the standards for food additives published on May 14 1979.
However, until now only a few work dealing with the analytical problems of these substances have been reported, mostly about their microscopical identification or some reactions of precipitation.

Later, the chemical identification of their constituents has been realized with difficulties because these thickeners contain nearly the same sugars like mannose, galactose and arabinose.
The problem of these analyses has been to complete the hydrolization and also to avoid the destruction of resulting sugars.

Such analyses have required too much time and the results were only acceptable in the case of single thickeners.
Mergenthaler and coworkers have been the firsts dealing with analytical problems in the field of polysaccharides. They have tried the fractional separation on DEAE cellulose (1) as well as the gaschromatographic separation of sugar constituents by transforming them into aldonitrilic acetates (2) or treating them by methanolysis (3). Friese described the separation of the sugars constituents, using the Zeisel - splitting method prior to gaschromatographic separation (4).
All these methods reported in the literature have had the disadvantage of requiring a very intensive analytical work.
Being aware of these problems, we have tried therefore to build a method for analyzing the thickeners without splitting or derivatisation process.
The fact that thickeners carry some negative charges due to the sulfuric acid groups or the uronic acids has given us the opportunity for a separation in the electric field by using an electrophoretic separation method. First the polyacrylamide gel and the agarose gel have been used for separation because of the high difference in their molecular weight.
The migration distance of the polysaccharides in these materials are short, owing to the great retention of the three dimensional structures of the thickeners during their migration trough the narrow porous material. This happens despite the use of boric acid buffer to modify the intact substances by formation of a charged complex. The identification has been made afterwards by using different dyes, as described later.
Starting with these experiences, we have tried to find another material able to allow a separation according to their charges.

We have chosen cellulose acetate membranes as a support material. This kind of membranes has been used also for the separation of thickeners by Padmoyo and Miserez in 1967 (5). The separation reported has been unsatisfactory, the shapes of some migrated substances have been distorted and therefore unqualified quantification.
On the basis of this work, we have developed a method for the

qualitative and quantitative determination of all common thickeners.
There were three different criterions :
> Migration in the electrical field
> Possibility of being dyed
> Shape of the migrated zone

APPARATUS AND CHEMICALS

The electrophoresis is done using a Beckman microzone cell type R-101. The power supply, type R-120, allows a voltage between 0 and 500 volts and a check of the current at any time. The membrane, 145 x 57 mm size is fixed in the cell bridge during application of the samples and during electrophoresis. A special applicator made of two metal lamina takes up a volume of 0.25 μl of sample when dipped onto the surface of a drop.
A such a small amount of sample is often insufficient for visualization , this process mostly must be repeated several times.
In order to avoid broad application spots, the membrane must be dried under a stream of cool air after 0.5 μl of sample has been applied at the same spot.
During the application, the dry membrane is fixed in the cell bridge and mounted under the cell cover which has 8 application grooves in 3 different slots.
After the application, the membrane is wetted in boric acid buffer (pH = 10) and installed in the cell. The electrophoresis is run at 300 V for 20 minutes, then the identification is started using different dyes:

- 1 / <u>Toluidine blue:</u> this reagent dyes polysaccharides containing acidic groups. The excess of dye is eliminated by washing with some portions of water. A small amount of denaturated alcohol added to the last portion of water converts the colour of spots having carrageenan, agar and furcellaran from blue into violett. This reaction offers an additional possibilty for identification of these substances.

- 2 / <u>Methylene blue:</u> staines the polysaccharides by precipitation. It reacts with the same thickeners dyed by toluidine blue.
Substances like carrageenan, furcellaran, pectin and carboxymethylcellulose develope characteristic shapes of the spots after the migration has finished. Different types of carrageenan (lambda, kappa, iota) and alginates may be identified in this way.

- 3 / <u>Amidoblack:</u> a special dye for proteins and therefore is used only in case of gelatine.

- 4 / <u>Fuchsin reagent:</u> (or Schiff's reagent), this reagent is well-known for its sensibility to the aldehyde groups. After electrophoresis, the cellulose acetate membrane has to be incubated in a periodic acid solution firstly. During this process, the C-C bond between vicinal OH-groups is ruptured, forming an aldehyde.

Only gum guar, carob and starch can be identified by this reagent; however alginic acid, alginates, gum arabic, tragacanth, pectin and gum ghatti can be dyed in the same way.
An excess of this dye is eliminated by denaturated alcohol. If it is necessary to keep the membrane for a long period, it has to be made transparent, using a mixture of methanol and acetic acid and dried afterwards at $105^{\circ}C$.
The quantitative determination of thickeners is made with wet membranes pressed between two glass plates and measured by a Zeiss TL scanner using H_2 or Xenon high pressure lamp.
The reproducibility for application of the same amount of sample on 8 possible positions is 3-10 %, mostly due to the incorrect localization of the measured spots.

ISOLATION

Prior to quantitative determination, the thickeners have to be isolated from the sample and transfered into a purified form. Therefore all other constituents in the food might be eliminated.
Fats and dyes can be removed by dioxane. Starch is degraded by the enzymes α- amylase and amyloglucosidase. This step has to be done very carefully, because the thickeners themselves contain α-1,4 bonds which can be ruptured also. If starch degradation is incomplete, there is no influence on the quantitative determination because starch is separated from all other thickeners. The removal of proteins is a little more complicated, especially in that case when such thickeners have to be isolated, which form a complex with proteins(for example carrageenans). Using dioxane to remove fat,the bonds between proteins and carregeenan are splitted. The proteins are then precipitated by trichloric acetic acid. However, some polysaccharides (like agar, alginate and methyl celluolose) might be possibly precipitated by this reagent also. Therefore all the proteins have to be dissoluted firstly by addition of urea before using trichloric acetic acid (6).
If all other components from the food are removed, the thickeners are precipitated by absolute ethanol standing overnight.
The percentage of thickeners recovered during the isolation depends on the numbers of isolation steps, type of thickeners and problems

arising during the precipitation of proteins.
For simple products like instant drinks and fruitice, the percentage of thickeners we have recovered is nearly 100 %.
For instant puddings 81-93 % of carregeenan and 100 % of carob have been recovered. Difficulties have been arising during the isolation of baby food based on milk powder. We have added alginate and carob as powders, but the recovery has been only 50 %.
The problems related with isolation of thickeners are dealt in a special research project. We plan to analyse the thickeners in different products like ketchup, milkicecream, precooked frozen foods, candies and others.
Being easy and quick, our method allows the identification and the quantitative determination of all common thickeners.

Acknowledgement
We are grateful to the "Forschungsförderungsfonds der gewerblichen Wirtschaft" for having sponsored this work by a grant.

REFERENCES

1   E. Mergenthaler and W. Schmolck: Z. Lebensm. Unters. Forsch. 155 (1974) 193-202
2   E. Mergenthaler and H. Scherz: Z. Lebensm. Unters. Forsch. 162 1976) 25-29
3   W. Schmolck and E. Mergenthaler: Z. Lebensm. Unters. Forsch. 152 (1973) 263-273
4   P. Friese: Z. Analytischen Chemie 303 (1980) 279-288
5   U. Glück and H.P. Thier: Z. Lebensm. Unters. Forsch. 174/4 (1980) 272-280

3 Element Analysis

Assets and Deficiencies in Elemental Analysis of Food-Stuffs

G. TÖLG

Max-Planck-Institut für Metallforschung, Institut für Werkstoffwissenschaften, Laboratorium für Reinststoffe, Seestrasse 75, D-7000 Stuttgart und Katharinenstrasse 17, D-7070 Schwäbisch Gmünd, F.R.G.

SUMMARY

The focal point of elemental analysis of food stuffs is centered on the determination of the concentration levels and types of bonding of trace elements that are yet physiologically active. In this, the accuracy of the analytical results is still very unsatisfactory for many elements as the levels to be determined lie in the ng/g- and, in part, already also in the pg/g range. The most important reasons for this are systematic errors, which can be recognized with difficulty only. Therefore, the most essential rules are summarized, how systematic errors (e.g. elemental cross interferences, contamination, adsorption effects, volatilizations), which distribute to a differing extent to the whole analytical procedure, can be substantially eliminated. Only if the traces of elements exist in an isolated form they can optimally be determined with high power of detection and reliability. In this connection special attention has to be directed to efficient decomposition and preconcentration methods.
The strategy followed in this is exemplified by the universal, very powerful and reliable determination of selenium in organic and inorganic materials. It propagates an extended systematic analytical basic research with respect to dependable and economic analytical procedures and a high cooperation between trace analysts and users of analytical data.

INTRODUCTION

Influence of Analytical Data on Quality of Life

The statement - without reliable data no quality of life - which is universally valid in an industrialized world, connotes also reliable information of chemical elements that - ubiquitous in nature - form the fundamental constituents of life. Their analytical detection is, in many fields of natural sciences and medicine - and consequently also in the field of human nutrition - an unavoidable necessity for tracing the essential and noxious concentrations. From this point of view, there is also an inseparable partnership between analytical chemistry on one side and many disciplines of vital sciences such as food chemistry, nutritional sciences, toxicology and medicine on the other side. In food stuff analysis elemental concentrations, ranging from major constituents to extreme traces, have to be determined. Today, special interest is centered on the trace elements, e.g. Hg, Pb, Cd, Mo, Se, As, Sb, Sn, Tl, Co, Ni, Cr, Mn, Be, F, I, which in ever very low concentrations act physiologically essential, toxic, oncogenic or mutagenic (1).

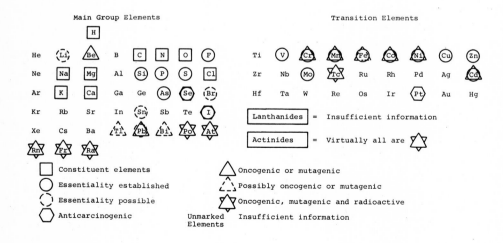

Figure 1 A periodic system of the elements indicating essentiality, oncogenicity, or mutagenicity after (1).

The concentration ranges, in which the elements exert a positive or negative effect to metabolism, lie very close together (2). The most impressive example is certainly the element selenium where the essential concentration and the yet tolerable limiting concentration differ by only about 1 order of magnitude. Generally speaking, if there is only a minimal deviation in either direction from the optimal concentrations fixed by nature, e.g., during production, storage or subsequent treatment, so we have to reckon with serious impacts on health. The concentrations must, therefore, be rigidly controlled. It is, however, much more important to elucidate by interdisciplinary investigation the complex ecological connections of the food chain - rocks, soils, water, atmosphere, plant, animal and man - which is fundamental for all legal regulations and controls. In this, particular problems arise as to the different types of bonding of the elements which are of quite different biological activity. Consider, for instance, Hg which exhibits, as is known, in form of its metal-organic compounds the highest toxicity so that a determination of the total concentration in pertinent samples yields only insufficient information (3).

Systematic Errors as a Cause of Wrong Data

Generally speaking, the analysis under discussion becomes only then problematic, considering the methodical point of view, if the concentrations to be determined fall below the $\mu g/g$ level. Then, we frequently have to be prepared for enormous difficulties which can only be portrayed very generally by means of some case studies.
If I, as an analyst, at this meeting have the honour to accomplish this task for which I want to express my thanks to the organizers at this opportunity, so I feel obliged not to withhold the reasons for an interdisciplinary cooperation from you.
My field of operation - the high-purity materials research - since many years faces especially difficult problems of the determination of very low levels of elemental impurities in ultra-pure materials. The requirements to be met here touch the limitations of what is possible to attain methodically today (4). Thus, we had to learn very early that the common analytical conception in high-purity materials analysis is valid to an ever lesser degree the more so the lower the levels of elements to be determined in a complex matrix.
Mainly, in the range of the extreme trace analysis the reliability of results was no longer in line with the familiar picture of a precision attainable at higher concentration levels. This perception soon showed us the difficulty to find not only speedy but also optimal pathways for the solution of problems with respect to reliability of results which have then to answer for scientific and economical purposes.

The paths to be pursued not only were very laborious but also required an absolutely new strategy which made perceive not only methodical but also more effective approaches to a collaboration of analysts with colleagues of other disciplines (5). In this context only the latter point is to be touched briefly. Meanwhile, extreme trace analysis is no longer only a domain of some few specialists. The determination of very low levels of elements gained attention in many other branches of research and supervision, too. In this, under stress of sudden emerging new analytical requirements more and more doctrinal instructions and proposals for a quick solution of problems arose which not seldom fall in conflict with our own experiences. The most dangerous doctrine was, and still is, to trust unprejudicedly an uncritical advertising of profit making manufacturers of analytical apparatus who promise rapid solution of problems and certainly unpremeditatedly in many cases make light of the importance of trace analysis. It was not until laboratory comparative studies, initiated by national and international institutions, made a wider section of data producers in an alarming way perceive, that analytical results in spite of a good reproducibility in the individual laboratories, may vary to a large extent and hence differ from the true value (6, 7, 8, 9, 10). Only slowly the knowledge gathers way that not only statistical but also systematic errors characterize the reliability of an analytical procedure and this the more so the lower the concentration levels to be determined. The difference of both types of errors is showed in Figure 2.

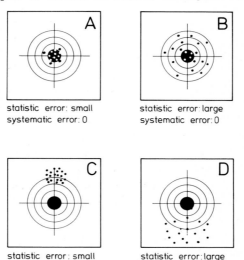

Figure 2   Distinction of statistic errors (precision) and systematic errors (accuracy)

Systematic errors which played by now in /ug/g analysis only a very inferior role, and can easily be recognized and eliminated by outlier tests, overlap in ng/g analysis greatly with statistical errors, thus becoming a serious problem. Their recognition and avoidance become, therefore, the most important task of trace analysis (6). Unfortunately, this knowledge only slowly gains ground in routine laboratories, which today are increasingly faced with ng/g analysis but must frequently follow very insufficient paths. I am very interested in underlining the word "must" because it is today less about rendering some help in finding arguments for an enforcement of better instrumental and personnel conditions in our laboratories. As of yet money lenders and commissioners on many levels do not sufficiently know that wrong results - and most of the results are at fault up to now in the mentioned concentration range - debit more to peoples fortune than sufficient operating material and qualified staff, both of which are preconditions for obtaining dependable data.

Perhaps, I even succeed to persuade some unteachables or novices in this field that trace analysis presupposes in great part attention of some unfamiliar rules, and considerable criticism to reduce the still prevailing strained relations between wishful thinking and reality in the field of accurate data.

Recognition and Causes of Systematic Errors

From what has been said so far follows that most problems in the ng/g analysis base upon systematic errors, which increase strongly and can then only be perceived with increasing difficulty as the absolute amount of an element to be determined decreases.

Of the great range of possibilites for their recognition (5) only an especially important one is to be mentioned: The application of at least two completely independent analytical procedures which display totally different sources of systematic errors. A verification has always to be applied also to such analytical prescription of the ng/g range which are taken from literature or from manuals of manufacturers of apparatus. Unfortunately, systematic errors are so complex and insidious that nobody can escape. Since they are not only method-specific but also may depend on smallest divergencies of parameters in a method, on the composition of the matrix and, largely, to a great extent, on the element to be determined, generalizations in this case are inadmissible. The various causes of errors, the most important of which are compiled in Table 1 distribute, to a different extent, to the various steps of an analytical procedure (Table 2).

Table 1   Causes of systematic errors

1. Improper sampling and storage of the sample
2. Contaminations: Surfaces of tools and vessels,
    of reagents, and auxiliary substances,
    laboratory air
3. Adsorption/Desorption:
    Surfaces of tools and vessel,
    interfaces (filtration, coprecipitation)
4. Volatilization: Elements (Hg, As, Se, u.a.)
    compounds (oxides, halogenides, hydrides)
5. Chemical Reactions:
    Change of valency of ions, precipitation
    exchange, ion exchange, complex formation,
    formation of volatile and non-volatile
    compounds
6. Interference of the signal:
    Matrix effects, target, overlapping of
    signals, signal background
7. Incorrect calibration and evaluation:
    Incorrect standards, instable standard
    solutions, blanks, errors during measure-
    ments, false calibration functions,
    inadmissible extrapolations, etc.

Table 2   Frequency of sources of systematic errors in trace analytical procedures; ± positive/negative error; +++/---: large; ++/--: medium; +/-: small

TYPE OF ERROR STEP OF OPERATION	CONTAMINATION REAGENTS, TOOLS LABORATORY AIR	ADSORPTION TOOLS, INTERFACES	VOLATILIZATION	CROSS INTERFERENCES BY ELEMENTS CHEMICAL REACTIONS, SIGNAL-INTERFERENCES, -COINCIDENCE, -BACKGROUND
SAMPLING	+ + +	- -	-	±
SAMPLE PREPARATION	+ + +	-	-	±
DISSOLUTION, DECOMPOSITION	+ + +	-	- - -	±
SEPARATION, PRECONCENTRATION	+ +	- - -	-	+ + + - - -
DETERMINATION: EXCITATION CALIBRATION EVALUATION	+	-	- -	+ + + - - -

Therefore, this synopsis allows only a very rough attachment which is
a result of long-years experience. We can only predicate unambiguously
that all instrumental direct procedures, e.g., SSMS, INAA, OES and XRFA
in which the samples are energetically excited to obtain directly an
analytical signal, are most susceptible to interferences, in large, by
energetical interaction of the elements to be excited and superimposi-
tion of signals during the detection. These errors can only be avoided, if for the calibration of these direct procedures standard reference materials are on hand which substantially match in composition
the sample under study. In consequence of this, in routine analysis the
most economic way of a multi-element determination with instrumental
automatic analyzers is successful in praxis only via dependable standard reference samples. These are, however, available only to a very
small extent. If we ask after the most promising universal, powerful
and dependable possibilities already on hand today, so we have to
answer clearly: multi-stage combined procedures.

Such procedures are, as a rule, essentially more complex and more cum-
bersome in implementation and are, therefore, reported to be not
economical.

In the first stage the sample has to be transferred into a homogeneous
liquid or gaseous phase. With organic matrices this includes always
a sample decomposition. In further steps the traces of elements to be
determined have to be separated as completely as possible from interfering concomitant elements and/or pre-concentrated to yield better
detection abilities. Finally, the this way isolated elements can be
determined substantially free from interferences with an optimal
detection ability. In such multi-stage procedures the actual step of
determination for which already a great number of methods are on hand
today, can relatively easily be calibrated and is, in general, least of
all subject to systematic errors. Hence, the causes for systematic
errors inherent to multi-stage procedures lie, in the first instance,
in all steps of sample preparation which have to precede the actual
determination. From Table 2 we can see that it is the main concern to
avoid contamination from tool materials, reagents, and the laboratory
air to keep the blank as low and constant as possible. Furthermore,
irreversible adsorption of traces of elements on container walls and
at interfaces as well as losses by volatilization are essential sources
of error which will be dealt with at some length later.

It is evident that starting from this optimal - in terms of accuracy
and precision of results - but also laborious solution, simplifications
have to be aimed at which then, however, in the pertinent case, pre-
suppose extensive investigations as regards additional systematic errors
which have to be expected.

General Rules to avoid Systematic Errors

The ng/g and much more the pg/g analysis are governed by laws, which can hardly be derived from the classical mode of thinking, prevailing in elemental analysis of higher concentration levels. In other words, certain borderline concentrations for each element or each ion species additional as of yet unfamiliar interfering effects emerge, which necessitate a more individual consideration of the respective elements (5).

The certainly greatest difficulty presents the problem of contamination which differs in its extent from one element to another and, therefore, may bias the actual detection ability of trace elemental analytical procedures. We may no longer disregard, that the detection ability for a given element depends to a lesser degree on the sensitivity of the detection system than rather on the ratio of the concentration of an element in the sample to its concentration in the environment of the analytical system.

All elements are omnipresent in our environment although in greatly differing concentration ranges which can ultimately be traced to different abundancies of the elements in the earth's crust. Thus, it is understandable that elements less frequent, e.g., Au, Pt, Co and Re can, in principle, be determined in lower concentrations than relatively frequently encountered elements, e.g., Si, Al, Ca, Mg, Fe, Na, K, and Ti. Elements such as Hg, Cu, Pb, Zn, Cr, which are already strongly preconcentrated in our environment or in our laboratories, respectively, are, therefore, as well subject to essentially higher interfering levels. The detection limits of analytical procedures for frequent elements, in the first instance, are determined by their blanks. It makes, therefore, in the present state of trace analysis, much more sence to look for new approaches to keep blanks low and constant rather than for new sensitive detectors. We have to change our minds also with respect to adsorption effects of elements and ion species, respectively, in solutions. They also are of negligible effect in classic elemental analysis. In extreme trace analysis, however, the relative part of the adsorbed elements and ions, respectively, at solid interfaces can strongly increase with a ratio of concentration of the solution to surface of the vessel becoming less favourable, that an element finds itself no longer in solution but is nearly completely adsorbed on the vessel surface.

Since adsorption effects are affected by many parameters, e.g., vessel material, surface conditions and structure, ion species, additional reactants, time, and temperature the conditions are soon becoming so complex that also here generalizations are no longer allowed.

We can state, at best, losses on the order of magnitude which lie in the range of 10^{-9}-10^{-12} mol/cm^2 (11).

Also in case of systematic errors engendered by volatilization of elements there are intricate and incalculable conditions as is best mirrored by a great range of conflicting literature. I want to demonstrate the conditions with the example of Hg. The losses occurring during storage of a weak nitric acidic solution, containing 2 ng/ml Hg^{2+}, in a PTFE-vessel at room temperature over a period of 10 h lie at about 25 % (12). Essentially higher Hg-losses occur naturally in the decomposition of organic matrices in open systems (11). They can, however, substantially be avoided if wet decomposition is used by boiling under reflux (e.g. Bethge-apparatus). In this, however, there is the risk of gaining high blank values by a large vessel surface if the apparatus is not carefully cleaned. Glass surfaces adsorb, e.g., from laboratory air about 0.5 ng Hg/cm^2 (13), resulting in blanks on the order of 50 ng if a 100 ml vessel is used.

The conditions are less extreme with other easily volatilized elements such as Se, As, Sb, Cr, B and I (11).

With these few examples in mind I would have you consider the following most important rules to reduce systematic errors in extreme trace analysis. They have been established in our laboratory in course of our experiments (4,6,11) and can be summarized in one sentence:

A multi-stage procedure is optimal for the ng/g and pg/g range, if all steps, - sample decomposition, separation of the traces or elements and their determination - proceed at as low a temperature as possible in a closed off system made of an substantially indifferent material and with a ratio of surface to element concentration as favourable as possible using only a minimum of easily purified reagents and/or auxiliary materials.

The following items are to illustrate this fundamental sentence.

1) Materials for vessels and tools should be made of quartz, glassy carbon, PTFE and PP which are very pure and have a relatively good thermal stability (14). Glass as a multi-component material is on account of its hydrolytic instability and its marked adsorption behaviour against ions in most cases not suitable. For surface cleaning and reduction of elemental losses by adsorption, special pretreatments of the surfaces, e.g., steaming with acids and water as the case may be, are inadmissible (14).

2) Reagents and auxiliary materials have to be confined to those which can easily be purified thus far that the remaining impurities (blanks and interferences) lie well below the element concentration to be determined (14). For most elements the blank in the required reagents can today be reduced to the pg/g level.

3) Since contamination by laboratory air - mainly by laboratory dust in which many elements are highly preconcentrated - carries ever greater weight in the determination of elements at low concentration levels, working in clean benches and/or clean rooms the detection limit of frequently encountered elements being improved by 1-3 orders of magnitude with the methods for determination available today (14).

4) The individual steps of operation (transfer, filtration, extraction etc.) ought to be confined to a minimum since each additional surface, reagent or operation introduces new systematic errors into the system. We should, therefore, alwys aim at techniques which are, if possible, easy to survey where all operations proceed in the same system - single-vessel system - best in a mechanized manner.

5) For dosing of solutions common techniques have to be supplanted by microchemical ones (15) which enable dispensing of μl volumes thus taking advantage of the use of higher concentrated solutions. This applies also for the preparation and handling of calibration solutions, which should best be delivered in as concentrated as possible forms with proportioning devices such as piston burettes and pipettes since in many cases element concentrations $\leq 0.1\ \mu g/g$ in solution are not stable.

6) By far the most frequent and serious errors occur already in sampling and sample preparation, which can not be dealt with in detail here. In the case of Hg they have been extensively investigated also in our laboratory (13).

7) In order to avoid the likewise heavy systematic errors in opening-out of samples, decomposition methods specially developed for the extreme trace analysis should unconditionally replace familiar ones such as ashing in a muffle furnace or applying oxydizing agents in solutions and fusion.

As long as there is no satisfying universal decomposition method for organic matrices which meets all requirements of trace analysis we have as the case may be to fall back on various decomposition principles (16-18). Wet opening-out procedures with oxydizing agents, e.g., HNO_3, $HClO_3$, $HClO_4$, H_2O_2 lend themselves best to both large sample throughputs and sample weights, entail, however, problems as regards contamination and losses of elements by volatilization. Strongly fatty matrices can only by completely decomposed with a mixture of HNO_3, $HClO_4$ and H_2SO_4. Aside from an explosion hazard the H_2SO_4 which is relatively difficult to purify limits the application of this method (11).

A mechanized wet decomposition* developed by Knapp is an optimal solution as to sample throughput (Figure 2).

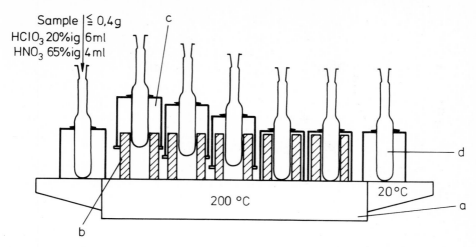

Figure 3 Apparatus for automatic wet-chemical decomposition after Knapp [19]; a) heating block; b) extension tubes to heating block; c) adjustable stop; d) Support for decomposition vessel; e) decomposition vessel.

Decompositions in PTFE pressure bombs** with oxydizing acids, e.g., HNO_3 which need only small acid volumes and exclude losses by easily volatilized elements, e.g., Hg, As, Se, as exemplified by a model developed in our laboratory (21) and in form of many variations helped in reducing numerous systematic errors in the decomposition of biological materials.

We can, however, in the first place, find fault with the sample weight which is, for reasons of safety, limited to 0,5 g which, contrary to numerous propositions for modifications of this technique should be rigidly adhered to. Frequently it is overlooked that under normal conditions no quantitative mineralization is achieved which is absolutely necessary for voltammetric determination methods (22,23). Using, e.g., HNO_3 sufficiently high oxydation potentials are obtained at decomposition temperatures $\geq 220\ °C$ in vessels made of glassy carbon (22).

For a determination of low concentrations of especially frequent elements, combustion in pure oxygen should be given priority over wet decomposition methods on account of a small risk of contamination. There are two alternatives:

---

* manufacturer: Anton Paar, K.G., A-8054 Graz, Postbox 58, Austria
** e.g. Forschungsinstitut Berghof, D-7400 Tübingen, Postbox 1523, Federal Republic of Germany

a) The combustion in an oxygen plasma excited by a HF (24,25) or UHF (24) field: Especially the second mode offers great advantages in extreme trace analysis. An universally applicable decomposition apparatus is commercially available ***.
b) The combustion in molecular oxygen: It can happen in an autoclave under pressure **** (Figure 4) (27-29) or according to the principle of a quasi-static combustion (30).

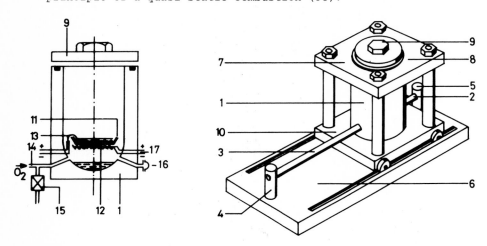

Figure 4 Principle of the Bioklav ® (27,28);
1 pressure vessel (stainless steel); 2 and 3 support bar; 4 and 5 mounting blocks; 6 bottom plate; 7 and 8 upper pressure bearing plate; 9 locker screw nut; 10 bottom pressure bearing plate; 11 sample tray; 12 Pt-heating wire; 13 Pt-ignition wire; 14 oxygen inlet; 15 pressure check valve for oxygen; 16 bursting disk; 17 Pt-heating wire to dry the sample.

In this, in a quartz apparatus with a cooled finger (liquid nitrogen) sample weights up to 1 g can be burnt completely in a combustion chamber of only about 75 cm^3 without losing even volatile elements, e.g., Hg, Se, As, I ***** (Figure 5). While the first method was designed for routine work in the μg/g-range only admitting sample weights in the g-range, the second one is laid out in the first instance for the extreme trace analysis.

***   e.g. Erbe Elektromedizin, D 7400 Tübingen, Postfach 1420, F.R.G.
**** Bioklav ®, Siemens AG, E 689, Postfach 211262, D-7500 Karlsruhe 21
***** Trace-O-Mat®, Anton Paar, K.G., A-8054 Graz, Österreich, Postbox 58

Both methods perfectly supplement each other and offer enormous improvements for destruction of biological materials.

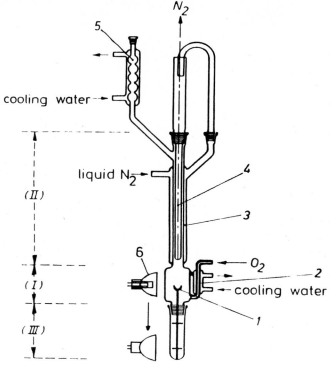

Figure 5 Combustion unit (quartz) of the Trace-O-Mat® (30)
(I) combustion chamber; (II) liquid nitrogen cooling system; (III) reagent tube of quartz; 1 sample holder; 2 capillary with cooler for oxygen supply; 3 cooling chamber; 4 cooling tube; 5 reflux cooler; 6 IR-lamp

8) The field of separation and preconcentration methods is too complex to show direct pathways for a solution of problems. Only some general points of view can briefly be broached. Decisive advantage is yielded for those elements, e.g., Hg, As, Sb, Se, Te that can be separated by volatilization from the mineral phase either directly or as easily formed volatile compounds, e.g., hydrides, halogenides or oxides. In this, the separation step can be directly tightly tied with ease to the decomposition, thus resulting in an efficient state of the art multi-stage procedure for the determination of Hg (13) and Se (31,32).

Similarly we can meet the new requirements, if conventional separation principles be so far modified, that they can jointlessly be included in a multi-stage system. Extreme examples of the high-purity materials research are precipitation exchange on thin layers (33) or electrolytical separation methods (12,34) which can in many modifications be transferred to trace analytical problems of food chemistry. Extraction methods, too, can essentially be optimized if their techniques are approximated to the "single-vessel principle" and the extraction doesn't happen in a separating funnel but directly in the decomposition flask, e.g., in a PTFE vessel which is used to separate the phases (35). This way we succeeded, for instance, in determining the distribution of traces of Be at pg/g levels in the individual protein fractions of blood serum (36). Using chelate GC and chelate HPLC (37-39) numerous - even simultaneous - separations of elements in the ng/g range are possible in which, on top of that, the separated species can especially powerful be determined. In this, the close coupling of the separation and determination step makes expect considerable strides with relatively simple means only.

9) The question, which method for determination is best, can generally be answered with the view to the aforementioned entities. If the elements to be determined exist substantially isolated after the preceding steps of a multi-stage procedure, many classical and instrumental determination methods succeed, if they have the necessary power of detection. Such methods are available to a large extent and can be selected following various points of view. Systematic errors play then, in general, only a very subordinate role. It is evident, that in the individual determination principles many general differences have to be taken note of. Thus, e.g., flame AAS, as a rule, is less prone to interference than the essentially more powerful ETA-AAS, which should only then be used if extremely good detection abilities are required. In simultaneous multi-element determination in optical emission spectrometry (OES), the ICP excitation is less interfered than other kinds of excitation, e.g.,CMP, MIP or hollow cathodes (39) which are, however, less lengthy. Similar pros and cons go for XRFA with its various kinds of excitation. But all differences carry, however, only then greater weight if, on account of insufficient preseparation of the elements ever more cross interferences occur in the excitation process. Then, however, minds begin to differ in that ones who prefer to incur a risk with respect to accuracy and in those who try to avoid risks. These different points of view lead latently to the certainly most serious problem of modern trace analysis. Either side sets forth economical arguments. One of them argues the lengthy methods being intolerable for reasons of being too costly, the other, uncertain

data being irresponsible and much more expensive at the end. Therefore, a compromise must be reached, to make both risks more calculable.

The scientific path I would like to demonstrate at least with the example of the determination of Se. As already mentioned hereinbefore, Se holds at present a special position based upon nutritional scientific points of view. On one side selenosis is certainly encountered if Se is present in the diet in concentrations above about 1 µg/g, on the other side, its essentiality is definitely proved, which eventually suggests even an anticancerogenic action (1). Many questions have still to be answered. One thing is, however, certain: Their elucidation presupposes a reliable trace analysis whose state of the art is impressively mirrored by recent interlaboratory comparative analyses. In Figure 6 we have an example of a still relatively high Se concentration of about 0.3 µg/g in animal muscle (16), a level suitable for activation analytical determinations being largely applied.

As a result, an analytical method has to be found that is favourable also for routine analysis, with which Se - and this applies equally to other biologically active trace elements - can, if possible, reliably be determined within a broad spectrum of matrices at ng/g levels and below.

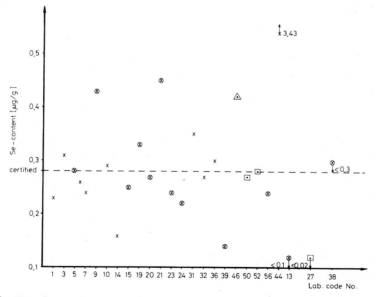

Figure 6   Results of an intercomparison study of the selenium-content in a sample of animal muscle by IAEA (10) methods: radiochemical or unspecified activation analysis (x); non-destructive instrumental activation analysis (⊗); emission spectroscopy (△); other methods (▫).

Many pathways promise to reach this goal, and are, therefore, followed since Se can be determined with high sensitivity by nearly all familiar principles of analysis. At present, INAA is the most used method for determination with high power of detection and sufficient reliability for biological materials (40), however, with the drawback of being tied to a reactor. Hydride-AAS (HG-AAS) (41), voltammetry (42), gas-chromatography (43), XRFA (44), and many spectrophotometric and fluorimetric methods for determination follow, which are, however, less selective. Literature citations on detection limits of analytical procedures for the determination of Se refer, in general, to pure solutions or are extrapolated from higher concentration levels. Similar conditions exist in the case of interferences which - if at all - have being investigated systematically for a few matrices only. Economical points of view are mostly debated in the circle of specialists, who survey only one method.

In consequence of this, the greatest difficulty lies in a sufficiently critical comparison of methods which has, therefore, to surpass the study of literature.

In a strategical future-oriented analysis an optimal solution, exempt from methodical and economical prestige-thinking should be aimed rather than additional, unsatisfactory partial solutions, that inundate literature already. This means, starting out from an ideal pattern that extended knowledge and experience on the limitations of all possibilities for determination must be available and digested, so that definitely the most powerful, reliable and economical solution for an interdisciplinary spectrum of matrices can be found.

We have been in the favourable situation to investigate and optimize at length in course of many years some of the most efficient determination principles for Se on many metallic, geochemical and organic materials (33, 44-48). In this, we often went astray, however, positive and negative experience increasingly gathered way.

Unfortunately, time is too short to point at recent investigations in the current HG-AAS dealing with systematic errors which soon made us recognize the limitations of this method (41,49).

Many concomitant elements, which have to be taken into account in each complex natural sample - that means also in food stuffs - may already strongly interfere with the Se-hydride formation, partly ever at very low concentration levels, that are on the order of the Se level to be determined (Figure 7). By means of some artifices, for a given matrix such interferences could be eliminated which, however, turned up again on changing the matrix for unknown reasons.

It could be clarified clearly that Hg-AAS is very powerful and also reliable for pure Se solutions.

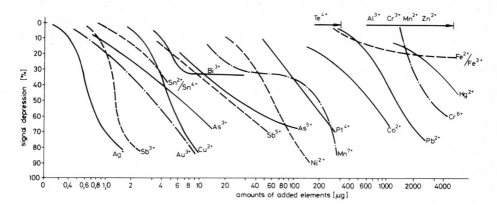

Figure 7   Cross interferences in the Se-determination by hydride AAS by concomitant elements (determination of 50 ng Se in 20 ml 0,3-N HCl using a reduction solution of 10 % $NaBH_4$ in 2,5 % NaOH) (44).

Only differential pulse cathodic stripping voltammetry (DPCSV) (45) and gas-chromatography with an ECD after separation of Se by way of piazselenol (46) still excel the detection ability, they are, however, less suitable in routine analysis for reasons of being too lengthy.

Similarly we investigated the most important aforementioned decomposition methods for inorganic and organic materials with respect to systematic errors (16,20,21,26,30), along with numerous possibilities, to separate Se, if possible selectively from elements interfering with HG-AAS (31,33,47). After long years investigation - in part by means of the radioactive isotope ^{75}Se - we found now a combination of methods approximating our ideal conception (32). It is made up of the combustion of the sample in oxygen in a quasi-static system (Trace-O-Mat®) (Fig. 5) and the HG-AAS. In the combustion which has, in part, to be carried out with additives, the Se is completely volatilized as SeO_2 at corresponding high temperature from each type of bonding, thus ensuring a separation from all interfering elements which remain in the combustion residue. The thus preisolated Se can now be transformed into the volatile Se-Hydride and reliably and economically be determined practically free from systematic errors by AAS in the concentration range ≥ 2 ng/g.

Table 3 shows some results (32) which verify the excellent agreement with both data of certified standard reference samples and those obtained using several independent procedures for a determination of Se in various matrices.

SAMPLE	SAMPLE WEIGHT [mg]	Se - CONTENT [µg/g]			
		HG-AAS		DPCSV	REFERENCE DATA
Orchard leaves (NBS 1571)	70 - 150	$\bar{x}$ $S_{rel.}$ n	0.076 3.0 % 12		0.08 ± 0.01 [a] 0.078 ± 0.008 [b]
Wheat flour (NBS 1567)	10 - 120	$\bar{x}$ $S_{rel.}$ n	1.03 4.2 % 9	$\bar{x}$ 1.00 n 2	1.1 ± 0.2 [a] 1.01 ± 0.06 [b] 1.10 ± 0.02 [31]
Rice flour (NBS 1568)	30 - 160	$\bar{x}$ $S_{rel.}$ n	0.38 3.1 % 11		0.4 ± 0.1 [a] 0.34 ± 0.034 [b] 0.40 ± 0.02 [31]
Soya groats	100 - 180	$\bar{x}$ $S_{rel.}$ n	0.50 3.5 % 4		0.51 ± 0.09 [b] 0.51 [d]
Bovine liver (NBS 1547)	20 - 110	$\bar{x}$ $S_{rel.}$ n	1.09 5 % 7	$\bar{x}$ 1.16 n 2	1.1 ± 0.1 [a] 1.12 ± 0.07 [b] 1.0 [c] 1.10 ± 0.02 [31]
Fish meal	20 - 160	$\bar{x}$ $S_{rel.}$ n	2.61 2.7 % 9	$\bar{x}$ 2.64 n 3	2.73 ± 0.1 [b]
Blood meal	50 - 150	$\bar{x}$ $S_{rel.}$ n	0.63 5.1 % 3		0.62 ± 0.03 [b] 0.71 [c]
Milk powder	100 - 150	$\bar{x}$ $S_{rel.}$ n	0.087 4.6 % 4		0.082 ± 0.016 [b] 0.09 [c] 0.096 [d]
Human hair	30 - 90	$\bar{x}$ $S_{rel.}$ n	1.73 1.5 % 4	$\bar{x}$ 1.77 n 2	1.73 ± 0.05 [b]

[a] NBS certified value; [b] Combustion with oxygen in the dynamic system and HG-AAS [31]; [c] INAA, Max-Planck-Institut of Metal Research, Schwäbisch Gmünd; [d] INAA, Hahn-Meitner-Institut, Berlin.

Table 3 Selenium contents in different biological matrices (hydride-AAS after decomposing the sample in oxygen in the Trace-O-Mat®).

Perhaps I succeeded with this example to somewhat acquaint you with our concept of obtaining reliable data in the region of trace elements. The strategy basing upon this approach can - we are aware of that - be consequently pursued in a few laboratories only. At a time with ever decreasing research funds but considerably rising demands on analytical chemistry only a few alternatives remain to offset growing differences between merits and deficiencies. The most sufficient alternatives are, to my mind, an improved analytical basic research and a more intensive partnership of analysts and all scientists of other disciplines, who have to rely on dependable analytical data. Perhaps this pathway requires somewhat higher investments and perseverence. I am, however, certainly convinced, that these investments are worthwhile at long term not only for research but also for our own benefit.

References

(1) G.N. Schrauzer, Trace Elements in Carcinogenesis; Advances in Nutritional Research 2, Ed.: H.H. Draper, Plenum Publ. Corp., (1979)

(2) G.H. Morrison, CRC Critical Reviews in Analytical Chemistry 8, (1979) 287

(3) G. Kaiser and G. Tölg, "Mercury" in: O. Hutzinger (Ed.): The Handbook of Environmental Chemstry, Vol. III, Part A, Springer-Verlag Berlin, Heidelberg, New York (1980).

(4) G. Tölg, Z. Anal. Chem. 294 (1979) 1

(5) G. Tölg, Strategy and Tactics in Micro Trace Analysis of Elements; 8th International Microchemical Symposium, Proceedings, Springer-Verlag Wien, in press

(6) G. Tölg, Naturwissenschaften 63 (1976) 99

(7) T.J. Murphy, Accuracy in Trace Analysis : Sampling, Sample Handling Analysis, Ed.: Ph. D. LaFleur: Proceedings of the 7th Materials Research Symposium; National Bureau of Standards Special Publication 422, Vol. I., U.S. Govrnment Printing Office, Waschington (1976)

(8) R. Dybczynski, A. Tugsavul and O. Suschny, Analyst 103 (1978) 733

(9) F. Ackermann, H. Bergmann und U. Scheichert, Z. Anal. Chem. 296 (1979) 270

(10) R.M. Parr, Report No. 2, Intercomparison of Minor and Trace Elements in IAEA Animal Muscle (H-4), International Atomic Energy Agency, Vienna, Austria, IAEA/RL/69, October 1980

(11) G. Tölg, Talanta 19 (1972) 1589

(12) G. Kaiser, D. Götz, P. Schoch and G. Tölg, Talanta 22 (1975) 889

(13) G. Kaiser, D. Götz, G. Tölg, G. Knapp, B. Maichin and H. Spitzy, Z. Anal. Chem. 291 (1978) 278

(14) P. Tschöpel, L. Kotz, W. Schulz, M. Veber and G. Tölg, Z. Anal. Chem. 302, (1980) 1

(15) G. Tölg, in Wilson and Wilson's Comprehensive Analytical Chemistry Ed.: G. Svehly, Vol. III, Elsevier Publ. Comp., Amsterdam, Oxford, New York, 1975

(16) R. Bock, A Handbook of Decomposition Methods in Analytical Chemistry; International Textbook Company, London 1979

(17) P. Tschöpel and G. Tölg, Vom Wasser 5, (1978) 247

(18) P. Tschöpel, Aufschlußmethoden, Ullmanny Encyclopädie der Technischen Chemie, Band 5, Verlag Chemie, Weinheim 1980

(19) G. Knapp, Z. Anal. Chem. 274 (1975) 271

(20) G. Knapp, B. Sadjadi and H. Spitzy, Z. Anal. Chem. 274 (1975) 275

(21) L. Kotz, G. Kaiser, P. Tschöpel and G. Tölg, Z. Anal. Chem. 260 (1972) 207

(22) L. Kotz, G. Henze, G. Kaiser, S. Pahlke, M. Veber and G. Tölg, Talanta 26 (1969) 681

(23) M. Stoeppler, K.P. Müller and F. Backhaus, Z. Anal. Chem. 297, (1979) 107

(24) C.E. Gleit and W.D. Holland, Anal. Chem. 34 (1962) 1454

(25) D. Behne and P.A. Matamba, Z. Anal. Chem. 274 (1975) 195.

(26) G. Kaiser, P. Tschöpel and G. Tölg, Z. Anal. Chem. 253 (1971) 177

(27) E. Scheubeck and A. Nielsen, Z. Anal. Chem. 294 (1979) 398

(28) E. Scheubeck, J. Gehring and M. Pickel, Z. Anal. Chem. 297 (1979) 113

(29) E. Scheubeck, Ch. Jörrens and H. Hoffmann, Z. Anal. Chem. 303 (1980) 257

(30) G. Knapp, S. Raptis , G. Kaiser , G. Tölg, P. Schramel and B. Schreiber, Z. Anal. Chem. , in press

(31) A. Meyer, Ch. Hofer, G. Knapp and G. Tölg, Z. Anal. Chem. 305 (1981) 1

(32) Heng-Bin Han, G. Kaiser and G. Tölg, Anal. Chim. Acta, in press

(33) A. Disam, P. Tschöpel and G. Tölg. Z. Anal. Chem. 295 (1979) 97

(34) G. Volland, P. Tschöpel and G. Tölg, Anal. Chim. Acta 90 (1977) 15

(35) G. Tölg, Pure and Applied Chemistry 50 (1978) 1075

(36) Th. Stiefel, K. Schulze, G. Tölg and H. Zorn, Z. Anal. Chem. 300 (198)) 189.

(37) G. Schwedt, Gas-chromatographische Trenn- und Bestimmungsmethoden in der anorganischen Spurenanalyse, Analytiker-Taschenbuch, Band 2, Springer-Verlag, Berlin, Heidelberg, New York, in press

(38) G. Schwedt, Chromatographia 12 (1979) 289

(39) N. Häring and K. Ballschmiter, Talanta 27 (1980) 873

(40) P. Tschöpel, in Wilson and Wilson's Comprehensive Analytical Chemistry, Ed. G. Svehla, Vol. IX, Elsevier Publ. Comp., Amsterdam, Oxford, New York, 1979

(41) H.E. Ganther, D.G. Hafeman, R.A. Lawrence, R.E. Serfass and W.G. Hockstin (Eds.), Selenium and glutathione peroxidase in health and disease - a review, Vol. II, Academic Press, New York, 1976

(42) G. Nordberg (Ed.). Factors influencing metabolism and toxicity of metals, Environmental Health Perspect. 25 (1978) 1

(43) J. Versieck and R. Cornelis, Anal. Chim. Acta 116 (1980) 217

(44) A. Meyer, Ch. Hofer, G. Tölg, S. Raptis and G. Knapp, Z. Anal. Chem. 296 (1979) 337

(45) G. Henze, P. Monks, G. Tölg, F. Umland and E. Weßling, Z. Anal. Chem. 295 (1979) 1

(46) A. Meyer, E. Grallath and G. Tölg, Z. Anal. Chem. 281 (1976) 201

(47) S. Raptis, W. Wegscheider, G. Knapp and G. Tölg, Anal. Chem. 52 (1980) 1292

(48) R. Engler and G. Tölg, Z. Anal. Chem. 235 (1968) 151

(49) S. Raptis, G. Knapp, A. Meyer and G. Tölg, Z. Anal. Chem. 300 (1980) 18

Acknowledgement

I would like to acknowledge the assistance of G. Kaiser in helping to translate the manuscript.

The Analysis of Inorganic Contaminants in Food

L.E. COLES
Mid Glamorgan County Public Health Laboratory, The Parade, Cardiff,
South Glamorgan, United Kingdom.

INTRODUCTION

The Commission on Food Chemistry of the Applied Chemistry Division of the International Union of Pure and Applied Chemistry has a Working Party on Inorganic Contaminants, of which I am the Chairman. It is my intention to speak generally of the problems involved in inorganic contamination rather than reiterate analytical details because I am certain that those of you who are expert analysts in the field of trace metal analysis do not wish to be reminded of the details of such analysis and those of you who are not analysts only require critical appraisal of the situation. In other words I have no intention of boring you with facts and figures but simply to outline the problems that exist.

Inorganic contamination is in the main concerned with toxic metals, although the analyst concerned with environmental contamination would also have to consider the presence of asbestos and of fluorides. Any practising analyst involved in the enforcement of legislation would also have to consider the presence of inorganic foreign bodies and these are often the basis of prosecutions brought under the food legislation of the United Kingdom (1). In addition my main concern is that the analytical methods should be precise and reproducible to the extent that if there is any subsequent dispute about the analysis then another independent expert analyst should obtain a similar result. Therefore, whatever I refer to in my paper, it must be in the hope that the methods have been, or should have been, subjected to the ultimate scrutiny of an analytical procedure, that is a properly conducted collaborative test (2).

ESSENTIAL AND NON-ESSENTIAL METALS

Trace elements in human nutrition have already been extensively reviewed by the World Health Organisation (3) and it is clear that classification of trace

metals into essential, non-essential and toxic groups can be inaccurate and
misleading. Indeed, all the essential elements become toxic at sufficiently
high intakes and the margin between levels that are beneficial and those that
are harmful may be small. The obvious examples of this is the role of fluorine
and selenium in the diet but more recently it has been inferred that there may
be essential physiological roles for the very toxic elements cadmium, arsenic
and lead. It has been further postulated that there is no functional relation
whatsoever between toxicity, on the one hand, and essential biological function,
on the other hand (4).

THE PROBLEMS OF SURVEILLANCE

In any programme for the monitoring of foodstuffs for inorganic contaminants it is essential to recognise that a single determination of the total amount of toxic metal is not sufficient in any meaningful examination. However, in terms of toxicity alone there may be some justification in making sure that the amounts present are extremely low or at concentrations near the limits of determination. Ideally, a simultaneous or sequential determination of many elements at all concentration levels in a single analytical technique, with accuracy and precision is required. Inductively Coupled Plasma with Atomic Emission Spectroscopy (ICP-AES) is obviously the technique of choice, which although expensive and only just being recognised will provide a vast amount of trace element information in the coming years (5). A review based on the factors influencing the toxicity of heavy metals (6) has been published illustrating the complexity of the situation. If there is no significant contamination then the food need not be examined any further. However, if it is present in an amount normally significant there are many questions that have to be answered. For example,

1. Is the contaminant in the elemental form? If so, is it as particulate matter or in solution? Although the lead in petroleum originates as lead tetraethyl the eventual form as deposition on crops and urban dust is as inorganic particulate matter (7). In canned foods it may be in the form of globules of solder which are soluble in fruit acids.

2. Is the contaminant combined with organic matter? What is likely to happen to the organo-metallic complexes when they enter the acid medium of the stomach? Speculation can be made about other chemical reactions further along the alimentary canal. It has been shown that arsenic is very firmly bound in an organic form (8) in many foods. Even for total arsenic determination very vigorous analytical procedures have to be adopted and separation of the organic form presents a formidable problem. It may be better perhaps to tackle the problem by determination of the inorganic species by sequestration using ligands. A number of studies (9) have been made into the relative merits of various procedures to remove metal species in simple solutions but only with

limited success.

3. What is the valency state of the inorganic contaminant? As is well known the toxicity varies enormously between one valency state and another. Procedures for the selective determination of various forms of arsenic have been reported (10).

4. Is the contaminant in a bioavailable form? Having considered the limited use of total elemental analysis and assuming a general picture of speciation knowledge, which in itself is a formidable task, the problem of bioavailability must then be studied in the context of the whole diet. Diets vary throughout the world, between countries and within countries and the permutation of all these variables produces an impossible situation in drawing up an analytical monitoring programme of food surveillance. In fact, any attempt to study possible factors in the diet which are responsible for disease must have its limitations. It has been shown that the bioavailability of dietary iron (11), for example, may be enhanced if there is more meat in the diet and less phytic acid because of a change from a primarily cereal diet to one containing more dairy and animal products. Further, the absorption of many trace elements is greatly influenced by the amount of phytic acid in the diet (12). It has also been shown that cadmium toxicity is decreased with dietary ascorbic acid supplements (13), and that the retention of lead varies inversely with dietary calcium content (14). In the case of chromium a determination of the total element bears no relationship to the glucose tolerance factor organic complex which is the important biologically active substance (15,16,17).

5. Has the contaminant a nutritional or toxic relationship with other inorganic species? For a number of years interdependence of trace elements in human nutrition has been recognised. The toxicity of methyl mercury is reduced in Japanese quail by a diet containing tuna which has a relatively high selenium content (18). Therefore, in establishing a maximum permissible level of mercury in fish the mercury-selenium ratio and interaction should be taken into account. Consequently, it is preferable to determine not only the mercury concentration in fish but also the selenium concentration (19). The additive effect of heavy metals has not received much attention, although studies have been made concerning the common action on many organs in the body, for example, lead and mercury (20), lead and cadmium (21), and mercury and cadmium (22).

A great deal of work remains to be carried out but there are already suggestions that protective and synergistic effects of trace metals are more common that hitherto realised. A great deal of evidence has accumulated (6) to show that the toxicity or otherwise of trace metals is modified considerably by the presence of other elements. Indeed it has been proposed that "safe" exposure standards should be based upon ratios and form of the element rather than relying solely upon the quantity of a particular element.

6. Is the foodstuff containing the contaminant a significant proportion of the diet? In fact, in any surveillance programme cognisance has to be taken of the fact that many diets of children contain large proportions of carbohydrate products, like potato crisps, fried potatoes, ice-cream, soft drinks, confectionery. Are these products contaminated with anything or lacking in anything?

FOOD SURVEILLANCE PROGRAMME

The need for a meaningful programme of food surveillance was realised after the discovery that methyl mercury compounds were more toxic than elemental mercury and present in the greatest proportion in certain fish, to mention one only of a number of similar examples. It is, therefore, essential that future food surveillance programmes have to be systematic and I am tempted to make a comparison with the approach of the forensic chemist in search for poisons and in his use of a systematic toxicological examination.

SYSTEMATIC TOXICOLOGICAL EXAMINATION

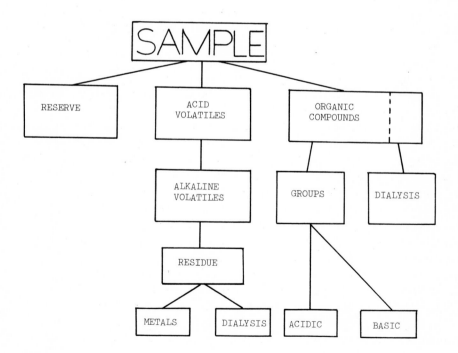

Such a system, however, is relatively easy because in the end it is a "one element at a time" classical technique having little regard to speciation properties, except possibly during the dialysis part of the procedure. At an International Conference in London in September, 1979 (23), on the Management and Control of Heavy Metals in the Environment, a paper was presented on Metal Forms and Speciation. The author rightly pointed out that the whole subject was an important challenge to analytical chemists. Another review on pollution pathway studies (24) concludes, amongst other things, that not enough use is made of neutron and/or proton activation analysis for "in vivo" tracer studies.

Any systematic scheme must have selective procedures, involving clear-cut separation of the various forms of the elements.

SYSTEMATIC SEPARATION SCHEME

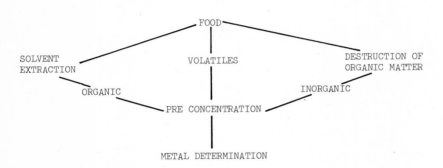

However, the above scheme would only separate inorganic forms from organo-metallic compounds. This would, of course, be a step in the right direction and some examples illustrate that, with sufficient investigation, an outline scheme can be prepared.

Methyl mercury may be extracted directly from fish by extraction with toluene (25). It has been shown that arsenic in marine organisms is present both as lipid soluble and water-soluble organo-arsenic compounds (26). The organo-arsenic compounds in aqueous extracts are very stable, even in strong acid conditions, whilst organo-arsenious acids and arsenious acid derivatives show no anion-like characteristics (27). Arsenic and selenium which are bound to peptides may be extractable by aqueous, acidic or alcoholic solutions. It has also been reported that the biological activity of chromium as the glucose tolerance factor is not related to the total chromium content of food but to that part of chromium which is extractable by ethanol (28). One way of distinguishing between small amounts of arsenate (As V) and arsenite (As III) in solution is based on the fact that arsenate forms an iso-amyl alcohol extractable molybdate complex while arsenite does not (29). Implications of further study (30) indicate that pentavalent arsenic can be separated from trivalent arsenic by using procedures analogous to well established phosphorus methodology (31).

Metals like lead are particularly prone to lability in biological systems and any analytical method should not destroy the quasi-equilibrium state in which the metal complexes are existing. It is possible that electro-analytical techniques have some unique advantages in this respect. Amongst these advantages are characterisation of species possibly with "in situ" measurements as well as high sensitivity. However, some workers have speculated that even with anodic stripping voltammetry (ASV) labile metal complexes may dissociate to some extent at the electrode surface.

A general procedure for the simultaneous determination of twelve elements by cathode ray polarography has recently been published (32). The method involves digestion with sulphuric acid and nitric acid, and bivalent mercury, copper, zinc, cadmium and lead are removed in the form of dithizone complexes by adjustment of the pH. The aqueous phase is then acidified, after which trivalent iron and antimony and quadrivalent tin are extracted as salts of cupferron. The pH of the resulting aqueous phase is adjusted with ammonia, after which hexavalent chromium is extracted with diethyl-dithiocarbamate in chloroform. In the remaining aqueous phase quadrivalent tellurium and selenium, pentavalent antimony and trivalent arsenic are determined.

However, it is still limited to analysis after wet oxidation when the speciation of the element will almost certainly have changed. An excellent paper has recently been published (36) on the sequential spectro-photometric determination of inorganic Arsenic (III) and Arsenic (V) species, but this was starting from a solution containing toxic elements which are not bound to an organic matrix.

SPECIATION ANALYSIS: THE DIFFICULTIES

In spite of the enormous amount of literature dealing with the biochemistry of toxic metals in living systems relatively little appears to have been done in relation to the composition or contamination of food. Investigations into speciation that have been attempted are in natural waters (33,34) and even in one of the most extensive reviews of metals in food that has appeared in recent years (35) very little reference is made to speciation. Nutritional experts are beginning to show interest but have been left behind in applying the analytical techniques and methods which have advanced so much in recent years. One of the reasons for this, in my opinion, is that many academic institutions plan their research programmes around their own facilities. For example, if a post graduate student happens to specialise in one particular instrument then the research will be orientated towards the capabilities of that instrument. The research should first recognise the need and then be designed to provide information to evaluate the situation.

OTHER INORGANIC CONTAMINANTS IN FOOD

FLUORIDE

Fluoride is present in most foods because it is ubiquitous in nature (37). Surveys have been made and it is significant that it has recently been found that in highly industrialised countries the fluorine intake from foods has increased owing to fluoride pollutants. The marked differences in the fluoride concentrations of various commonly used foods and in the food habits due to climatic, national, individual and environmental factors cause distinct differences in the average daily intake of fluorine in individual members of the population (38).

There are a number of procedures for the determination of fluorine and as usual the main problem remains to distinguish between that which is naturally present and that which is a contaminant. The preparation of the sample is all important.

As with other inorganic contaminants before determination usually the fluoride has to be separated from the organic material by ashing and extraction or by distillation. The method of choice (39) depends upon the facilities available at the time, for example, oxygen flask combustion (40) is commonly used although distillation (41), diffusion (37,42), ion-selective electrodes (39), colorimetric methods (43), radioactivation and isotope dilution analysis (44) have all been used and comparative work has been carried out. A comparison of results of total fluorine analysis by the closed bomb technique and open ashing (using calcium phosphate as a fixative) suggests that open ashing values are lower (45). In recent years gas/liquid chromatography has been successfully used after extraction of the fluorosilanes formed by reaction with alkyl-chlorosilanes.

Work of the Inorganic Contaminants Working Party of the Commission on Food Chemistry (46) has confirmed that the gas/liquid chromatography method is satisfactory, especially at low levels.

ASBESTOS IN RELATION TO FOOD AND DRINK

A U.K. Government Committee undertook a wide ranging review (47) of the health risks to workers and members of the public which could arise from exposure to asbestos.

It was concluded that there is a need to investigate sources and levels of asbestos in food and drink, for example, filters containing asbestos. Having considered the technological need for asbestos filters and published results the Committee came to the conclusion that such filters were not a significant source of contamination - although alternative filtration material should be sought.

It was also considered whether there should be a statutory control for asbestos in materials and articles in contact with foods. In the UK and EEC countries it is possible for enforcement authorities to take action against any person responsible for materials and articles from which asbestos fibres pass into food in a quantity which is held to endanger health or to make the food unacceptable.

It was further considered whether there should be statutory maximum permitted levels for asbestos in food. If there were a limit it would be inconsistent with the recommendation that levels should be kept as low as possible.

As asbestos is present in the environment it is impracticable to require that no residues whatsoever should find their way into food. There are many reasons for not introducing statutory control, for example,

1. There is no available data on which a technologically and toxicologically acceptable limit could be based.
2. The setting of any limit is unlikely to lead to any reduction in actual levels.
3. Asbestos is difficult to detect in food.
4. There are already adequate enforcement powers.

LEGISLATION

As far as I can find out there has been very little advance on the directives of the EEC of many years ago, which were concerned about the general purity criteria of colours, preservatives and antioxidants. In fact, the general standard at the time was that "They must not contain any measurable traces of elements dangerous from a toxicological point of view" (48). These criteria need revising having regard to the advances in analytical techniques. In fact, future limits of trace metals may have to take into consideration the total toxicological burden rather than the individual toxicological effect of trace metals. In this respect the EEC Directives (48) are well ahead of their time - but how they are going to be enforced eventually is another matter.

CONCLUSION

Over the years it has been gradually recognised that there is need to control the burden of heavy metal contamination and efforts in the developed countries have been made. For example, the use of the two-piece can for canned products will inevitably result in less contamination from lead because the solder used in the three piece can will no longer be present. Gross inorganic contamination of foods must surely decrease with increased awareness of the dangers and the sources of contamination.

All the WHO recommendations (49) are directed towards reducing contamination particularly from the environment, such as lead from petrol.

It is desirable to establish an international monitoring system to ensure that inorganic contamination is kept to a minimum. But the problems are so complicated that a great deal of international cooperation is needed not only to avoid duplication of effort but to provide the necessary resources and expertise required.

The only sensible attitude to take at the present time is to reiterate some of the conclusions of the FAO/WHO Reports (49) that there must be a systematic collection of national data of contaminants in foods (based on agreed methods of sampling and analysis): that food consumption patterns should be studied and total diet investigations should be continued. However, in addition there should be some effort to direct such monitoring programmes towards meaingful speciation studies.

In fact, I endorse in full the recommendations of Huisingh (6) that a food monitoring network must also include research into the human reaction to heavy metals and how it is related to the population of different areas in the world. In addition, if in the future the increasing population of the world may be partially fed, at lease in some areas, with recycled and processed animal wastes it will become vitally necessary to increase our knowledge of inorganic contamination. If only a few things are remembered from this conference I hope that one of them will be the concept of a meaningful international monitoring programme. The most important word being meaningful, because whether or not action is taken on any of the recommendations it should be emphasised that protracted investigations must lead to conclusions that are soundly based, otherwise we are all wasting our time.

REFERENCES

1. Food and Drugs Act, 1955 (Section 2). Eliz. 2. C. 16.
2. Steiner, E.H. The Planning and Analysis of Results of Collaborative Tests. Published by British Food Manufacturing Industries Research Association, Leatherhead, Surrey. England (1974).
3. Trace Elements in Human Nutrition. World Health Organisation Technical Report Series No. 532 (Geneva), (1973).
4. Schwarz, K. Clinical Chemistry and Chemical Toxicology (ed. S.S. Brown) Elsevier, Amsterdam, The Netherlands (1977).
5. Fassel, V.A. Anal. Chem. $\underline{51}$, No. 13, p. 1290A (Nov. 1979).
6. Huisingh, D. et. al. Ecology of Food and Nutrition, $\underline{3}$, p. 263 (1974).
7. Chemistry International I.U.P.A.C. No. 2, p. 9 (1980).
8. Braham, R.S. and Foreback, C.C. Science, $\underline{182}$, p. 1247 (1973).

9. Pevin, D.D. C.R.C. Critical Reviews in Analytical Chemistry, 5, 85 (1975).
10. Yasui, A. et. al. Agric. & Biol. Chem. 42, (11), p. 2139 (1978).
11. Bremner, I. et. al. Effects of Diet on the Toxicology of Heavy Metals p. 139 - Management and Control of Heavy Metals in the Environment. London, September 1979. (E.P. Consultants Ltd., 26 Albany Street, Edinburgh, EH1 3QH UK).
12. Davies, N.T. et. al. Br. J. Nutr. 34, 243 (1975).
13. Fox, M.R.S. et. al. Science, Cadmium toxicity decreased by dietary ascorbic acid supplements. 169, p. 989 (1970).
14. Moore, M.R. Proc. Nutr. Soc. Diet and lead toxicity. 38, 243, (1979).
15. Hambridge, K.M. J. Hum. Nutr. 32, 99, (1978).
16. Anderson, R.A. J. Agric. Fd. Chem. 26, 1219, (1978).
17. Gursen, C.T. Adv. Nutr. Res. 1, 23, (1977).
18. Ganther, H.E. and Sunde, M.L. J. Food Sci. Progress Report. 39, p. 1 (1974).
19. Luten, J.B. et. al. J. Food Sci. 45, p. 416 (1980).
20. Fridlyand, S.A. Gig. Sanit. 30, 20, (1965).
21. Ferm, V.H. Experimentia, 25, 56, (1969): Challop, R.S. New Eng. J. Med. 285, 970, (1971).
22. Gale, T.F. Environ. Res. 6, 95, (1973).
23. Management and Control of Heavy Metals in the Environment. London, September 1979, p. 439. (E.P. Consultants Ltd., 26 Albany Street, Edinburgh EH1 3QH UK).
24. Morgan, G.B. and Bretthauer, E.W. Anal. Chem. Metals in Bioenvironmental Systems. 49, No. 14, p. 1210A (Dec. 1977).
25. Analytical Methods Committee. Determination of mercury and methyl mercury in fish. Analyst, 102, No. 1219, p. 769 (1977).
26. LeBlanc, P.J. et. al. Marine Pollution Bulletin, 4, 88 (1973).
27. Lunde, G. Nature, Lond. 224, 186, (1969).
28. Toepfer, E.W. et. al. J. Agric. Fd. Chem. 21, 69, (1973).
29. Johnson, D.L. Environ. Sci. & Technol. 5, 411, (1971).
30. Brown, E.J. et. al. Bull. Environ. Contam. & Toxicol. 21, p. 37 (1979).
31. Strickland, J.D.H. et. al. Fish. Res. Board. Can. Bull. 125, 43, (1965).
32. Kapel. M. and Komaitis, M.E. Analyst. Polarographic Determination of Trace Elements in Food from a Single Digest. 104, p. 124 (1979).
33. Skogerboe, R.K. et. al. Anal. Chem. Exchange of Comments on Scheme for Classification of Heavy Metal Species in Natural Water. 52, p. 1960 (1980).
34. Aggett, J. et. al. Analyst. The Determination of Arsenic (III) and Total Arsenic by Atomic Absorption Spectroscopy. 101, p. 341 (1979).
35. Crosby, N.T. Analyst. The Determination of Metals in Foods. 102, No. 1213, p. 225 (1977).
36. Howard, A.G. et. al. Analyst, 105, p. 338 (April 1980).
37. Hall, R.J. Analyst. Observations on the Distribution and Determination of Fluorine Compounds in Biological Materials. 93, p. 461 (1968).
38. Kumpulainen, J. et. al. Residue Reviews. Fluorine in Foods. 68, p. 37 (1977).

39. Cooke, J.A. et. al. Environ. Poll. Determination of Fluoride in Vegetation. A Review of Modern Techniques. 11, p. 257 (1976).
40. Vandeputte, M. et. al. Z. Analyt. Chem. Comparison of Methods for the Determination of Fluorine in Plants. 282, p. 215 (1976).
41. A.O.A.C. Official Methods of Analysis, 11th Ed. Sec. 25.029-25.035, (1979).
42. Dabeka, R.W. et. al. J.A.O.A.C. Microdiffusion and Fluorine Specific Electrode. Determination of Fluoride in Foods. 62, p. 1065 (1979).
43. Tusl, J. Coll. Czechoslav. Chem. Commun. Spectrophotometric Determination of Fluorine in Biological Materials after Diffusion. 35, p. 1001 F.S.T.A. 3, 2A 79.
44. Tomura K. et. al. J. Radioanal. Chem. The Determination of Fluorine in Bones by Non-destructive Neutron Activation Analysis. 34, p. 375 (1976).
45. Venkateswarla, P. Z. Analyt. Chem. Comparison of Methods for the Determination of Fluorine in Plants. 282, p. 215 (1976).
46. Beswick, G. et. al. The Gas Chromatographic Determination of Fluoride in Milk and Water. (1978). Dept. of Applied Biology and Food Science, Polytechnic of South Bank, Borough Road, London SE1 OAA (UK).
47. An Examination of Asbestos in Relation to Food and Drink (FAC/REP/30) M.A.F.F. (HMSO) UK (1979).
48. European Communities Secondary Legislation. Part 26. Food Standards. H.M.S.O. London (1972) and subsequent amendments.
49. FAO/WHO (1972) Evaluation of Certain Food Additives and the Contaminants, Mercury, Lead and Cadmium; Joint Expert Committee on Food Additives Sixteenth Report (1972). W.H.O. Tech. Report Series No. 505, Geneva.

A version of this lecture has also been published in Ernährung 5 (1981) by agreement of the publishers.

Determination of Arsenic in Muscles and Liver of Some Fresh-Water Fish

I. PETROVIĆ[1], F. MIHELIĆ[2], AND T. MAŠINA[2]

[1] Institute of Public Health of SR Croatia, Zagreb
[2] Faculty of Technology, Zagreb University.

SUMMARY

Arsenic determination findings in the muscle tissue and spleen of certain fresh-water fishes are given in this paper.
While arsenic levels in muscle tissue of trout and spleen of carp were the lowest amounting to only 0,00 µg/g, in barbel samples the highest levels of arsenic were found in the muscle tissue and varied between 1.45-3.15 µg/g.

INTRODUCTION

Arsenic belongs to a group of elements that are nearly ubiquitous: it is present in our environment, biosphere as well as foodstuffs drinking water included.
Although arsenic compounds have been known, even as drugs, for quite some time (since before new era), their toxicity is of topical concern primarily because arsenic's toxicity is of the cumulative type whose effects manifest themselves in conditions of exhaustion or bodily illness (1). The effects of arsenic compounds take the following sequence:
$AsH_3 > As^{3+} > As^{5+} > RAsX$, wherease the excretion from body flows in the exactly opposite direction (2).
World Health Organization (WHO) has devoted special attention to the arsenic content of foodstuffs ranking it second among foodstuffs for

toxicity and this particularly because of a possible cancerogenic effect of its compounds (3). It is also quite understandable that such great attention should be paid to arsenic as human body actually is endangered by this contaminant and the sources from which arsenic derives are multiple. Primarily it passes through air, water and foodstuffs and is also released into the air when metals are obtained from sulphide ores, through coal burning, and equally it enters the vegetable foodstuffs through plant protection agents, various chemicals used in the manufacture of foods and additives, and drugs, and especially through additives for animal feeding which contain arsenic compounds together with other growth stimulating agents.

It should be particularly emphasized that tragic cases of human poisoning by food raw materials or additives that were contaminated with arsenic are known. To mention a case in England in 1900 with more than 4,000 human poisoninings, 300 deaths were beer was incriminated as the cause of poisoning. Glucose contaminated with arsenic from sulphuric acid (4) was used in the manufacture of the beer. Or, a case in Japan in 1955, when 130 children died and another 12,000 manifested the symptoms of arsenic poisoning when children were nourished with powder milk to which dipotassium monophosphate was added which contained between 8-25 mg of arsenic oxide (5).

According to literature data the amounts of arsenic in individual foodstuffs vary within a wide range. So it was found that fruit, vegetables and cereals contained between 0.5 mg/kg of arsenic and rarely in excess of 1 mg/kg, the meat of slaughter animals rarely above 0.5 mg/kg, while in milk the amounts ranged between 0.03-0.06 mg/kg. The amounts of arsenic rang between 0.07-1.5 mg/kg in the milk of New Zealand cows which were fed with the grass from contaminated areas. Foodstuffs from sea-fish, molluscs, crabs and shells contained far greater amounts of arsenic. Thus fish contained between 2-8 mg/kg oysters 3-10, conger-eel 1-15 mg/kg, while mussels, shrimps, and prawns contained 0.5-174 mg/kg of arsenic (2,6,7,8 and 9).

The stated reasons, significant amounts of arsenic in seafish, shrimps, shells and molluscs were the incentive for determining the level of arsenic in freshwater fish from fish ponds and rivers as such investigations, according to the literature available to us, have not been undertaken, particularly not in this country.

Material and determination method

Determining the amounts of arsenic in muscle tissue and liver of fresh-water fish from fishponds and rivers at given time intervals was set as the objective for the present work.

30 fish samples were taken, i.e.:
- 10 specimens of trout (Trutta fario)
- 10 specimens of carp (Cyprinus Caprio)
- 10 specimens of barbel (Barbus barbus)

The trout specimens came from the Bregana fishpond, carp specimens from the Crna Mlaka fishpond, while barbel specimens came from the Danube.

Arsenic level determinations were carried out by using the method of diethylethiocarbamate (Ag-DDTC) in pyridine, while absorbance measurements were done on a Beckmann DU model spectrophotometer.

The lower threshold of method's sensitivity is 0.1 $\mu g/g$

Results and discussion

It may be concluded rom the enclosed table that the levels of arsenic in the liver of trout are higher than the levels of arsenic in the muscle tissue of trout, while the levels of arsenic in the muscle tissue of carp and barbel are higher than those in the liver of the same specimens.

The lowest level of arsenic in the liver of the examined fish samples was 0,00, the highest 1.50 $\mu g/g$, while the lowest and highest levels in the muscle tissue were 0.00 and 3.15 $\mu g/g$ respectively.

The levels of arsenic in the muscle tissue of trout and liver of carp were the lowest amounting to 0.00, while in barbel samples the highest established levels were those in the muscle tissue and ranged from 1.45-3.15 $\mu g/g$.

Table 1 PRESENTATION OF THE LEVELS OF ARSENIC IN THE MUSCLE TISSUE AND LIVER OF THE TESTED FISH SAMPLES

Samples	Amount of arsenic (µg/g)								
	in muscle tissue					in liver			
	Minimum	Maximum	$\bar{X}$	s	C.V.%	Minimum	Maximum	$\bar{X}$	s	C.V.%
Trout	0.00	0.60	0.06	–	–	0.00	1.50	0.501	0.46	91.82
Carp	0.26	0.75	0.50	0.15	30.00	0.00	0.00	–	–	–
Barbel	1.45	3.15	2.48	0.47	22.48	0.50	1.48	1.1	0.32	29.09

CONCLUSION

The completed study allows of the following conclusions:
The lowest levels of arsenic were found in the liver of carp and muscle tissue of trout, whereas the highest levels of arsenic were found in the liver of trout and muscle tissue of barbel. As evidenced by the table the results mutually differ significantly in that the result of the type of nutrition and metabolism for the above kind of fish, and to the highest degree so the levels of arsenic are due to the extent of contamination of their living environment, for undoubtedly the Danube in Yugoslavia is to a considerable extent contaminated with organic and inorganic pollutants.

REFERENCES

1 K. Bauer, Kemija u industriji 9, 235 (1960)
2 H. Woidich, W. Pfannhauser, Deutsche Lebensmittel Rundschau 6, 190 (1979)
3 WHO Techn. Report Series No 532, p. 49, Genf 1973
4 M. Mokranjac, Toksikološka hemija, Naučna knjiga, Beograd, (1963)
5 D.G. Chapman, L.I. Pugsley, Public Health Aspects of the Use of Phosphates in Foods, Symposium-Phosphates in Food Processing, Westport, Connecticut (1971)
6 H.A. Schroeder, The Poisons Around us. Toxic metals in Food, Air and Water, Indiana Univ. Press (1974)
7 E.J. Underwood, Trace Elements in Human and Animal Nutrition, University of Western Australia, Academic Press, New York, London, 3rd Ed (1971)
8 E.J. Underwood, Toxicants Occuring Naturally in Foods. Trace Elements, National Academy of Sciences, Washington, D.C. (1973)
9 I. Petrović, Arsen u morskim ribama, rakovima, školjkama i mekušcima, Rukopis pripremljen za štampu (1981).

Quality Assessment of the Edible Part of Mussels: Determination of Lead and Cadmium by Atomic Absorption Spectroscopy and Voltammetry

L. G. FAVRETTO, G. P. MARLETTA AND L. FAVRETTO

Institute of Commodity Science, University of Trieste, Italy

SUMMARY

The concentration of lead in the edible part of mussels (*Mytilus gallo-provincialis* Lamarck) is $\log_{10}$-normally distributed in the sample extracted from a population living in a polluted site, where mussels collect particulate lead from sea water. In the sample from an unpolluted site the concentration of lead is normally distributed and, in this case, lower range and mode are observed. In all samples cadmium appears to be log-normally distributed.

INTRODUCTION

The evaluation of empirical frequency distribution of the content of toxic heavy metals is a simple statistical approach in quality assessment of foods from marine origin. Some tentative researches have been already performed on the edible part of the common mussel (*Mytilus galloprovincialis* Lamarck), which is extensively harvested along the shores of the gulf of Trieste (Northern Adriatic Sea) (1,2). In this paper the interest is focused on the concentration of lead and cadmium. The significance of the presence of these metals in mussels has been often discussed (3,4,5).

In coastal sea waters polluted by urban outfalls, lead is mainly present in < 0.8 μm suspended particles (6) and it appears to accumulate in mussel flesh (7). The relationships between the concentration of lead in sea water and in the edible portion have been also studied in mussels grown in artificial sea water environments containing a definite concentration of metal added as a soluble salt (3). In sea water at pH $\simeq$ 8, lead added at trace levels (less than 1 µg/kg) is likely converted to sparingly water soluble oxo- and hydroxo-compounds. Therefore this metal seems to travel through a natural sea water environment mostly as a component of insoluble chemical species. The particulate fraction of the suspended matter comprises also lead containing atmospheric dust from automobile exhaust gases and other combustion smokes.

SAMPLING, METHODS AND DATA PROCESSING

Mussels have been sampled on the wall of a pier in the harbour of Trieste about 0.5 m below the mean sea level. This site is about 400 m from the outfall of an urban sewer, which carries untreated liquid wastes to sea. Commercial purified mussels have been also obtained from an industrial hatchery (Grado). All samples have been weekly collected during 1975-76 years.

Lead and cadmium have been determined in the mussel ash prepared and solubilized as previously indicated (7). Voltammetric methods have been already described (7,8). Voltammetric curves have been traced with an AMEL 448 instrument connected to an X-Y Hewlett-Packard 7041 recorder.

A Perkin Elmer HGA graphite furnace mounted in a 372 atomic absorption spectrophotometer with D_2-lamp background corrector has been used. Absorbance peaks have been recorded with an X-Y AMEL 862/A time-basis instrument using a Perkin Elmer Intensitron lead hollow cathode lamp at 283.3 nm (current 10 mA). Details of the method for lead analysis in the solution of ashed mussel samples have been already published (9). The method utilizes a curved calibration line, $A' = abq/(1 + aq)$, where A' (mm) is the absorbance peak height corrected for the reagent blank, q (ng) is the amount of lead atomized. This equation fits the points of the calibration graph obtained in a wide range of lead vaporized (0-4 ng), and tends to linearity when approaches to 0. This curve has been used in the linearized form by plotting $1/A'$ *versus* $1/q$. Advantages of linear over non-linear regression have been already discussed (9).

Atomic absorption spectroscopic analyses have been triplicated, taking the arithmetic mean for further processing. All concentration data are referred to dry weight.

All statistical calculations have been performed on an Olivetti 1040 desk processor. The adequacy of the calculated normal (or log-normal) frequency distribution to fit the observed histogram has been evaluated with the χ^2-test.

RESULTS AND DISCUSSION

Table 1 summarizes the results obtained in processing the concentration data of lead and cadmium in the edible part of mussels grown in differently polluted sites. The concentrations of lead in polluted mussels were determined by two different methods.

Table 1. Frequency distributions calculated for mussel samples from polluted and unpolluted environments. V, voltammetry; AS-MS, atomic absorption spectroscopy with a microsampling cup; AAS-GF, atomic absorption spectroscopy with graphite furnace. n, number of observations; RV, random variable; $\bar{x} \pm s$, arithmetic mean and root mean square deviation (geometric mean appears in brackets). All concentrations are given in μg/g dry weight.

Metal	Sampling site	Method	n	Range	RV	$\bar{x} \pm s$
Pb	Polluted hatchery	V	38	3.9–14.5	$x_i = \log c_i$	0.827 ± 0.134 (6.71)
Pb	Polluted hatchery	AAS-MS	38	4.6–15.9	$x_i = \log c_i$	0.901 ± 0.128 (7.96)
Pb	Unpolluted hatchery	AAS-GF	36	0.8–5.6	$x_i = c_i$	2.70 ± 1.34
Cd	Polluted hatchery	V	35	0.10–0.65	$x_i = \log c_i$	$\bar{1}$.517 ± 0.216 (0.329)
Cd	Unpolluted hatchery	V	32	0.17–0.73	$x_i = \log c_i$	$\bar{1}$.537 ± 0.215 (0.344)

A polluted environment affects the lead level not only by increasing the range in comparison with that of commercial mussels, but also by changing the frequency distribution. Figure 1 compares the histograms of lead concentration in unpolluted and polluted mussels. The distribution of lead concentration in unpolluted mussels is acceptably normal with an arithmetic mean of 2.70 µg/g on dry weight, whereas that of polluted mussels appears to be skewed and acceptably $\log_{10}$-normal with a geometric mean of 7.97 µg/g.

There is no significative difference between the environments when cadmium is considered. Ranges are nearly the same and a log-normal distribution in both samples is approximately observed.

In the polluted site of the harbour of Trieste, lead in mussels has an evident urban origin. Lead pollution is mainly due to particulate matter suspended in sea water (7) and this particulate lead decreases with the distance from the urban sewer (10).

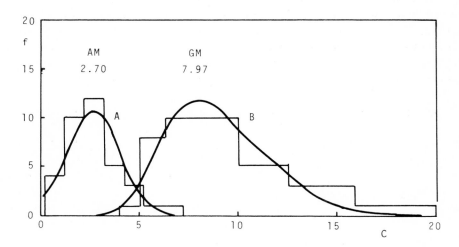

Fig.1. Frequency (f) distribution of lead concentration (c, µg/g dry weight) in the flesh of mussels grown in an unpolluted hatchery (A) and in a polluted environment (B). Histograms show observed (grouped) frequencies, continuous lines indicate the calculated distribution, assuming a normal (A) and log-normal (B) model. AM, arithmetic mean; GM, geometric mean.

To conclude, the most frequent lead content in polluted mussels is about 8 µg/g on dry weight. However, this content can rise up to 20 µg/g as a consequence of rare but severe events of sea water pollution. As a matter of fact, a normal frequency distribution (arithmetic average 2.7 µg/g dry weight) of lead content seems to characterize the mussels sampled in an industrial hatchery free from pollution by urban areas.

REFERENCES

1. L. Favretto and F. Tunis, Rev. Intern. Océanogr.Méd. 33 (1974), 67
2. L. Favretto and F. Tunis, Rass. Chim. 27 (1975), 179
3. M. Schulz-Baldes, Marine Biol. 16 (1972), 226
4. M. Schulz-Baldes, Marine Biol. 21 (1973), 98
5. M. Piscator, Proc.Int. Conf. on Managment and Control of Heavy Metals in the Environment, London (1979)
6. L. Favretto and L. Favretto Gabrielli, Rev. Intern. Océanogr. Méd. 33 (1974), 61
7. G. Pertoldi Marletta, L. Favretto Gabrielli and L. Favretto, Z. Lebensm. Unters.-Forsch. 168 (1979), 181
8. G. Pertoldi Marletta, Riv. Merceologia 18 (1979), 97
9. L. Favretto, G.Pertoldi Marletta and L. Gabrielli Favretto, Mikrochim. Acta (1981) (in press)
10. L. Favretto, G.Pertoldi Marletta and L. Favretto Gabrielli, Proc. 3es Journées Etude Pollutions, CIESM, Split (Jugoslavia) (1976), 51

A Statistical Approach to the Balance of Lead in Milk and Some Dairy Products

G. P. MARLETTA, L. G. FAVRETTO and C. CALZOLARI

Institute of Commodity Science, University of Trieste, Italy

SUMMARY

A systematic study on milk and its primary by-products (cream and skimmed milk, both obtained from the same milk batch) is considered in order to define a tentative balance of lead. Lead was determined by atomic absorption spectroscopy with graphite furnace on a solution of the ashed sample. As a consequence of the appreciable interference due to the matrix effects, the method of standard additions was adopted to determine the metal in some samples with a statistically acceptable degree of accuracy and precision. The solution of these ashed samples was then used as a reference standard for the rapid determination of lead in other samples having a comparable metal level.

INTRODUCTION

Many studies have been recently devoted to the determination of lead in milk in view of the toxicity of this heavy metal and the importance of milk in the diet of infants and children. As milk and its by-products are basal constituents of their diet, it is essential to ensure that they are free from contamination with this metal and a rapid and reproducible method of determination is therefore necessary (1-5).

In view of the enrichment of lead in fatty dairy products that has been sometimes noted (6,7), a preliminary study on milk and its primary by-

-products (cream and skimmed milk, both obtained from the same milk batch was performed in order to define a tentative balance of lead in these products.

Many difficulties arise in the determination of lead in biological materials at trace level, which often is the normal level of this heavy metal in foods. Furthermore other difficulties, that should not be overlooked, are the interference of the matrix involved and the contamination risk (3,8).

One of the most sensitive and widely used techniques for the detection of lead in milk is flameless absorption spectroscopy, using a graphite furnace as atomizer; however if one wishes to reduce the appreciable interference of this particular matrix, direct analysis of the sample is not feasible and the ashing procedure is necessary. Therefore in this research, the preliminary ashing of the milk samples has been performed and possible interference has been checked by the standard additions method. The aim of this investigation was to verify the analytical method when applied to the determination of lead in milk, skimmed milk and cream and to make a survey of the content of this metal in the above products.

EXPERIMENTAL

The samples of pasteurized whole milk, skimmed milk and cream, obtained from the same milk batch, were supplied by a Central Dairy of the Friuli-Venezia Giulia region.

An HGA graphite furnace mounted on a Perkin Elmer 372 atomic absorption spectrophotometer connected with an X-Y AMEL 862/A time-basis recorder was used with D_2-lamp background corrector. The samples were atomized in normal graphite tubes under nitrogen atmosphere. Absorbance peaks were traced at 283.3 nm using a Perkin Elmer Intensitron lead hollow cathode lamp (current 10 mA) with a slit width 0.7 nm. The optimized operating conditions were as follows : drying step- 100°C for 30 s, N_2 flow 250 ml/min; charring step- 500°C for 30 s, N_2 flow 200 ml/min; atomizing step- 2100°C for 5 s, N_2 flow 20 ml/min. An instantaneous ramp (1600°C/s) has been used in order to obtain the maximum sensitivity of absorbance peak height as well as to avoid pre-vaporization of lead during ramp, which enlarges the peak in accordance with a previous literature report (9).

5-10 g of each sample were weighed in a small platinum crucible, then charred under an infra-red lamp and ashed at 500°C in a muffle overnight. The ash was treated with 1 ml of 2 M nitric acid near the boiling point for 1 min; 2 ml of 0.1 M nitric acid were added, filtering the solution through a 5-cm diameter filter paper (Schleicher and Schüll 589[1]), collecting the filtrate in a 10-ml calibrated flask and then washing the filter with 0.1 M nitric acid up to the mark.

RESULTS AND DISCUSSION

Calibration curves were obtained in the range 0-4 ng of lead vaporized. Absorbances at 283.3 nm were read directly as peak heights for increasing amounts of lead from standard solutions of lead nitrate in 0.1 M nitric acid.

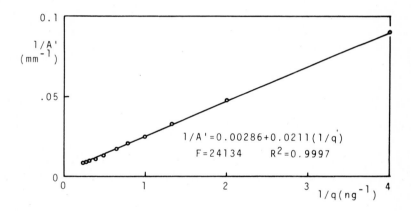

Fig.1. Linearized calibration line corresponding to the hyperbole obtained by non-linear regression. The continuous line has been calculated by regression of $1/A'$ on $1/q$ data. The adequacy of fitting has been tested by the analysis of the variance and expressed as F and R^2.

The calibration graph was obtained by non-linear regression, assuming that the empirical model function is a hyperbole having the general formula $A' = abq/(1 + aq)$. This function has been chosen (10) because it passes through the origin and it has a *quasi*-linear trend near the origin. In the experimental conditions adopted, the resulting equation is $A' = 48.3q/(1 + 0.164q)$. The curve fits the observations at least up to $q = 4$ ng, with a small root mean square deviation (s = ±2.1 mm). If A' (mm) is the peak height corrected for the reagent blank (2 mm), the working linear calibration graph is obtained by plotting $1/A'$ (mm^{-1}) $versus$ $1/q$ (ng^{-1}). Figure 1 shows the linearized equation $1/A' = 1/b + (1/ab)(1/q)$, where the goodness of fit appears from the index ratios F and R^2 estimated from the analysis of the variance.

Advantages of linear over non-linear regression are as follows: i) the verification of a linear calibration graph requires only a few points instead of 10-12 points at least involved in computing with a non-linear program; ii) a desk processor is able to run a linear regression program; iii) simple index ratios F and R^2 test the adequacy of the model function. The graph at every run of the analyses must be checked with at least 4 points.

The graph of figure 1 can be utilized for direct determination of lead in the analyzed solutions only if the matrix effects are absent. The method of standard additions was adopted for checking that $1/A'$-values corresponding to the additions lie on the linearized graph. As an example, figure 2 shows that no matrix interference on the calibration graph is detectable with solutions of ashed samples of whole milk, skimmed milk and cream (11). In all samples the 3 additions lie on the linearized calibration graph.

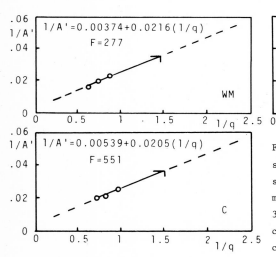
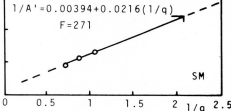

Fig.2. Linearized plots of the standard additions of lead on samples of whole milk (WM), skimmed milk (SM) and cream (C). The 3 additions, corresponding to the circles, lie on the linearized calibration graph.

The precision of the method is discussed below. Recoveries vary from 96 to 108% in some analyses repeated after addition of lead from a standard solution to the samples before charring. The relative difference (in %)

of recoveries is largely explained by the coefficient of variation observed in duplicated analyses.

Taking into account that the whole milk examined has a mean fat content of 3.5%, the skimmed milk 0.5% and the cream 33%, an average value for cream recovery is about 10% by weight. Therefore, the content of lead in whole milk (c'_{WM}) is easily calculated from the contents of skimmed milk (c_{SM}) and of cream (c_C): $c'_{WM} = 0.9\, c_{SM} + 0.1\, c_C$.

Table 1 shows some preliminary analytical data of lead in whole and skimmed milk and in cream. The root mean square deviation, calculated from duplicated analyses, indicates that the best intra-laboratory reproducibility is observed with skimmed milk: in this product, a coefficient of variation of about 6% at a lead level of 20 µg/kg is observed. Whole milk shows a lower reproducibility, that is about 12% at 20 µg/kg level. The lowest reproducibility is observed with cream, which is about 15% at 35 µg/kg lead level. This decrease in precision with the increase of fat content is probably due to the difficulty of ashing without production of fumes in fat-containing materials.

Table 1. Lead content (c = µg/kg) in whole milk, skimmed milk and cream. The arithmetic mean is underlined. s, root mean square deviation calculated from 4 duplicated analyses.

Whole milk c_{WM}		Skimmed milk c_{SM}		Cream c_C		Calculated c'_{WM}
18.1		19.1		42.2		
21.0	19.6	17.8	18.5	30.2	36.2	20.2
10.3		13.5		40.8		
14.2	12.2	15.5	14.5	32.5	36.6	16.7
25.5		25.2		35.6		
29.8	27.6	24.7	24.9	36.2	35.9	26.0
17.3		12.6		33.2		
16.0	16.6	14.5	13.5	35.0	34.1	15.6
s = ±2.3		s = ±1.1		s = ±5.2		

Table 1 indicates also that cream has an higher lead content than whole milk and skimmed milk. This is in accordance with some early results on the lead analysis of butter, which contains about 4-5 times more lead than milk (7). Lead in cream seems to be, at least in part, bound to the fat content of this food. In fact, skimmed milk seems to have a slightly appreciably lower lead content than whole milk. As further confirmation for the samples considered so far, the calculated values of lead concentration in whole milk correspond to those found directly in whole milk.

REFERENCES

1 P.J. Barlow, J. Dairy Research 44 (1977), 377
2 D.C. Manning, Amer.Lab. 5 (1973), 37
3 H. Jönsson, Z.Lebensm. Untersuch.-Forsch. 160 (1976), 1
4 J.A. Fiorino, R.A. Moffitt, A.L. Woodson, R.J. Gajan, G.E. Huskey and R.G. Scholtz, J.Ass.Off.Anal.Chem. 56 (1973), 1246
5 J. Koops and D. Westerbeek, Neth. Milk Dairy J. 32 (1978), 149
6 G. Velghe, M. Verloo and A. Cottenie, Z. Lebensm. Unters.-Forsch. 156 (1974), 77
7 G.Pertoldi Marletta, Ind. Alim. (Pinerolo, Italy) 110 (1974), 113
8 J.A. Krasowski and T.R. Copeland, Anal. Chem. 51 (1979), 1843
9 C. Hendrikx-Jongerius and L. De Galan, Anal.Chim.Acta 87 (1976), 259
10 L. Favretto, G. Pertoldi Marletta and L. Gabrielli Favretto, Mikrochim.Acta (1981) (in press)
11 W.Wegscheider, G.Knapp and H. Spitzy, Z.Anal.Chem. 283 (1977), 183

Study on the Presence of Heavy Metals in Cereals

M. Baldini, M. Centi, C. Micco, A. Stacchini

Istituto Superiore di Sanità, viale Regina Elena, 299, Roma
Italia

SUMMARY

In this work are reported the results of a research conducted on samples of wheat and of rice from various regions in Italy, with the purpose to value the conditions of contamination from lead, cadmium, chromium, zinc and copper.
The determination have been performed employing the technique of Atomic Absorption Spectrophotometry.
The samples are representative of about the 60% of the Italian production.
The distribution of the values found in function of the areas of origin is described and the levels of concentration of each metal are discussed.

INTRODUCTION

We have planned a sistemic study on the determination of Pb, Cd, Cr and also Cu, Zn in cereals.
These pollutant agents are nowaday spread éverywhere and an aim of this research was also the evaluation of the local contamination level in relation on the originary enviroonmental conditions.
In the present comunication are reported data concerning the '78 - wheat-crop, obtained on samples of different origins, representing more than 60% of the whole national production, are also presented data on rice producted in some northern area. Wheat samples were examined coming from the following areas: Veneto, Lombardia, Piemonte, Emilia Romagna, Toscana, Lazio, Basilicata, Puglie.

Analytical method

The determinations were carried out by A.A. spectrometry on meal ashes.
The results have been given in $\mu g/g$ and they referred to dry substance.

It was necessary to verify, the loss - percentage in each step of the analytical procedure in particular for the loss due to volatilization during charring.

Operative conditions of Atomic Absorption spectrometry

Pb   λ :   283,3 nm
     Furnace:
     Charring temperature   400 °C
     Atomization Temperature   2.700 °C

Cd   λ :   228,8 nm
     Furnace:
     Charring temperature   250 °C
     Atomization Temperature   2.100 °C

Cr   λ :   357,9 nm
     Furnace:
     Charring temperature   1.000 °C
     Atomization Temperature   2.700 °C

Zn   λ :   213,8 nm
     Flame: air/acetylene

Cu   λ :   324,7 nm
     Flame: air/acetylene

Results discussion of wheat

By the concentration levels on each pollutant metal one can draw the following conclusions.
Lead is omogeneously spread in wheat at an extent comparable with the environmental levels.
About 90% of the examined samples display a lead content of 0 - 0,4 µg/g, most of the values being close to 0,2 µg/g. (Fig.1)
Higher concentration levels have been found only in 3 places: Emilia Romagna, Marche, and Lazio. (Fig. 2).

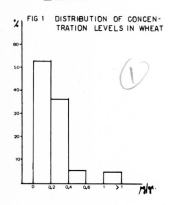

FIG 1 DISTRIBUTION OF CONCENTRATION LEVELS IN WHEAT

LEAD

FIG 2 Pb LEVELS IN VARIOUS AREAS

● 0 - 0.2 ppm
□ 0.2 - 0.5 ppm
△ > 0.5 ppm

LEAD

Cd levels display a % distribution profile parallel to that of lead. About 90% of the samples had a Cd content of 0 - 0,3 μg/g with values crowding at 0,2 μg/g (Fig. 3).

CADMIUM

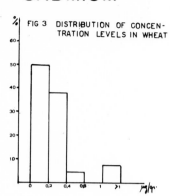

CADMIUM

Nevertheless, with respect to lead, for Cd have been found more samples containing more than 1 μg/g, and also the top values were in some cases respectively much higher (conc. Cd 3 μg/g). For some area (Marche - Toscana) these maximal levels represent singular and anomalous situations with respect to average of other values; a different meaning has to be attributed to high contamination values found in Emilia Romagna, where are many industries, an area in which the average Cd - contamination is higher. (Fig. 4).

Cr - concentrations are averagly higher than those of Pb and Cd. (Fig. 5)

CHROMIUM

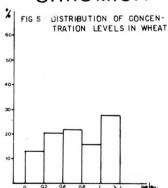

Higher concentration levels have been found (conc. Cr. 1 μg/g) in Lombardia, Piemonte and Emilia Romagna.
Peaks higher than 1 μg/g are distributed also in other areas, in which, anyhow, a wide dispersion of values is found (Fig. 6).

CHROMIUM

FIG 6 Cr LEVELS IN VARIOUS AREAS

- • 0 - 0.2 ppm
- □ 0.2 - 0.5 ppm
- △ > 0.5 ppm

In Lazio there is a gathering of the concentrations around values higher than the average.
Zn and Cu contents are generally high. The values for Zn are in the range 20 - 40 µg/g, most of the data being around 30 µg/g, (Fig. 7) and are generally omogeneously distributed. (Fig. 8).

ZINC

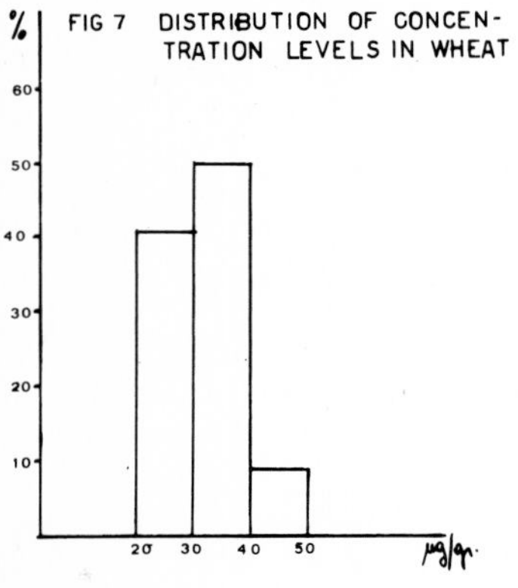

FIG 7 DISTRIBUTION OF CONCENTRATION LEVELS IN WHEAT

ZINC

FIG 8 ZINC LEVELS IN VARIOUS AREAS

- • < 30 ppm
- □ 30-40 ppm
- △ > 40 ppm

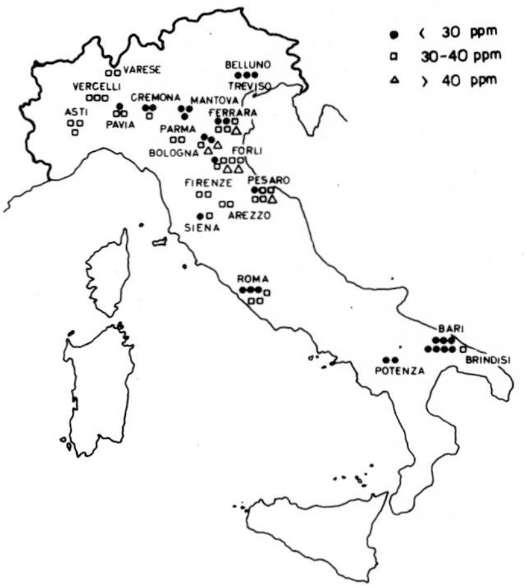

Cu concentrations are generally lower, 0 - 26 µg/g (Fig. 9).
Particularly high levels have been found in some parts of Emilia Romagna, in Lazio and Marche (Fig. 10).

COPPER

FIG 9 DISTRIBUTION OF CONCENTRATION LEVELS IN WHEAT

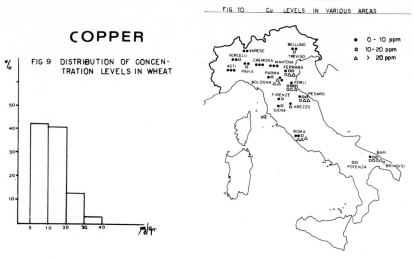

COPPER

FIG 10 Cu LEVELS IN VARIOUS AREAS

As for a correlation between contamination and origin of the sample, one may observe that the highest metal contents have been obviously found in samples coming from the area of higher industrialization, f.e. Lombardia, Piemonte, Emilia Romagna.

Results discussion of rice

As for rice, the study has still to be completed. Only about 1/3 of the available samples were examined, and the data concerne samples coming from Piemonte. Most of them display a lead content of 0,2 - 0,6 µg/g, with values crowding at 0,4 µg/g. (Fig. 11).
No samples had high values of contamination.
Cd - contents fall substantially around two ranges: about 45% of the samples had a content of less than 0,2 µg/g, and about 45% of the samples had a content ranging from 0,4 to 0,9 µg/g. The top value is 0,86 µg/g. (Fig. 12)

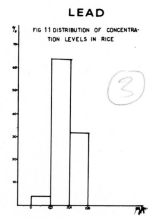

LEAD
FIG 11 DISTRIBUTION OF CONCENTRATION LEVELS IN RICE

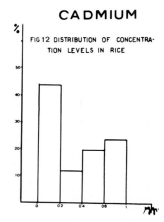

CADMIUM
FIG 12 DISTRIBUTION OF CONCENTRATION LEVELS IN RICE

About 90% of the examined samples had a chromium content of 0,15 - 0,40 µg/g
The top value is 0,97 µg/g (Fig. 13).
Zn values are in the range 10 - 30 µg/g most of them being around 20 µg/g, and Cu content is generally low, 90% of the values being less than 4 µg/g. (Figg. 14 - 15).

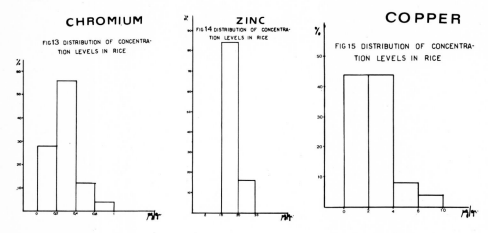

A comparison has been made between Pb - Cd - Cr levels in wheat and rice samples coming from the same geografic area, in order to verify how the similar enviromental conditions affect the absorption of the metals in that two kinds of cereals.
In wheat Pb and Cd levels are considerably lower than those found in rice. Values in wheat are always lower than 0,2 µg/g, while in rice levels are ranging 0,30 - 0,50 µg/g for lead and 0,30 - 0,70 µg/g for Cadmium.
On the contrary, in wheat samples Cr content reaches higher levels than those found in rice, so that wheat-values are close to 1 µg/g, while rice values are generally lower than 0,5 µg/g.
We plan to continue with our researches studying other cereals growing in the same area, with the aim of correlating heavy metals in the environment, and in cereals-coltivations.

REFERENCES

1. Garcia Wy., Lessin Cw., Inglet Ge. "Heavy metals in food products from corn". Cereal Chemistry, 51, 6, 1974.
2. Santoprete G., "Il contenuto in oligoelementi (Cr-Mn-Co-Cu-Zn-Cd-Hg-Pb) del riso commerciale e relativa assunzione da parte del consumatore italiano" Riso - Quaderni di Merceologia, 14, 2, 1975.
3. Masironi R., Koirtyohann S.R., Pierce J.O., "Zinc, Copper, Cadmium and Chromium in polished and unpolished rice" The science of the Total Environment, 7, 1977.
4. Yamagata N. and Shigematsu I., Bull. Inst. Publ. Health (Tokyo), 19, 1970.

Trace Elements (Cr, Mn, Cu, Zn, Cd, Hg, Pb) Content in Italian Foods and in the Meals Consumed in Other European Countries

G.C. SANTOPRETE[1],[2] and N. WOLKENSTEIN[2],[3]

[1] Istituto di Merceologia, Università, Bologna (Italy)
[2] Istituto di Ricerche Aziendali, Università, Pisa (Italy)
[3] Istituto di Elaborazione dell'Informazione del C.N.R., Pisa (Italy)

SUMMARY

Taking into account: a) the contents in trace elements of the basic alimentary products consumed in Italy, and b) the official Balance Sheets of seventeen European Countries, we have made an approximate computed determination of the average daily pro caput amounts of Cr, Mn, Cu, Zn, Cd, Hg and Pb contained in the products used in the diets of the populations taken into consideration, giving thus sufficiently reliable and practically useful elements for a first conclusion for sanitary and nutritional purposes.

INTRODUCTION

In order to give a real contribution to the problem of ascertaining whether the average ingestion of a given trace element by the population of a certain area is to be considered as satisfactory, balanced, insufficient or dangerous, we believe one should determine the quantities of this element in the alimentary products actually consumed by the population, independently from the places where these products are obtained: in fact, while there is a direct relation between a plant and the soil on which it grows, the same thing is no longer true for man, owing to the variety of the sources of his food.

One of us has conducted his research in accordance with this principle, inserting it in the pluriennal programme of his Institute (1, 2). Systematic studies were made to determine the quantities of Cr, Mn, Cu, Zn, Hg, and Pb in the alimentary products that form the basis of the diet of the Italian population (bread, paste products, rice, flour, vegetables, etc.)(3).

This work is based on the above mentioned experimental studies, as well as on further research made chiefly in the "Istituto di Merceologia" in Bologna (literature to be found in (2)). By combining these experimental data with statistical information about the average diets in various European Countries (4), we have then calculated the average quantities (pro caput per day) of trace elements to be reasonably expected in those diets; by comparing these amounts with: a) the safety limits for toxicity indicated by FAO/WHO (5), and b) the requirements given in the literature - and especially in (6,7) - for healthy subjects following balanced diets comprising derivatives of cereals, meat, fish, pulses, greens and fruits, we have attempted to reach a first - indicative - conclusion about the adequacy of the amounts of Cr, Mn, Cu and Zn, and about the eventual excesses of Cd, Hg and Pb to be found in the foods consumed in AUSTRIA, BELGIUM-LUXEMBOURG, BULGARIA, CZECHOSLOVAKIA, DENMARK, FINLAND, FRANCE, WESTERN GERMANY, GREECE, IRELAND, ITALY, THE NETHERLANDS, SPAIN, SWITZERLAND, THE UNITED KINGDOM and YOUGOSLAVIA.

Before giving and examining the results of our experiments and computations, we feel it necessary to express our opinion that, in the present state of scientific knowledge, it would be impossible to reach conclusions more analytical and accurate than those we have been aiming at, especially because various aspects of the problem have not yet been sufficiently investigated. Such are, for instance, the questions concerning:
- the metabolic functions of the single elements, the amount of each element required for each function, the priorities between the functions involving a given element;
- the chemical forms in which the elements can be used by the human organism, and those in which they are present in the alimentary products before and after the latter are variously processed (and/or cooked, etc.) prior to final consumption;
- the reciprocal interactions of the various trace elements.

RESULTS AND CONCLUSIONS

Unless there is some explicit different indication, all the "amounts" quoted hereafter are to be understood as expressed in μg/person/day. It should furthermore be remarked that - but for the evident exception of Hg - it seems that the processing or cooking of the different alimentary products does not affect appreciably their contents in trace elements, at least for the aim of this work, which is - let us stress it again - to arrive at a preliminary sanitary and nutritional conclusion.

CHROMIUM (excluding the hexavalent form): the average amounts vary from 170 (Finland) to 280 (Bulgaria), and are therefore well within the limits (80 - 400) determined by various Authors by studying the balanced diets of healthy subjects. On the other hand, they are far below the levels considered as toxic (50 µg/g in the dry matter for rats, vs. some .3 - .5 µg/g in the European diets). We can therefore conclude that the amounts of this element actually present in our food: a) can be considered as sufficient for the metabolic functions of Cr (chiefly, normal metabolism of glucose), and b) are such as to make it possible to exclude that there may be classes of population submitted to any danger, even if one takes into account the Cr eventually contained in beverages.

MANGANESE: the average amounts vary from 1,660 (Denmark) to 3,090 (Yougoslavia), while the amounts found in the balanced diets of healthy subjects vary from 2,200 to 2,700. The higher amounts observed in some Countries can be considered as quite safe, even if one takes into account the contribution of beverages, because they are still far below the safety limits (1,000 to 2,000 µg/g on the dry basis for rats, vs .4 to .6 µg/g in the European diets). The quantities ingested in Denmark (1,660), Finland (1,675), the Netherlands (1,790) and the United Kingdom (1,850) are probably insufficient for an optimal metabolic action, especially for the wealthier classes, which have a tendency to replace the products derived from cereals with other foods (meat). The quantity of Mn ingested can indeed be very strongly influenced by the quality of the diet, reaching up to 6,000 - 12,000 µg/day for consumers of not-too-refined derivatives of cereals, and remaining far below these values in diets with a preponderance of meat, vegetables and fruits.

COPPER: the average amounts vary from 1,400 (Denmark) to 2,160 (Spain). Taking into account the additional contribution of beverages (eg. wine), these amounts can be considered as close enough to the value of some 2,000 µg/person/day esteemed to be sufficient for a satisfactory functional action. Some doubts may, however, reasonably arise for some classes in Denmark (1,400), Austria (1,510) and the Netherlands (1,540). Diets containing pulses, un-refined cereals, and especially seafood, can raise the level of ingestion up to 5,000 - 7,000. Even these higher values are still far below the safety limit (500 µg/g for rats, vs .3 - .5 µg/g in the European diets).

ZINC: for this element, we have average amounts from 8,450 (Denmark) to 12,100 (Czechoslovakia), vs. requirements ranging from 6,000 to 12,000. One of the biological functions of Zn is the control of bodily growth. Now, there are areas in the Middle East affected by d w a r f i s m , notwithstanding an average daily supply of up to 15,000 µg/person; the cause of this anomaly lies in an insufficiently balanced diet, combined with the presence of intestinal parasites, which prevent a perfect absorption as well of this element as of other nutritive substances (8).

CADMIUM: the average amounts vary from 34.0 (the United Kingdom) to 46.5 (Italy), and are therefore below the safety limits (60 - 70). This fact should eliminate any danger of anaemia, hypertension, injury to the kidneys, etc. connected with an accumulation of this element.

MERCURY: the average amounts vary from 39 (the Netherlands) to 67 (Spain). But for the Netherlands, Austria, Switzerland, Germany and Finland, they are a b o v e the safety limit of 43 µg/person/day. It should however be remarked that: a) the "safety limits" were determined by applying a "safety factor" of approximately 100 to an amount of Hg not yet really dangerous and mostly in its more active (alchylic) form; and b) we have personally ascertained that the amounts of Hg found in alimentary products when cooked and ready for final consumption, are smaller than the amounts contained in the same products in their raw forms. Also, when the samples were collected by us, the usage of p e s t i c i d e s containing Hg for the desinfection of silos and plants for the preparation of derivatives of cereals, had not yet been completely banned in Italy ((9) and literature listed in (2)). One can therefore assume that the amounts of Hg to be found nowadays in the European diets comply with the safety limits prescribed by FAO/WHO (5). This conclusion should hold even for Spain and some other Countries characterized by a large consumption of fish.

LEAD: the average amounts contained in the diets vary from 230 (Belgium-Luxembourg) to 390 (Ireland), and are therefore below the safety limit (430) indicated by FAO/WHO - the more so, as a certain amount of this element disappears during cooking. However, for some classes in some Countries, the contribution of beverages may well raise the amounts ingested above the safety limit. But we do not think that this fact should cause any serious pathological effect.

TO SUM UP

Within the limits set up by: a) the present state of our knowledge concerning metabolism in general and the functions, proportions, chemical forms and correlations of the trace elements, etc.; and b) our assumption that the contents in trace elements of the alimentary products consumed outside of Italy are similar to those determined by us for our Country, it seems that one can reasonably conclude that - for the Countries and elements taken into consideration - there are no particular motives of preoccupation, but for a possible insufficiency of manganese and of copper in the diets of the wealthier classes in some Countries, as a consequence of a tendency of these classes to consume smaller quantities of derivatives of cereals (and in excessively refined qualities) vs. greater quantities of meat.

REFERENCES

1. G.C. Santoprete, Rivista di Merceologia, 17 (1978), 431
2. G.C. Santoprete, Rivista di Merceologia, 18 (1979), 149
3. ISTAT, Annuario Statistico Italiano, Roma
4. FAO, Provisional Food Balance Sheets (1972 - 1974 average), Roma, 1977
5. FAO/WHO, Sixteenth Report of the Joint FAO/WHO Expert Committee on Food Additives, 1972
6. E.J. Underwood, Trace Elements in Human and Animal Nutrition, Academic Press, New York and London, 1971, p. 2
7. Proceedings of WAAP/IBP International Symposium, Aberdeen (Scotland), July 1969, Trace Metabolism in Animals, Edimburgh and London, 1970
8. A.S. Prasad et al., Archives of Internal Medecine, 111 (1963), 407
9. W. Ciusa, M. Giaccio, F. Di Donato, L. Lucianetti, Boll. Laboratorio Chimico Prov., XXIII (1972), 580

4 Chemical Reactions and Interactions

Carbonisation and Caramelisation Measurements by Near Infrared Reflectance

M. MEURENS and M. VANBELLE

Laboratory of Biochemistry of Nutrition, Louvain University, B-1348 Louvain-la-Neuve. Supported by I.R.S.I.A., Rue de Crayer, 6   B - 1050 Bruxelles (Belgium)

SUMMARY

   Carbonisation and caramelisation are overheating effects produced during the industrial drying of sugar beet pulps. All reflectance measurements made on ground pulps packed in sample cells were recorded as log (I/R) values, where R is the sample reflectance at each wavelength from 1100 to 2500 nm. The spectral curves are composite and represent optical density summations of the absorption caused by the various components present in the samples. These components such as the carbonisation and caramelisation products were not directly measured but prediction equations were calculated from samples with known chemical or physical parameters. A special sample preparation procedure has been developed. Instead of analysing lots of blindly collected samples, we have prepared a limited number of standard mixtures with different proportions of overheated and unheated representative dried pulps. In a model system part of the pulp was carbonised by flame action and part caramelised by oven drying. Mathematical treatments were applied to the data in conjunction with multiple linear regression to find according to a least square criterion the best equation to relate the reflectance data to the percentages of incorporated carbonised and caramelised pulps. The carbonisation and caramelisation levels of unknown sugar beet pulps samples can now be measured within a few minutes of the basis near infrared reflectance spectrum and prediction equations.

INTRODUCTION

The recent introduction of quantitative near infrared analysis provides interesting characteristics in many applications. In general, it has been shown that these techniques allow unskilled workers to make measurements equal in accuracy and precision to the reference chemical methods. The advantage of using near infrared reflectance spectroscopy is that these measurements take only a few seconds without requiring special sample preparation or destroying the sample. This paper provides a new application of this technology: the measurement of carbonisation and caramelisation on dry sugar beet pulps.

The pulps are the fibrous residues of the beet after sucrose diffusion in the beet sugar factories. Wet, pressed or dried sugar beet pulps serve as an important base product for cattle feeding in Europa. If the energy consumption is not considered to be too expensive, the pressed pulps coming from the diffusion step are industrialy dried by means of large tubular rotary air dryers and commercialy distributed in the practical form of pellets.

Carbonisation is the surface blackening of the pulps due to the very high temperatures of the flame surrounding gases in the beginning of the dryer. Caramelisation is the browning in depth due to the stay of the pulps during a few minutes in an atmosphere whose temperature is between three hundred, and one hundred degrees in the last part of the dryer.

MATERIALS

Studies on the influence of carbonisation and caramelisation effects of the industrial drying on the nutritional value of dry pulps for cattle feeding have been indertaken with a research composition analyser NEOTEC 6350. This instrument is a multipurpose spectro/computer wich we used for this work only in reflectance mode of operation and in the near infrared part of the light spectrum.

The dried pulps were ground with a cyclone mill CYCLOTEC of the Swedich TECATOR Company. The ground samples were packed in a special three parts cell with quartz window. The samples were illuminated through the quartz window. In the monochromator the light goes from a 100 Watt tungsten halogen lamp to the sample cell through lenses and a rotary infrared filter and is reflected on a concave holographic oscillating grating mirror. The monochromatic radiation illuminating the sample cell is diffusely reflected and collected by four lead sulfide detectors equaly spaced around the incident beam.

Coupled to the spectrophotometer, a NOVA IV mini-computer of DATA GENERAL worked with NEOTEC programs wich assess video display of spectra and execution of mathematical operations on the optical data and the corresponding laboratory informations (analytical results).

METHODS

The most common example for use of near infrared reflectance in quantitative measurements is the moisture determination in a product.

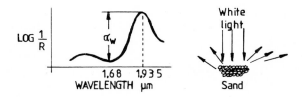

Figure 1. Near infrared spectrum of light reflected by wet sand. More energy light is reflected at 1.68 micrometer than at 1.935 micrometer because the water is absorbing the light at 1.935 micrometer.

The figure 1 presents the well known near infrared absorption curve for water. Light is strongly absorbed at thousand nine hundred thirty five nanometers; conversely at thousand six hundred eigthy, this measurement is rather insensitive to the presence of water. Thus, by measuring the relative light absorption at these two wavelenghts, we can obtain a direct measure of water content. For example, if it is desired to measure the percentage of moisture in beach sand, it could be accomplished as follow: % water = $K_o + K_1 \alpha_w$ where K_o and K_1 are specific proportional constants for the material being measured and α_w is the ratio of the light energy reflected at the absorption point (1.935 μm) to the light energy reflected at the reference point (1.680 μm).

On a chemically complex product - for example soybeans - where oil and protein affect moisture measurements the equation : $K_o + K_1 \alpha_w$ wil not give accurate results.

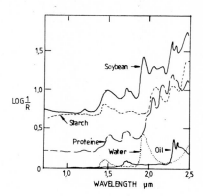

Figure 2. Near infrared absorption curves of soybean and his constituents.

By reviewing the figure 2, it can be easily seen that oil, protein, starch all have influences at the moisture measurement wavelengths (1.935 μm). The soybean absorption curve is composite and represent the summation of the absorption of all the sample constituents at each wavelength. In this case, % water = $K_o + K_1 \alpha_w + K_2 \alpha_{oil} + K_3 \alpha_{pro} \ldots$.

The typical method used to determine K values is to obtain a representative number of accurately analysed calibration samples, usualy more than 50. The optical proporties of these samples are then measured (i. e., α_w, α_{oil}, α_{pro}) and the K values calculated using regression techniques in a digital computer. Once the K's are determined, the near infrared spectro/computer system can then directly measure the constituent percentages (persent moisture, oil and protein).

A special preparation procedure for the calibration samples has been developed by us in order to measure directly carbonisation and caramelisation on dry sugar beet pulps. Instead of analyzing lots of blindly collected samples like in the above-mentioned analytical method, in our new synthetic procedure we made the samples by mixing different proportions of overheated and unheated representative dried pulps.

In a modelisation system, a part of the basical overheated pulps, the carbonised pulps were totaly blackened by a flame treatment and the other part, the caramelised pulps were browned by an oven treatment during one hour at two hundred degrees. With our synthetic preparation procedure there is no difficulty to have an uniform distribution of the calibration samples throughout the range of constituent variance.

RESULTS

The light absorption increases with the level of carbonised pulps in the calibration samples such as indicated by the figure 3.

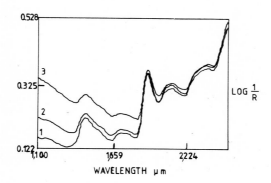

Figure 3. Near infrared absorption curves of 3 samples of one beet pulps lot. Curve 1 = 0 % carbonised pulps; curve 2 = 2 % carbonised pulps; curve 3 = 6 % carbonised pulps.

The light absorption increases also with the level of caramelised pulps in the calibration samples but on a smaller part of the near infrared spectrum such as indicated by the figure 4.

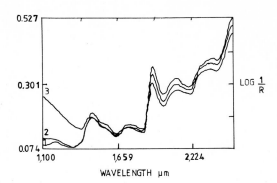

Figure 4. Near infrared absorption curves of 3 samples of one beet pulps lot. Curve 1 = 0 % caramelised pulps; curve 2 = 10 % caramelised pulps; curve 3 = 100 % caramelised pulps.

First derivative transformation of the near infrared spectra was choosen as the algorithm wich allows the best quantitative analysis of the beet pulps. Multiple linear regression was applied to the synthetic and optical data to find the best equations to relate the reflectance results to the percentage of carbonisation and caramelisation. Even now, these equations supply a very good accuracy and precision to the spectro/computer system in runtine measurements of carbonisation and caramelisation on beet pulps samples of the Belgian market.

CONCLUSION

By the aid of the near infrared reflectance and a powerful spectro computer system we can evaluate the carbonisation and caramelisation effects of the heat drying on the composition of the sugar beet pulps. It is obvious that similar application of the near infrared quantitative analysis can now be directly undertaken in the measurement of heat effects on other food products as for example malt, cacao, potato chips, milk powder, ...

REFERENCES

R.D. ROSENTHAL, Neotec Instruments, Inc. (1978), An introduction to near infrared quantitative analysis.
I. LANDA, Neotec Instruments, Inc. (1978), Model 6100 Fast Scan High Energy Spectrophotometer for the range of 0.25 to 2.40 μm.

Importance and Quality of Lipoproteins from Wheat Flour

H. STACHELBERGER and E. SCHÖNWALD

Institute for Applied Botanics, Technical Microscopy and Organic Raw Materials, Technical University of Vienna, Austria

SUMMARY

The removal of wheat flour lipids by extraction with petroleum ether (b.p. 40-60°) or chloroform-methanol (1:1, v/v) has a marked influence on the glutenin subunit distribution obtained after sodium dodecyl sulfate-polyacrylamide gel electrophoresis (SDS-PAGE). There are qualitative and quantitative differences among the electrophoretic patterns. First of all the decreasing portion of high-molecular subunits (103,000, 120,000, 140,000 mol wt.) as a consequence of delipidation has to be mentioned thus indicating that the interactions between the low-molecular flour lipids (plain fatty substances, proteolipids) and the wheat storage proteins leading to the high-molecular lipoprotein complex of gluten are an essential precondition for extracting gluten proteins to the largest possible extent when tenside-containing extraction media such as acetic acid-urea-hexadecyltrimethylammonium bromide (AUC) shall be used. To clarify the question if the removal of perhaps originally existing high-molecular lipoproteins during delipidation could also have contributed to the decreasing yield of high-molecular glutenin subunits the protein fractions of the petroleum ether extract have been isolated and characterized by their component and subunit distributions obtained after SDS-PAGE as well as their amino acid compositions. The two fractions obtained consisted predominantly of subunits of 16,500 and 29,000 mol wt. respectively, subunits of 117,000 mol wt. being present only in one fraction to a very small extent; a direct influence on the yield of high-molecular glutenin subunits should therefore be negligible. On the basis of their high lysine, arginine, and cystine contents a close relationship of the petroleum ether-soluble protein fractions to the purothionins has been established.

INTRODUCTION

The dependence of the functional properties of gluten proteins on intra- and intermolecular disulfide bonds as well as on interaction phenomena due to aggregation and association on the basis of polar and hydrophobic bonding is undisputed. Nevertheless attempts to explain rheological behaviour of doughs by means of chemical parameters and therefore to predict the technological properties of different wheat flours using methods of biochemical analysis have failed so far. For the purpose of characterizing wheat flours by their glutenin subunit distribution, as proposed by ORTH & BUSHUK (1), an extent of extraction for the gluten proteins as large as possible is imperative. According to SIMMONDS & WRIGLEY (2) the extent of extraction can be increased or decreased by the presence of lipids in dependence on the composition of the extraction media. In the course of our investigations, the aim of which has to be regarded mainly as basic research on the properties of wheat proteins, it seemed therefore of interest to evaluate the influence of the free lipids of wheat flour on the subunit distribution of glutenins.

MATERIALS AND METHODS

Glutenins were prepared from a brand of commercial available wheat flour (Anker W 700, Vonwiller & Schöller KG, Austria) by the alcohol-pH precipitation method of ORTH & BUSHUK (3) using the AUC-extraction medium (0.1M acetic acid, 3.0M urea, 0,01M hexadecyltrimethylammonium bromide) of MEREDITH & WREN (4). Defatting of flour samples was carried out by exhaustive extraction with petroleum ether (b.p. 40-60°) and chloroform-methanol (1:1, v/v) respectively. It should be stated here that the formation of a gluten ball out of defatted flours has been practically impossible. In this case the flour samples were treated directly with the AUC-solvent removing unsoluble components like starch etc. in a second step by centrifugation. For the isolation of petroleum ether-soluble proteins a fractionation procedure given by HOSENEY, POMERANZ & FINNEY (5) was used. SDS-PAGE of reduced glutenins as well as unreduced and reduced petroleum ether-soluble proteins (2-mercaptoethanol as reducing agent) was done according to KHAN & BUSHUK (6). Densitometry of electropherograms was carried out by photographing the stained gels and measuring the diffuse densities of the electrophoretic bands with a Beckman Analytrol densitometer. Amino acids were analyzed as n-propyl-N-acetyl esters in a Perkin-Elmer GC 900 gas-chromatograph (7). Cystine was estimated by amperometric titration (8, 9).

RESULTS AND DISCUSSION

Comparing the glutenin subunit distributions of untreated and defatted flour qualitative differences between the electrophoretic patterns as a consequence of delipidation can be observed. These differences are expressed mainly by the occurrence of additional bands in the middle-range of the molecular weights, as this can be clearly seen from Figure 1.

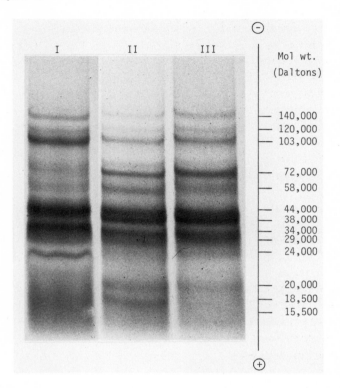

Figure 1. Subunit distributions of glutenins from untreated (I), petroleum ether (II), and chloroform-methanol defatted (III) flour

Furthermore there is a marked decrease of the intensities of the bands of the high-molecular subunits (103,000, 120,000, 140,000 mol wt.) (Table 1) indicating a decreasing yield in the extraction of high-molecular protein dependent on the extent of delipidation; extractable protein amounted to 4.9 mg per 100 g flour with petroleum ether and 32.5 mg per 100 g flour with chloroform-methanol. It should therefore be concluded that the interaction products between the low-molecular flour lipids (plain fatty substances, proteolipids) and the wheat storage proteins are highly essential for dissolving or dispersing gluten proteins in tenside-containing extractants like the AUC-solvent. This conclusion agrees very well with the findings of KOBREHEL & BUSHUK (10) according to which it is possible to dissolve glutenins completely in aqueous solutions of sodium stearate, thus indicating that the insolubility of wheat glutenin in aqueous solvents is caused mainly by hydrophobic interactions.

Table 1. Intensity of the electrophoretic bands (area-%) of wheat glutenin subunit distributions obtained after SDS-PAGE (I untreated flour, II petroleum ether, and III chloroform-methanol defatted flour)

Mol wt. (Daltons)	Intensity of electrophoretic bands		
	I	II	III
140,000	3.2	2.3	1.4
120,000	3.1	2.2	1.1
103,000	8.8	4.2	3.5
72,000	3.2	8.8	9.0
58,000	3.7	6.8	4.9
44,000	27.7	22.2	18.9
38,000		17.9	30,7
34,000	20.7	22.8	25.2
29,000	10.4	2.8	2.1
24,000	4.6		
20,000		5.0	3.2
18,500	3.2	5.0	
15,500	11.4		

Considering this dependence it seemed of utmost interest if there could be high-molecular lipoproteins originally existing in wheat flour the removal of which in the course of delipidation could also contribute to the decrease of the yield of high-molecular subunits. This rather sensible assumption was supported by findings of BEKES (11) who has been able to detect components of 116,000 and 130,000 mol wt. in petroleum ether-soluble lipoprotein fractions of wheat flour. To clarify this question we have isolated two protein fractions from the petroleum ether extract following the procedure of HOSENEY et al. (5). The amino acid composition of these fractions shows up (Table 2) a close relationship to the purothionins already described by BALLS & HALE (12) on the basis of their high lysine, arginine, and cystine contents. The high proline content of fraction II and the rather low protein portions of fraction I and II (27.6 % and 22.4 % respectively) seem unusual but it should be kept in mind that data concerning purothionins often are very controversial. The close relationship to purothionin is further supported by the complete absence of cysteine (13); according to BALLS, HALE & HARRIS (14) purothionin represents the oxidized form of a strong redox system and its presence in the disulfide form implies its properties as inhibitor against papain.

The investigation of fractions I and II by means of SDS-PAGE shows up differences in the molecular weight range of their components as well as in their subunit distributions after reduction with 2-mercaptoethanol. As a matter of fact fraction I

Table 2. Portions of some selected amino acids (g amino acid per 100 g protein) in proteins isolated from wheat flour

	Glutenin	Gliadin	Fraction I	Fraction II
Proline	8.8	14.9	3.2	20.9
Glutamic acid	23.6	38.3	7.2	1.5
Lysine	1.8	0.3	3.5	4.9
Arginine	4.2	0.6	21.3	13.7
Histidine	1.8	0.3	6.5	1.7
Cystine	2.1	1.8	14.1	20.0

contains subunits of 117,000 and 80,000 mol wt. The comparison between component and subunit distribution allows the conclusion that fraction I consists mainly of subunits of 16,500 mol wt. and fraction II of subunits of 29,000 mol wt. linked to higher-molecular compounds by intermolecular disulfide bonds (Table 3). A ratio between

Table 3. Comparison between component (1) and subunit distribution (2) of lipoprotein fractions I and II obtained after SDS-PAGE (intensity of electrophoretic bands expressed as area-%)

Mol wt. (Daltons)	Intensity of electrophoretic bands			
	Fraction I		Fraction II	
	1)	2)	1)	2)
117,000	1.8	1.2		
80,000	2.5	2.6		
56,000			5.0	
42,000			80.3	0.5
41,000	9.9	7.6		
34,000	7.2	13.2		
29,000				86.6
24,000	56.0	14.4		
22,000			13.6	12.9
16,500	17.4	59.8		
13,500			1.1	

intra- and intermolecular disulfide bonds cannot be derived from these results. The question at issue if the removal of high-molecular petroleum ether-soluble lipoproteins originally existing in wheat flour contributes to the decrease of the yield of high-molecular protein in the course of AUC-extraction has to be answered in the way that this sort of influence should be negligible since the high-molecular subunits of the petroleum ether-soluble fraction I represent minute portions within a protein fraction which itself amounts to only 4.9 mg per 100 g of flour. The conclusion is therefore affirmed that the interaction products between the free flour lipids and the wheat storage proteins in the course of gluten formation are mainly responsible for the extremely high extent of protein extraction in the case that tenside-containing extractants are used.

REFERENCES

1. R.A. Orth and W. Bushuk, Cereal Chem. 50 (1973), 191
2. D.H. Simmonds and C.W. Wrigley, Cereal Chem. 49 (1972), 317
3. R.A. Orth and W. Bushuk, Cereal Chem. 50 (1973), 106
4. O.B. Meredith and J.J. Wren, Cereal Chem. 43 (1966), 169
5. R.C. Hoseney, Y. Pomeranz and K.F. Finney, Cereal Chem. 47 (1970), 153
6. K. Khan and W. Bushuk, Cereal Chem. 54 (1977), 588
7. R.F. Adams, J. Chromatog. 95 (1974), 182
8. R. Hamm and K. Hofmann, Z. Lebensm.-Unters. Forsch. 130 (1966), 133
9. H. Stachelberger, Lebensm.-Wiss. u. -Technol. 11 (1978), 45
10. K. Kobrehel and W. Bushuk, Cereal Chem. 54 (1977), 833
11. F. Békés, Acta Alimentaria 6 (1977), 39
12. A.K. Balls and W.S. Hale, Cereal Chem. 17 (1940), 243
13. L.S. Stuart and T.H. Harris, Cereal Chem. 19 (1942), 288
14. A.K. Balls, W.S. Hale and T.H. Harris, Cereal Chem. 19 (1942), 279

Comparison of the Antioxidative Activity of Maillard and Caramelisation Reaction Products

A. HUYGHEBAERT, L. VANDEWALLE, G. VAN LANDSCHOOT

Laboratory of Food Chemistry and -microbiology, Faculty of Agricultural Sciences, State University of Ghent, Coupure 533, B-9000 Ghent, Belgium

SUMMARY

The antioxidant activity of nonenzymatic browning in sugar and sugar-amino acid mixtures at various stages of the reaction was determined. Maillard reaction products showed a high activity in ascorbic acid degradation and fat oxidation, while caramelization is only active in fat oxidation. It is demonstrated that especially colourless products, with a low molecular weight act as antioxidants.

1. INTRODUCTION

The brown discoloration due to nonenzymatic browning is the most evident result of interactions in sugar and in sugar amine systems. This type of browning is reviewed in a number of papers (1-4). From a food point of view, it is very important that browning reaction products can positively influence the stability of oxidation sensitive compounds. The formation of antioxidative substances from carbonyl compounds and amino acids is well known (5-10) but little progress has been made in the structure determination of these substances. In this work the antioxidative properties of Maillard and caramelization reaction products are compared in different systems composed of different sugars and glutamic acid. In order to further identify, the antioxidative products are separated according to their molecular weight by ultrafiltration and columnchromatography.

2. EXPERIMENTAL

2.1. Sample preparation

Maillard reaction products are obtained by heating 0,5 M sugar and 0,5 M glutamic acid at pH 7 and 100° C for different periods of time. A solution, heated for 120 minutes is separated on Sephadex G 10. Caramelization products are obtained by heating 1 M sugar at pH 7 and 100° C.

2.2. Oxidation stability tests

10 mg ascorbic acid is dissolved in 20 ml KH_2PO_4 / Na_2HPO_4 buffer (pH 7), with 25 ppm Cu^{++}-ions added as a catalyst. The ascorbic acid degradation was followed at 22° C upon addition of 5 ml of the browning reaction products.
10 ml reaction solutions are concentrated under vacuum to a viscous solution and extracted with 2 x 10 ml acetone. The extracts are blended with fat and acetone is evaporated under vacuum. TBA-values and POV-values were determined upon storage at 50° C.

2.3. Methods

The following general methods are applied : TBA-value, Sidwell et al. (11); POV-value, A.O.A.C. method (12); Ascorbic acid determination, A.O.A.C. method (13). Maillard reaction products are fractionated on a column of 20 cm X 2 cm packed with Sephadex G 10.
Ultrafiltration conditions : in a filtration apparatus (Model 202 Amicon Corporation) 180 ml Maillard reaction solution, heated for 120 minutes is filtered on diaflo ultrafiltration membranes (PM 10 and DM 5).
Determination of Amadori compound : the dried Amadori compounds are treated with hydroxylaminehydrochloride and the resulting oximes are converted to trimethylsilyl ethers by the addition of silylating reagents. These derivatives are determined with high resolution gas liquid chromatography by using a 25 meter column (ID ; 0,3 mm), packed with OV-1.
HMF is determined spectrometrically at 284 nm; the brown pigments are measured at 420 nm.

3. RESULTS

3.1. Ascorbic acid systems

In the first system the antioxidative activity of the Maillard and the caramelization reaction is compared. Browning reaction products, obtained by heating for several periods of time, were added to an ascorbic acid solution. The ascorbic acid content was estimated periodically. Generally degradation reactions like vitamines are first order reactions (14); the reaction constant was calculated.

The antioxidative activity is expressed as k_B/k_A : the ratio of the reactionconstant of the reference k_B (vitamine C solution) and the reactionconstant of the sample k_A (vitamine C solution and browning products). In figure 1 relative reactionconstants for different solutions at different heating periods are given.

Figure 1

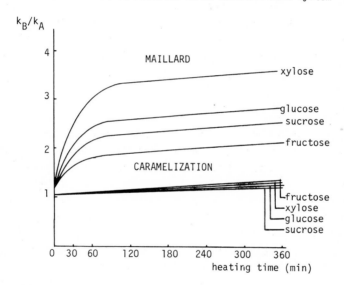

Comparative antioxidation activity of Maillard- and caramilization reaction in an ascorbic acid system

From this results it is clear that caramelization reaction products have a very poor or no antioxidative activity. Maillard products on the other hand show a well pronounced activity. Especially in the initial heating period there is an increase in the activity with a rather limited rise with longer reaction times. There is a clear distinction between the activity of sugars in the Maillard reaction. Xylose is the most reactive sugar ; it is very well known that xylose reacts to a greater extend in browning than hexoses and disaccharides ; this reactivity is also reflected in the antioxidative properties. Glucose shows an intermediate activity, while the reaction of sucrose is due to hydrolysis ; this sugar has an activity somewhat between glucose and fructose. Rather unexpected fructose gives the lowest effect although fructose is very sensitive to browning. On the other hand it is known that upon heating fructose easily gives 5-hydroxymethylfuraldehyde, some difructosides and similar compounds which may not have the same effect on ascorbic degradation.

The chelate formation with metal ions (in this case Cu^{++}) by initial stage reaction products has been proposed as the active principle for those antioxidative effects, specially the Schiff'sbase of a sugar-amine browning system is very active (15). Another possibility proposed by El-Zeany et al. (15) and Laing et al. (16) is the interaction with peroxides. Indeed ascorbic degradation gives dehydroascorbic acid and H_2O_2 ; this H_2O_2 reacts further with ascorbic acid or another compound in our case brown products. The competitive effect of ascorbic acid and of brown products for H_2O_2 could explain some activity. Indeed the bleaching effect of H_2O_2 was easily observed. This competitive effect can only apply for the limited effect at longer heating periods.

3.2. Fractionation by ultrafiltration and gel chromatography

In an attempt to further identify the active principles, two fractionation methods have been applied, ultrafiltration and Sephadex separations.
The reaction mixtures was separated by ultrafiltration. In a first filtration molecules superior to 10.000 moleculair weight are filtered off, and in a second filtration the molecules superior to 5.000. The antioxidative activity of both fractions was tested in an ascorbic acid solution. There is no decrease in antioxidative activity after removal of molecules superior to a molecular weight of 10.000 or 5.000. As expected low moleculair weight fractions showed the highest activity.

The results of the fractionation on a Sephadex G 10 are illustrated in figure 2 ; the relative degradation constant of ascorbic acid, the amount of brown pigments, the content of 5-hydroxymethylfuraldehyde and the amount of Amadori compounds are represented as a function of the elution pattern of a heated glucose-glutamic acid solution.
It is concluded that here is a very close relationship between antioxidation activity as measured by ascorbic acid degradation and the amount of Amadori compounds in Maillard type reaction.

Figure 2

Sephadex separation of Maillard reaction products and antioxidative activity

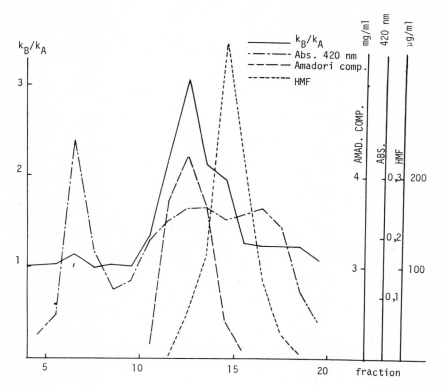

3.3. Lipid systems

In a second series of tests, acetone extracts of both browning reactions have been added to butterfat. The activity of these acetone extracts has been compared in a fat stability test at 50° C. Oxydation has been expressed in terms of peroxyde- and TBA-values.

Some results are given in figure 3 ; unlike the first system caramelization products also showed a good inhibitory activity in this lipid system. In this figure peroxide values are plotted after a given incubation period of the fat.

It can be seen that in this particular case the reference was heavily oxidized ; upon addition of reaction mixtures to the fat the oxidation rate decreases as well for caramelization products as for Maillard products. A similar pattern was obtained when oxidation was expressed in terms of TBA-values, as measured at 530 nm. The antioxidative properties increase sharply in the first stage with a reduced color formation ; the effects become less important in the further stages where a pronounced browning is observed. In general the oxidative stability was better for Maillard than for caramilization products.

Figure 3

Peroxyde value of fats with acetone extracts of caramelization after 21 days storage at 50° C

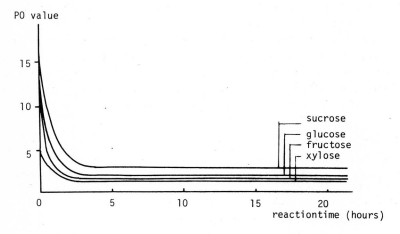

Peroxyde value of fats with acetone extracts of Maillard reaction after 24 days storage at 50° C

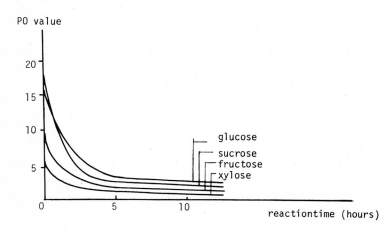

Also in this system the activity of Maillard products has been attributed to initial enediol (Maillard and caramelization) and enaminol (Maillard) structures, particularly in lipids (17). These intermediates act as free radical acceptors in the oxidizing system.

4. CONCLUSION

By heating sugar or sugar-amino solutions antioxidant active substances are obtained. These components are formed in the initial phase of the Maillard reaction. An increased oxidation stability has been observed before any visible discoloration. This means that especially products formed in the initial phase of the browning, i.e. the colourless products, act as antioxidants.

The antioxidant activity of caramelization reaction products depend on the system studied ; these compounds have a very low activity in the ascorbic acid degradation but have approximately the same activity as Maillard reaction products in a fat system. By ultrafiltration and column chromatography, it is demonstrated that especially reaction products with a low molecular weight show the highest activity.

REFERENCES

1. W. Baltes, Ernährungs-Umschau 20 (1973), 35
2. N.A.M. Eskin, H.M. Henderson and R.J. Townsend, Biology of Foods, Academic Press New York and London (1971), 69
3. T.M. Reynolds, Adv. in Food Research 12 (1963), 1
4. T.M. Reynolds, Adv. in Food Research 14 (1965), 167
5. T. Griffith and J.A. Johnson, J.A. Cereal Chemistry 34 (1957), 159
6. C. Hwang and D. Kim, Korean J. Food Sci. Technol. 5 (1973), 84
7. K. Kawashima, H. Itoh and I. Chibata, J. Agric. Food Chem. 25 (1977), 202
8. N. Kirigaya, H. Kato and M. Fujimaki, Agr. Biol. Chem. 32 (1968), 287
9. M. Morita, T. Aonuma and N. Inaba, Agr. Biol. Chem. 40 (1976), 2491
10. J. Velisek, J. Davidek, B. El-Zeany, J. Pokorny and G. Janicek, Z. Lebnsm. Unters. Forsch. 154 (1974), 151
11. C.G. Sidwell, H. Salwin and J.H. Mitchell, J. Am. Oil Chemists' Soc. 32 (1955),13
12. Association of Official Analytical Chemists, Washington (1975), 484
13. Association of Official Analytical Chemists, Washington (1975), 829
14. T.P. Labuza, Critical Reviews in Food Technology 3 (1972), 217
15. B.A. Elzeany, J. Pokorny, J. Velisek and G. Janicek, Z. Lebensm. Unters. -Forsch. 153 (1973), 316
16. B.M. Laing, D.L. Schlueter and T.B. Labuza, J. Fd Sci. 43 (1978) 1440
17. K.S. Rhee, Y.A. Ziprin and K.C. Rhee, J. Fd Sci. 44 (1979), 1132.

Model Studies on the Heating of Food Proteins: Influence of Water on the Alterations of Proteins Caused by Thermal Processing

J.K.P. WEDER and U. SCHARF

Institut für Lebensmittelchemie, Technische Universität München, Lichtenbergstr. 4, D-8046 Garching, FRG

SUMMARY

Two pure proteins, lysozyme and ribonuclease, were heated from 1-24 h at 80-160°C with water contents of 10-30 %. Colour and consistency already indicate more severe changes than in dry-heated proteins. Weight losses up to 30 % and protein losses up to 60 % were measured. Losses in some amino acids, including some of the essential ones, rise to values as known for lysine during the Maillard reaction. Heating at 160°C with increasing heating time and water content results in the complete destruction of cysteine and in an over 90 % destruction of serine, threonine, and aspartic acid. Lysine, methionine, and isoleucine are 50-75 % damaged. TNBS-reactive lysine drops to 10 % of the original content. At the same time the formation of atypical amino acids, lysinoalanine, lanthionine, ornithine, allo-isoleucine, and, probably, α-amino butyric acid could be demonstrated. Some preliminary experiments with water contents up to 70 % indicate maximal damaging of proteins in the range of 30-50 % water content. Furthermore, wet heating results in considerable changes in molecular weights. Crosslinking and splitting occur simultaneously. At low temperatures and/or shorter heating times crosslinking predominates incorporating hydrolysis products into the network. In the beginning crosslinking is due to formation of disulphide bridges by oxidation and/or disulphide interchange while, later on, crosslinks are detectable which are stable to reduction. They were identified as lysinoalanine, lanthionine, and N^{ϵ}-(ß-L-aspartyl)-L-lysine with a maximum content of 1.45, 0.63, and 1.29 residues per mole of monomer, respectively. In contrast to this, at higher temperatures and/or longer heating times hydrolysis reactions predominate. Molecular weights decrease with temperature, time, and water content. Finally, up to 10 % of the protein content estimated by amino acid analysis consists of free amino acids.

Foods are repeatedly heated at various temperatures for different times during processing and preparation. Some papers deal with the alterations in proteins in complex foods during these processes, but they do not permit generally valid predictions. We have therefore attempted to compose a model food stepwise, starting with pure well-known proteins in order to study the influence of each component on the behaviour of this composite food during the heating process. Out of this complex only the protein/amino acid plus carbohydrate system (the Maillard reaction) has been studied very frequently and nevertheless even this problem has not yet been fully clarified.

After having studied the alterations in selected model proteins during dry heating [1], we investigated the influence of water content on these alterations with the aid of two proteins, lysozyme (LSH) and ribonuclease (RNase). Treatment was performed with water contents of 10, 20, and 30 % at 80, 120, and 160°C for 1, 2, 4, 8, 16, and 24 hours in sealed ampoules. Even the appearance of the water-containing samples heated indicates more severe damage (figure not given). While the dry samples become light brown within 24 h at 160°C, water-containing samples become dark brown to black within the same time. Consistency of the samples is also quite different after treatment (chewy or lava-like to dry granular). Weight losses range up to 30 %, while protein losses (calculated from amino acid analyses) reach 60 %.

	Amino acid loss (%) with water contents of			
	0 %	10 %	20 %	30 %
Isoleucine	6.5	42.2	45.5	50.0
Leucine	12.1	40.4	44.6	49.2
Lysine (TNBS)	30.5 (76)	56.0 (94)	60.7 (91)	61.8 (88)
Methionine	40.4	31.3	43.5	45.7
Phenylalanine	5.5	30.4	38.2	41.8
Threonine	20.6	93.0	95.8	95.8
Valine	13.1	28.5	35.3	35.2
Protein loss (%)	17	53	57	59

Table 1: Ribonuclease heated for 24 h at 160°C

The compilation of amino acid losses after heating at 160°C for 24 h clearly demonstrates the influence of water content (Table 1, only essential amino acids are given). The losses of water-containing samples are twice to seven times as high as in the dry-heated sample, methionine excepted, where water seems to serve as a protectant. With increasing water content losses rise slowly. Preliminary studies with higher water contents indicate maximum damage between 30 and 50 % of water. TNBS-reactive lysine (values given in brackets), an index for biologically available lysine, exhibits only minor differences compared with the dry-heated product, as it is always severely damaged after dry heating. After less severe thermal stress, e.g. after 2 h at 120°C, most of the amino acids remain unaltered. Only lysine and methionine start to be damaged, nevertheless resulting in a loss of about one third for TNBS-reactive lysine with 30 % of water. With lysozyme similar results are generally obtained.

Of the atypical amino acids discussed as indicators for damage during processing ('hot spots'), lysinoalanine, lanthionine, ornithine, and allo-isoleucine are demonstrated to be present, furthermore probably α-amino butyric acid (Table 2). At 160°C the first two compounds go through a maximum, indicating an instability under these conditions. This means that they cannot be used as certain indicators in all cases. The remaining compounds increase within the range examined.

	Max.det.contents % of heated protein	Conditions			
		H_2O	Temp.	Time	
Lysinoalanine	1.4 (RNase)	10	160	8	Maximum
Lanthionine	0.6 (LSH)	30	160	1	Maximum
Ornithine	3.5 (LSH)	30	160	24	Rising
allo-Isoleucine	2.3 (LSH)	30	160	24	Rising
α-Amino butyric acid	+ (both)	from 160		4	Rising

Table 2: Atypical amino acids ('Hot spots')

The alterations which remain undetected by the studies outlined above, that is to say alterations in molecular size, will now be treated in a little more detail. They already begin to occur after significantly lower stress and exhibit interesting transitions within the range examined.

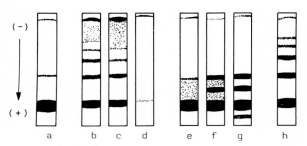

Fig. 1: SDS/PAGE of lysozyme treated for 4 h at 80°C; acrylamide 10 %, sodium phosphate buffer pH 7.2, e-g: with ME, staining with Coomassie brilliant blue R-250, 17.5 μg of a: LSH, b-g: LSH 80°C/4 h with 10 (b,e), 20 (c,f), and 30 % (d,g) water, h: DEPC-LSH

After heating at 80°C for 4 h a considerable oligomerization and polymerization can already be demonstrated by polyacrylamide gel electrophoresis with sodium dodecyl sulphate (SDS/PAGE, Figure 1). Lysozyme with 10 and 20 % of water clearly exhibits stoichiometric oligomers from the dimer up to the pentamer and a high-molecular fraction only just able to penetrate into the gel. Chemically oligomerised lysozyme (DEPC-LSH) is given as a comparison. The sample containing 30 % of water is polymerised to such a degree that only parts of it can penetrate into the gel. These oligomers and polymers are mainly linked by disulphide bonds, as can be demonstrated by SDS/PAGE after reduction (with mercaptoethanol, ME, right hand side of Fig. 1). However, oligomers that are stable against reduction can already be seen in the sample containing 20 % of water and in particular in that containing 30 %. These oligomers must be explained by another kind of crosslinking. Even here products with a molecular weight of one and a half times that of the original protein as well as minor parts with a molecular weight of less than that of lysozyme are remarkable.

A schematic representation of the reactions leading to these alterations is given in Figure 2. Thiols (1) are formed by hydrolytic cleavage of disulphide bonds from the monomer [2]. Hydrolytic cleavage of the peptide chain [3, 4] yields two polypeptides linked together only by a disulphide bridge (2), while destruction of both cysteine residues [4-6] leads to open polypeptides free of thiol groups (3). Low molecular-weight peptides can be formed by further cleavage reactions. Thiol and monomer react to form the dimer by disulphide interchange and, furthermore, to form higher stoichiometric oligomers (4). Thiol and disulphide 2 yield non-stoichiometric oligomers and new low-molecular thiols (1'). Products of the same molecular size can also be formed from this thiol and the monomer. At higher stresses other intermolecular linkages can also be formed between the same primary products, yielding mixed SS/CX-polymers (5) or even mere CX-polymers (6), respectively. These polymers can be formed even from smaller fragments, as indicated in the right hand lower part of the figure.

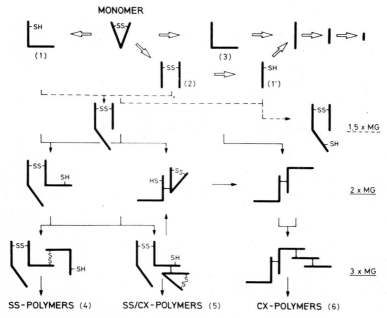

Fig. 2: Reaction pathways from monomer to polymers

Within the range following 80°C/4 h the proteins form polymers with molecular weights of over 100 000. Nevertheless the polypeptide chains are mainly linked by disulphide bridges. A change in the kind of linkage is demonstrated in Figure 3. After 4 h at 120°C all oligomers are still detectable under reductive conditions, shifting to higher oligomers with increasing water contents. This demonstrates a considerable amount of non-disulphide reduction-stable crosslinks. After 8 h at the same temperature these oligomers can only be seen clearly in the sample containing 10 % of water, the other samples exhibiting only traces of oligomers and minor parts penetrating the gel surface. This demonstrates molecular weights of 100 000 and more for reduction-stable oligomers after such treatment.

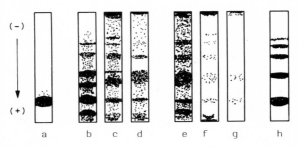

Fig. 3: SDS/PAGE with ME of lysozyme treated at 120°C as in Fig.1; a: LSH, b-d: LSH 120°C/4 h, e-g: LSH 120°C/8 h with 10 (b,e), 20 (c,f), and 30 % (d,g) water, h: DEPC-LSH

Figure 4 shows the reverse of these trends. After 4 h at 160°C products within the range of 7000 to 100 000 are again detectable. Definite protein bands are only visible with difficulty because the various, different fragmentation and linking reactions can yield all molecular weights. After 16 h molecular weights decrease below that of the monomer.

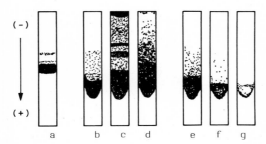

Fig. 4: SDS/PAGE with ME of lysozyme treated at 160°C as in Fig.1; a: LSH, b-d: LSH 160°C/4 h, e-g: LSH 160°C/16 h with 10 (b,e), 20 (c,f), and 30 % (d,g) water

Parallel to this, free amino acids can be demonstrated in these samples. They can already be detected in traces in ribonuclease containing 30 % of water after 24 h at 80°C (Table 3) and increase with increasing stress. With lower water contents they are only detected, as expected, at higher stresses. Maximum content of free amino acids is about 10 % of the protein content estimated by amino acid analysis in the sample containing 30 % of water and heated for 24 h at 160°C.

Temp.-Time	80-24	120-8	120-24	160-2	160-24
30 % H_2O	Traces	0.4	1.2	1.5	4.3
20 % H_2O			0.5	1.1	2.5
10 % H_2O			0.3	0.3	0.9

Table 3: Total free amino acids in wet heated ribonuclease

As crosslinks that are stable against reduction, the atypical amino acids lanthionine and lysinoalanine (cf. Table 2), which are formed from dehydroalanine and cysteine or lysine, must be taken into account. Furthermore, a maximum of 1.29 residues of N^{ε}-(β-L-aspartyl)-L-lysine per mole of monomer could be demonstrated in wet heated samples by a modified amino acid analysis after enzymic total hydrolysis [7]. The corresponding N^{ε}-(γ-L-glutamyl)-L-lysine is only formed in traces. The high amounts of crosslinks, in total about three linkages per mole of monomer, are to be expected as it must be taken into account that with decreasing molecular size of the single polypeptide chains the amount of crosslinks per original monomer must of course increase.

Summarizing the results concerning the changes in molecular weights it can be stated that the reactions which take place tend in two directions within the range examined. On the one hand oligomerization occurs, with different reactions dominating according to stress conditions. On the other hand fragmentation occurs. While oligomerization predominates at first, with increasing stress it is fragmentation which does so. Conclusions from these results should be taken into account for relevant procedures, e.g. for steaming legume meals in order to inactivate antinutritional factors or for texturizing protein dispersions.

REFERENCES

1 J.K.P. Weder and S. Sohns, Mitteilungsbl. GDCh-Fachgruppe Lebensmittelchem. Gerichtl.Chem. 30 (1976), 9
2 A. Schöberl and H. Eck, Justus Liebigs Ann.Chem. 522 (1936), 97
3 A.Patchornik, M. Sokolovsky and T. Sadeh, Proc. 5th Int. Congr.Biochem., Moscow 1961, Vol. 9 (1963), 69
4 J. Bjarnason and K.J. Carpenter, Br.J.Nutr. 24 (1970), 313
5 M. Fujimaki, S. Kato and T. Kurata, Agric.Biol.Chem. 33 (1969), 1144
6 J. Kisza, Z.Zbikowski and P. Przybylowski, Z.Ernaehrungswiss. 10 (1970), 115
7 J.K.P. Weder and U. Scharf, Z.Lebensm.Unters.Forsch. 172 (1981), 9

Fractionation and Sensory Evaluation of the Reaction Products of Reducing Sugars and Their Degradation Products with L-Lysine

J. DAVÍDEK, J. POKORNÝ, H. BULANTOVÁ, A. MARCÍN, J. PAVLIŠ AND G. JANÍČEK

Department of Food Chemistry, Prague Institute of Chemical Technology, CS-166 28 Prague, Czechoslovakia

SUMMARY

Reducing sugars /glucose, fructose, galactose, lactose/ were heated with L-lysine monohydrochloride either in dry state or in buffered solutions to temperatures between 80 and 140 °C. Reactions with sugar degradation products /D-arabino-hexosulose and 2-furaldehyde/ were studied under analogous conditions. Reaction mixtures and their fractions were evaluated by sensory profile analysis. The sensory profiles were affected both by temperature and heating time, and by concentration of reactants, changing from baked, caramel odours to rancid and burnt odours. The reactions of lysine with sugar degradation products were more rapid than those with reducing sugars, and fruity, rancid, and burnt odour notes caused by Strecker degradation were more intensive. The reaction products were fractionated by gel chromatography on Sephadex LH-20 and by liquid partition chromatography on silica gel. The evaluation of separate fractions instead of the total reaction mixture increased the sensitivity and accuracy of sensory analysis.

INTRODUCTION

The aroma of Maillard products is due to various precursors, e. g. deoxyhexosulose, hexosulosene or 2-furaldehyde and its derivatives. In presence of oxygen various other compounds may be formed and act as precursors of flavour compounds, e. g. hexosulose /1, 2/. The contribution of various possible precursors may be estimated by model experiments with mixtures containing pure sugar degradation products.

Maillard reactions result in very complicated mixtures of substances the aroma of which is difficult to evaluate sensorically. It may be easier evaluated after prefractionation into simpler mixtures of odour active substances.

EXPERIMENTAL

The reactants were heated in dry mixtures or dissolved in 0.2 M sodium acetate buffer /pH = 5.6/, and 0.5 M to saturated solutions were heated to the reaction temperature in a thermostated water bath. The sensory /odour/ profiles of the products were determined by intensity scoring partial odour notes /3/. A 30-member panel, 28 or 10 odour descriptors, and a 9-point intensity scale were used /contribution of a partial odour note to the overall odour: 1 = absent, 2 = detectable only in traces, 3 = slightly modifying the overall odour, 4 = distinctly modifying the overall odour, 5 = present as an important minor note, 6 = present as a main odour note, 7 = the most important odour note, the prevailing odour note was scored 8, and 9 = the sole odour not present/.

The reaction mixture was fractionated by gel chromatography using a 1 m x 10 mm or 1 m x 50 mm columns packed with Sephadex LH-20^R /Pharmacia, Uppsala, Sweden/ and 0.2 M sodium acetate buffer as a solvent; flow-rate 83 $\mu l \cdot s^{-1}$; UV detector at 254 and at 440 nm. For the subfractionation, Sephadex G-10 and G-15 /Pharmacia, Uppsala/ and silica gel Silpearl^R /Kavalier, Votice, Czechoslovakia/ and 600 mm x 25 mm columns were used, flow-rate 35 $\mu l \cdot s^{-1}$.

Reaction mixtures were characterized by means of their UV and visible spectra, determination of residual lysine /enzymic method/, and by IR, UV, and NMR spectra of chloroform extracts.

RESULTS AND DISCUSSION

When D-fructose or other reducing sugars were heated with L-lysine to 100 °C for 1 h, the odour was nearly imperceptible if the reaction was carried out in dry state, and only after the reaction dissolved in the buffer. Even at higher temperatures /up to 130 °C/ solutions with a very weak odour were obtained in this way, in spite of intensive coloration of the mixture and low residual lysine content.

The odour intensity was distinctly greater when sugars were heated with L-lysine in buffer solutions but even then the aroma was weak if the reaction temperature was 100 °C and the reaction time shorter than 2 h. On the contrary sugar degradation products /D-arabino-hexosulose or 2-furaldehyde/, being very active even at 100 °C, produced brown mixtures where roasted, spicy, fruity, and ethereal odour notes prevailed. Sensory profiles of the last two reaction mixtures were very similar /Fig. 1/.

Fig. 1 Sensory profiles of Maillard reaction products
Reaction temperature 100 °C: ———— mixture of D-arabino-hexosulose and L-lysine; ------- mixture of 2-furaldehyde and L-lysine; reaction temperature 130 °C: ———————— mixture of D-glucose and L-lysine; ---------- mixture of 2-furaldehyde and L-lysine

Odour note

Burnt, sulphuric
Roasted, bread crust
Caramel, gingerbread
Spicy, heavy, sweet
Fruity, flowery
Ethereal, light
Acidic, pungent
Buttery, rancid
Cheese, mushroom, meat
Moldy, stale, rotten

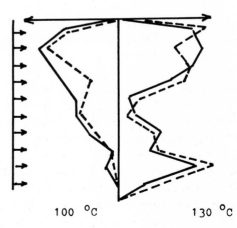

Aroma intensity

100 °C   130 °C

When sugars are heated with L-lysine to 130 °C in buffer solutions /similar results are obtained at 140 °C/ the sensory profiles of the reaction mixtures are widely different /Fig. 1/. The sensory profiles of mixtures of L-lysine with D-glucose or L-lysine with 2-furaldehyde were similar in this case /those of L-lysine heated with D-fructose or D-galactose were also nearly the same/. Roasted and caramel odours still prevailed in the mixture but odours reminding cheese, mushroom, meat, heated oil appeared contrary to lower reaction temperatures. Fruity and ethereal odours were only weak in mixtures heated to 130 °C. Odours resembling roasted material or bread crust were reported in the literature /4/. Contrary to D-glucose and even to 2-furaldehyde, D-arabino-hexosulose is extremely reactive /5/, especially as a substance possessing high Strecker-activity /6/, so that the reaction mixture had a strong burnt odour when heated to 130 °C which masked other odours; therefore, the profile is not given in Fig. 1.

It has been demonstrated in the experiments carried out in the temperature range of 100 - 130 °C that the increasing temperature stimulated the formation of burnt, roasted, meaty and moldy odours while it suppressed gingerbread, fruity and flowery notes produced mainly at lower temperatures. The effect of increasing reaction time and that of increasing concentration of reactants was similar to the effect of increasing temperature.

The sensory evaluation of mixtures subjected to Maillard reactions was found difficult because of the complexity of odour character. Less intensive odours which contribute to the fullness and pleasantness of odour and modify it, nevertheless, cannot be easily distinguished in presence of strong, intensive odour notes of roasted, caramel, and burnt character.

The chloroform-soluble substances were fractionated by gel chromatography on a Sephadex LH-20 column. The fractions were further subfractionated by liquid partition chromatography on silica gel. More polar fractions possessed odours resembling meat, cheese or mushrooms while less polar fractions possessed spicy, sweet, heavy and ethereal odour notes. The least polar fraction resembled 2-furaldehyde in its ultraviolet, infrared, and NMR spectra but differed in its colour and both odour and flavour character, it contained thus obviously a small amount of a sensorically very active substance.

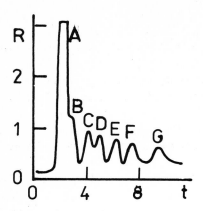

Fig. 2 Fractionation of the chloroform-insoluble fraction on Sephadex LH-20 R - detector response /at 254 nm/; t - elution time /h/; main odour notes of the respective peaks: A - mushroom, wet dog; B - sweat, roasted, buttery; C - sweat, rancid, mushroom; D - burnt, rubber, sweat, wet dog; G sulphuric, burnt, roasted

Similarly the chloroform-insoluble fraction was fractionated by gel chromatography on Sephadex LH-20. The fractionation of a mixture of 2-furaldehyde and L-lysine heated to 80 °C for 2 h is given here as an example /Fig. 2/. All fractions were analyzed by the method of sensory profiles. The fraction in the front consisted of melanoidins and had a caramel odour and flavour. The following large fraction contained unreacted L-lysine but also some ninhydrin-positive and ninhydrin-negative reaction products which imparted the fraction a mushroom odour. The series of smaller fractions which followed in the chromatogramme had odour characteristics which were not particularly agreeable /Fig. 2/ when presented separately as they mostly resembled rancid oils, rubber, sweat, wet dog or burnt products. Nevertheless, they suitably modified the odour of chloroform-soluble compounds, making it richer and fuller. Most assessors /94 %/ preferred the odour of the combined chloroform-soluble and chloroform-insoluble fractions to that of the chloroform-soluble fraction alone.

The fractions obtained by the first fractionation /Fig. 2/ could be further subfractionated by chromatography on Sephadex G-10 or G-15 /7/, each fraction being resolved into 2 - 4 subfractions, however, such small amounts of efluent were not satisfactory for the sensory evaluation, with the exception of the main L-lysine-containing fraction, which was separated into a fraction containing bitter compounds and another with a mushroom odour.

REFERENCES

1 S. Umemoto, S. Ishiie, Y. Irie and T. Imai, Nippon Nogei Kagaku Kaishi 44 /1970/, 64
2 B. M. White and R. Carubelli, Carbohydr. Res. 33 /1974/, 366
3 J. Pokorný, L. Dvořáková, A. Marcín, H. Bulantová and J. Davídek, Nahrung 23 /1979/, 921
4 J. Adrian, Ind. Alim. 90 /1973/, 559
5 J. Pokorný, N.-T. Luan, H. Bulantová and J. Davídek, Nahrung 23 /1979/, 693
6 A. Schönberg, R. Moubacher and A. Mostafa, J. Chem. Soc. /London/ /1948/, 176
7 J. Pokorný, H. Bulantová and J. Davídek, Nahrung, in press

Interactions of Artificial Sweeteners with Food Additives

G. KROYER, and J. WASHÜTTL

Institute of Food Chemistry and Food Technology, Technical University Vienna, Austria

SUMMARY

A possible interaction of the artificial sweeteners sodium cyclamate with ingredients of food, particularly with watersoluble vitamins and essential amino acids, in model-experiments and at the preparation and storage of food is indicated.

INTRODUCTION

Today the synthetic substances sodium saccharin and sodium cyclamate number among the most used artificial sweeteners. Refering to new scientific results, they caused heavy discussions about their advantage as food additives in nutrition, but also about the sanitary disadvantages by using them. In the last years there were made a lot of examinations regarding these artificial sweeteners, especially their toxity, carcinogenity, embryotoxity and mutagenity, succeeding in the prohibition or at least restriction of the use and sale of these sweeting agents in some countries. On the contrary only few studies are dealing with the interaction of artificial sweeteners with food ingredients, especially at the preparation and storage of food, although this aspect should be not unnoticed from the food chemical and nutritional point of view.

RESULTS and DISCUSSION

In our study about the interaction of artificial sweeteners with food ingredients we analysed at first in model-experiments the influence of sodium saccharin and sodium cyclamate on watersoluble vitamins and essential amino acids at room temperature and increased temperature (80°C). The natural sweetener saccharose was submitted to the same conditions for comparison. By doing all these experiments, it was taken care to use all substances in such quantities, which are appropriate to their practical application.

The comparative model-experiments with aqueos solutions of sodium saccharin, sodium cyclamate respectively saccharose and essential amino acids or watersoluble vitamins at room temperature and increased temperature (80°C) effected in most cases practically no influence of these sweeteners on the mentioned food ingredients. Remarkable losses could be proved only with phenylalanine and above all with tryptophane at the presence of sodium cyclamate after storage of some weeks at room temperature (table 1)

Table 1: %-reduction of the quantity of tryptophane and phenylalanine.

storage time	%-reduction of tryptophane			%-reduction of phenylalanine		
	s.saccharin	s.cyclamate	saccharose	s.saccharin	s.cyclamate	saccharose
1h	∅	∅	∅	∅	∅	∅
1d	∅	∅	∅	1	1	∅
5d	7	18	∅	2	4	1
10d	5	25	2	5	11	2
14d	9	42	3	10	16	3
24d	13	47	3	-	-	-
28d	-	-	-	10	17	3

Mixtures of watersoluble vitamins with sodium saccharin, sodium cyclamate and saccharose in solid form indicated no significant losses of the vitamins B_2, B_6, B_{12} and nicotinamide by heating them one hour at a temperature of 120°C and 150°C, however distinct changes of the vitamins C and B_1, as is shown in table 2. (particularly at the presence of sodium saccharin)

Table 2: %-reduction of the quantity of vitamin C and vitamin B_1.

Sample	%-reduction of vitamin	
	120°C	150°C
vitamin C-standard	∅	∅
vitamin C + sodium saccharin	14	93
vitamin C + sodium cyclamate	8	11
vitamin C + saccharose	3	9
vitamin B_1-standard	∅	∅
vitamin B_1 + sodium saccharin	6	33
vitamin B_1 + sodium cyclamate	8	7
vitamin B_1 + saccharose	4	3

The effect on the essential amino acids isoleucine, leucine, lysine, methionine, phenylalanine, theonine, tryptophane and valine using the same conditions is not as remarkable, although some changes with certain amino acids could not be overseen. (table 3).

Table 3: %-reduction of the quantity of amino acids at a temperature of 150°C

Sample	%-reduction of the quantity of the amino acids
phenylalanine + sodium cyclamate	16
phenylalanine + sodium saccharin	9
tryptophane + sodium cyclamate	10
methionine + sodium cyclamate	8

After these studies, which indicate the influence of artificial sweeteners - in comparison to natural ones - on essential food ingredients, experiments were made to examine on the other hand the effect of these food components on the stability of the artificial sweeteners themselves.

Pure sodium saccharin was hardly destroyed when heatened at 250°C for one hour, while comparatively with the addition of the appropriate quantity of watersoluble vitamins or essential amino acids, the decomposition of sodium saccharin was significantly accelerated and occured already at relatively low temperatures (about 150°C; table 4).

Table 4: %-reduction of sodium saccharin

heat-treatment	sodium saccharin pure	sodium saccharin-vitamin-mixture	sodium saccharin-amino acid-mixture
100°C	Ø	Ø	Ø
150°C	Ø	5	2
200°C	Ø	5	10
250°C	10	23	27

The thermal decomposition of sodium cyclamate was noticed already at a temperature about 200°C, and by heating for one hour to a temperature of 250°C no sodium cyclamate could be detected at all. At the same time the sample got a brown-black colour, combined with an intensive sulphurous gas generation. As a result of the addition of watersoluble vitamins or essential amino acids, sodium cyclamate was completely destroyed already at temperatures about 200°C (table 5).

In connection with the decomposition of sodium cyclamate, with the temperature rising quantities of cyclohexylamine, a toxic decomposition product of sodium cyclamate, could be noticed . (table 6)

Table 5: %-reduction of sodium cyclamate

heat treatment	sodium cyclamate pure	sodium cyclamate- vitamin-mixture	sodium cyclamate- amino acid-mixture
100°C	∅	3	3
150°C	∅	37	14
200°C	20	100	100
250°C	100	100	100

Table 6: cyclohexylamine-content of the sample in % of sodium cyclamate

heat treatment	sodium cyclamate pure	sodium cyclamate vitamin-mixture	sodium cyclamate amino acid-mixture
100°C	–	–	–
150°C	–	14	–
200°C	2	37	20
250°C	43	37	31

Although some authors cited a relatively good stability of sodium cyclamate at high temperatures, our studies indicated, that the decomposition already took place at relatively low temperatures and that moreover the beginning of the decomposition shifted even to lower temperatures at the presence of food ingredients like vitamins and amino acids, increased with the rising of the temperature and led to the complete destruction of the sweetener. At the same time the quantity of cyclohexylamine as a toxic decomposition product increased according to the disappearance of sodium cyclamate.

The human organism will hardly take up artificial sweeteners directly, furthermore primarily they will be added to foodstuffs. In this case there could take place a mutual influence of the artificial sweeteners on food ingredients, which may succeed in a changing of the biological value of the foodstuffs as well as in the sweeting intensity of the sweeteners, yet before the food is supplied to the organism.
In the last part of our study the results obtained from the above mentioned model-experiments should be proved on some foodstuffs, which we produced ourselves, in order to examine, inhowfar a noticeable interaction could take place between artificial sweeteners and essential food ingredients at the preparation and storage of these foodstuffs. The analysed food was prepared according to original recipes, edited from the producer of the artificial sweeteners, using strictly the recommended sweetening agents, which are combination products of sodium saccharin and sodium cyclamate in solid or liquid form. The same foodstuffs prepared with the appropriate quantity of saccharose respectively without sweeteners acted for comparison.

The following foodstuffs were prepared:
as representative of vitamin-containing food: cake (vitamine B_1),
jam (vitamine C)

as representative of alkaloid-containing food: coffee (coffein)
as representative of alcoholic food: punch (ethanol)
and as representative of acid-containing food: vinegar-dressing (acetic acid)
The analysis of the vitamine B_1-content of cakes, which were baked by using artificial as well as natural and also without sweeteners, brought about minimal vitamin losses of the sweetened cakes compared with the unsweetened one, but it was not possible to distinguish, whether they were sweetened artificially or naturally. The analysis of the vitamin C-content of artificial and natural sweetened jam showed, that significant higher vitamin losses occured by adding artificial sweeting agents compared to saccharose. (table 7)

Table 7: Analysis of cake and jam

sweetening agent	relative losses (%) of	
	vitamin B_1 in cake	vitamin C in cake
artificial	13	11
saccharose	11	6

In further experiments it was tested, if the coffein-content of coffee was influenced by adding artificial sweeteners. The results indicated no interaction between the investigated substances.

Regarding the stability of artificial sweeteners during the preparation of these foodstuffs, the most remarkable result was obtained by baking cakes. At this process the content of sodium saccharin was reduced to nearly one third of the original used quantity (table 8).

Table 8: Stability of artificial sweeteners

foodstuff	relative reduction of the content of	
	sodium saccharin	sodium cyclamate
	%	%
cake	64	16
jam	1	6
punch	∅	1
coffee	2	2
vinegar dressing	7	17

Because of a likewise significant loss in the sodium cyclamate content of the cake, an examination of a possible cyclohexylamine-content was carried out, which actually could be found in small quantity. The average content of cyclohexylamine in the cake was 0,55 percent of the used quantity of sodium cyclamate, which meant, that 1,1 percent of sodium cyclamate was reduced to cyclohexylamine.

Only very small losses of artificial sweeteners - if at all - were noticed at the preparation of jam, an alcoholic hot beverage (punch) and coffee. Cyclohexylamine was not found in these foodstuffs.

Actually interesting results could be obtanied by analysing an artificially sweetened vinegar dressing for fish-dishes, which was stored eight days at room temperature and practically was a 3,75% solution of acetic acid in water. In accordance with results obtained by japanese authors there could be proved an average reduction of 7 % of the sodium saccharin and an average reduction of 17 % of the sodium cyclamate content. Cyclohexylamine was also found, although in very small quantities, namely 0,13 % of the used sodium cyclamate quantity, corresponding to 0,26 % reduced sweetener.

As conclusion of our experiments about the interaction of artificial sweeteners with food ingredients at the preparation respectively storage of food can be stated, that there could be exerted a decreasing effect on the stability of the artificial sweeteners on the one hand and some essential food ingredients on the other hand, when sufficiently high temperatures are used, e.g. at baking processes, or when the substances are allowed to react for a longer time, e.g. at tne storage of food. If these suppositions take place, the using of artificial sweeteners should be thought over with regard to the possibility of a reduction or partly destruction of essential food components and also of the formation of new toxic substances.

5 Poster Session

Determination of Carbamate Pesticides

G. BLAICHER, W. PFANNHAUSER, H. WOIDICH

Forschungsinstitut der Ernährungswirtschaft
Blaaßstraße 29, A 1190 Wien

SUMMARY

This paper describes the determination of carbamates and dithiocarbamates used as pesticides for protection of fruits and vegetables. Determination of carbamate pesticides is done by extraction of the material with ethyl acetate, followed by liquid-liquid partition and clean up on a florisil column. Final determination is by hplc and glc respectively, depending on the detection limits of the substances to be determined.
Analysis of dithiocarbamates is carried out by decomposition of the substances with hot hydrochloric acid/stannous chloride and determination of resulting carbon disulfide by headspace gas-liquid-chromatography with a special thermionic detector doted with potassium sulfate, for quantitative analysis.
A quick semiquantitative method allows determination of dithiocarbamates within 20-30 minutes: Decomposition of the substances occurs by using hot hydrochloric acid/stannous chloride. The determination of the resulting carbon disulfide is achieved by a test tube, filled with copper acetate/diethanolamine on basic alumina as reagent. Presence of carbon disulfide is detected by a brownish colouration of the reagent layer. Compairing the colouration with that, resulting from standards, allows a semiquantitative evaluation of carbon disulfide in the sample. The method also works well outside of the laboratory as a field method for screening purposes.

INTRODUCTION

Carbamate pesticides (1,2) are esters of carbamic acids or in the case of dithiocarbamates metals salts of N-alkyldithiocarbamic acids.
Carbamate pesticides are of great significance in pest control and are increasingly used instead of organochlorine and organophosphorus pesticides. Dithiocarbamates are widely used as fungicides and there are no other substances, which can fully substitute these compounds. The great disadvantage of dithiocarbamates (especially ethylene dithiocarbamates) is the possibility of degradation to cancerogenic ethylenethiourea. Therefore valid methods for controlling these pesticides are to be worked out.
A special problem is the control of dithiocarbamates on fresh fruits and vegetables. Values for the pesticides often are obtained only when the goods have been sold. A quick method therefore is of great advantage.

DETERMINATION OF CARBAMATE PESTICIDES (3)

Determination of carbamate pesticides (except the dithiocarbamates) is carried out by extraction of 250 g sample three times by use of ethyl acetate. After addition of 30 ml water to the combined ethyl acetate extracts, ethyl acetate is evaporated. 70 ml methanol are added to the resulting water phase followed by extraction with n-pentane to remove chlorophylls. The methanol-water phase still contains carbamates and the yellow and orange plant colours (carotenes, carotenoids, xanthophylls). Removal of yellow and orange plant colours is carried out by extraction of the methanol-water phase by chloroform and clean up of the chloroform layer by a florisil column (using dichloromethane + 0,5 % methanol and after an elution volume of 100 ml dichloromethane + 1 % methanol for further 500 ml as eluent). Column eluents are evaporated to dryness and filled up to 0,5 ml with ethyl acetate. Final determination is carried out by hplc or glc respectively.

HPLC Determination of Carbamate Pesticides:

Apparatus and Materials
pump : Waters M 6000 A
injector : Waters U6 K
detector : Waters M 440 uv-detector; 254 nm
column : 30 cm stainless steel (1/4" o.d.)
stationary phase: LiChrosorb SI 60 (10 µm)
mobile phases : A : 2,2,4-trimethylpentane (iso-octane)
+ dioxane = 97+3
B : 2,2,4-trimethylpentane (iso-octane)+dioxane = 76+24
injection volume: 50 µl

GLC Determination of Carbamate Pesticides:

In the case of too low sensitivity in hplc, glc is a good alternative, because these compounds, not detectable in uv are stable enough for glc.
Apparatus and Materials
GC: Varian 2100
detector : AFID (Varian)
column : 1,8 m glass (1/4" o.d.)
stationary phase : 5 % OV 17 on Chromosorb W/AW-DMCS
carrier gas: N_2; 20 ml/min.
temperatures : detector: 270°C
injector: 210°C
column: program 110°C - 195°C; 4°C/min.
injection volume: 1 µl
The following carbamate pesticides may be determined by these methods:

HPLC, mobile phase A :
Diallate, triallate, chlorobufam, chloropropham, propham (naphthole-1, the degradation product of carbaryl), barbane, promecarb, mercaptodimetur, propoxur, carbaryl, aldicarb.

HPLC, mobile phase B:
Phenmedipham, aldicarb, methomyl, benomyl, formetanate.

GLC:
Phenmedipham, aldicarb sulfone, butylate, tillam, propham, ordram, cycloate, chloropropham, diallate, promecarb, chlorobufam, triallate.

The method described are valid to determine all carbamate pesticides mentioned in the Austrian regulations for pesticide residues and some other carbamates used in the EEC.
Only benzimidazolamine-2 (2-AB) (also mentioned in the Austrian regulations), the degradation product of benomyl and carbendazime cannot be determined in this way. Therefore determination of benomyl, carbendazime and 2-AB is run together in the following way: The sample is extracted with ethyl acetate, benomyl is hydrolysed by hydro - chloric acid to give carbendazime and this is hydrolysed by sodium hydroxide to give 2-AB (4-7). Liquid-liquid extraction yields in an extract suitable for hplc determination on µ-Bondapak C_{18} with methanol + water = 80+20 + 1 % acetic acid as mobile phase (2 ml/min.) Detection is at 254 nm.

ANALYSIS OF DITHIOCARBAMATES

GLC-METHOD(8)

Gaschromatographic determination of dithiocarbamates is carried out using a thermionic detector sensitive to sulfur compounds. This detector consists of an usual flame ionisation detector with a clay pellet covered with molten potassium sulfate mounted on the flame tip. Sensitivity of the detector to sulfur compounds depends on the sulfur content of the molecule and significantly on hydrogen and air flow. Sulfur dioxide is also detectable by this system. Dithiocarbamate analysis is done by decomposition of 10 g sample with 50 ml hydrochloric acid (4 N) and 0,75 g stannous chloride in a 100 ml flask sealed with a septum at $80^{\circ}C$ for two hours (9). To make sure to have a representative result, 2-4 parts of the sample are treated in the same manner. Resulting carbon disulfide is determined by head space analysis with thiophene as internal standard. Detection limit of the method is 0,1 mg/kg carbon disulfide. The duration of an analytical run is 7 minutes.

GLC-Conditions:
GC: Varian 2800
column: 3,8 m stainless steel (1/8" o.d.)
stationary phase: 30 % DC 200 on Chromosorb W/AW-DMCS
temperature: 80° isotherm
carriergas: N_2: 8,5 ml/min.
detectorgases: H_2: 300 ml/min.
air: 240 ml/min.
injection volume: 0,5 ml from headspace

QUICK SEMIQUANTITATIVE METHOD (10)

The quick semiquantitative determination of dithiocarbamate fungicides is done by decomposition of 40 g sample with 150 ml hydrochloric acid (4 N) + 2,5 g stannous chloride at 80°C immediately followed by the detection of the resulting carbon disulfide with a test tube. The reaction is run in a round bottom flask (250 ml) with a T-shaped connector mounted on the top. A tube (preferable a pipet) is dipped through the connector near to the bottom of the flask. At the horizontal outlet of the connector, the test tube is fixed.
A gas detector pump (Dräger) is connected to the outlet of the testtube. Air is pumped through the reaction mixture when actuating the pump and flows through the test tube. Usually 600 ml of air, carrying carbon disulfide resulted from the reaction are suckled through the test tube. A yellow brownish colour of the reaction layer shows presence of carbon disulfide.
Colour intensity depends on the amount of dithiocarbamates present in the sample.
The colour intensity of the test tube is due to copper dithiocarbamate resulted from reaction between diethanolamine, copper acetate and carbon disulfide. Use of basic alumina as support is required to get the sensitivity necessary for determination. A layer consisting of lead acetate on celite preceding the reaction layer will absorb hydrogen sulfide which could give rise to erroneous results when reacting with copper acetate.
For semiquantitative determination dithiocarbamate standards have to be analysed prior or after the sample. The method is valid to determine differences of 0,5 mg/kg carbon disulfide on a 2 mg/kg level. Time necessary for the whole determination is 20 to 30 minutes.

ACKNOWLEDGEMENT

We wish to thank the Forschungsförderungsfonds der gewerblichen Wirtschaft for sponsoring the work about carbamate pesticides. The work about the quick semiquantitative determination of dithiocarbamate fungicides was sponsored by the Bundesministerium für Gesundheit und Umweltschutz of Austria.

REFERENCES

1 N.N. Melnikov, edited by F.A. Gunther and J.D. Gunther: "Chemistry of Pesticides", Springer Verlag, New York-Heidelberg-Berlin (1979)

2 R. Wegler (ed.): "Chemie der Pflanzenschutz- und Schädlingsbekämpfungsmittel", 2 Volumes, Springer Verlag, Berlin-Heidelberg-New York (1970)

3 G. Blaicher, W. Pfannhauser, H. Woidich: Chromatographia, 13 (1980), 438

4 H.L. Pease, R.F. Holt: J.Ass. Off. Anal. Chem. 54 (1971), 1399

5 J. Vogel, C. Corvi, G. Veyrat: Mitt. Lebensm. Unters. u. Hygiene 63 (1972), 453

6 K. Polzhofer: Z. Lebensm. Unters.-Forsch. 163 (1977), 109

7 J.P. Calmon, D.R. Sayag: J. Agr. Food Chem. 24 (1976), 311

8 G. Blaicher, H. Woidich, W. Pfannhauser: Ernährung 4 (1980), 440

9 H.A. Mc Leod, K.A. Mc Cully: J.Ass. Off. Anal. Chem. 52 (1969), 1226

10 H. Woidich, G. Blaicher, W. Pfannhauser: Berichte (ed. by Bundesministerium für Gesundheit und Umweltschutz) 5 (1979), 1

Evaluation of the Lipid Composition of Some Infant Formula

G. BELLOMONTE, B. CARRATU[1], R. DOMMARCO, S. GIAMMARIOLI, E. SANZINI

Food Laboratories - Istituto Superiore di Sanità - Rome - Italy

SUMMARY

Lipid, fatty acid, sterol and Position 2 triglyceride fatty acid contents were assayed in 13 nationally and internationally widely used adapted powdered milks for infant. Milk mineral contents had already been previously studied (1). Main purpose of this investigation was to compare experimentally obtained results with figures recommended for a more correct nursing diet by authoritative national and international pediatricians.

INTRODUCTION

Commercial products were obtained from the Abbot, Dieterba, Guigoz, Mellin, Milupa, Nestlè and Plasmon companies. Sample 9 consisted of unadapted powdered milk alone and was used as a baseline for other samples whose lipid content had been altered by the addition of vegetal oils and use of skimmed milk.

EXPERIMENTAL

<u>Determination of total fats</u> - Samples were subjected to acid reflux hydrolysis with 3 N HCl for one hour. Residues were filtered. Once dried, they were placed in a Soxhlet and then extracted with petroleum ether for six hours. Total fat contents were determined by weighing after solvent removal.
<u>Determination of fatty acids</u> - The lipid fraction was collected with a known volume of solvent and an exactly measured aliquot removed. After evaporation, the lipid fraction was reflux esterified with 5% sulphuric acid ethyl alcohol for three hours. Ethyl esters were extracted with ethyl ether. The solvent was partially removed. A quantity of the thus-prepared sample was injected into a 3.5 m GC steel column packed

with 15% BDS Chromosorb WHP (80-100 mesh). Temperature increments were programmed to 6°C/min from 80 a 205°C. Injector and detector temperatures were 250°C. Nitrogen was used as carrier gas at a flow rate of 25 ml/min.

Determination of sterols - The samples were saponified with Potassium hydroxide in alcohol at room temperature for one night and then under reflux at 80°C for 30 minutes. The unsaponifiable fraction was extracted with a 1:1 ethyl ether:hexane mixture. Further handling was according to the official Italian procedure for oils and fats (2). The silalized sample was theninjected into a 3 m GC glass column packed with 3% OV 17 Chromosorb WHP (80-100 mesh). Internal standard was dotriacontane. Column temperature was 255°C. Injector and detector temperatures were 300°C. Nitrogen was used as carrier gas.

Determination of Position 2 fatty acids - Lipid extraction from the sample was according to the Rose-Gottlieb method (3). Further handling was according to the official Italian procedure for oils and fats (4). Lipase digestion was modified nd extended to 15 minutes. Esterification was according to official Italian procedure (5). GC conditions were as for fatty acid determination.

RESULTS AND DISCUSSION

The lipid contents in the 100 g samples of powdered milk assayed (Tab. 2), found to correspond to the contents declared (by manufacturers), were used to estimate infant lipid intake in grammes for 100 Cal (Tab. 3). It may be noted that, although some samples had a lower-than-minimum lipid content, none exceeded the maximum recommended value.

Tab. 2 - Lipid content of dried milk: g % g

1	2	3	4	5	6	7	8	9	10	11	12	13
19,6	14,6	24,2	22,5	11,3	20,6	14,7	17,7	14,4	26,0	27,9	22,5	27,8

Tab. 3 - Grams of lipid / 100 Cal
Recommended : 4-6 g / 100 Cal (ESPGAN)

1	2	3	4	5	6	7	8	9	10	11	12	13
4,08	3,24	4,85	4,63	2,56	4,35	3,37	3,84	3,43	5,14	5,38	4,62	5,55

Tab. 4 - Fatty acid composition of dried milk: g % lipid

Fatty acid	1	2	3	4	5	6	7	8	9	10	11	12	13
4:0	1,52	2,36	0,42	0,61	1,56	1,87	1,89	1,27	2,84	1,99	1,64	1,28	--
6:0	1,13	2,15	0,47	0,61	1,35	1,49	1,56	1,00	2,09	1,91	1,22	1,22	0,43
8:0	0,70	1,32	0,86	0,83	0,82	0,92	1,01	0,71	1,26	1,23	1,11	1,40	5,17
10:0	1,46	3,50	1,13	1,32	2,33	2,06	2,07	1,70	2,86	2,76	1,69	2,18	4,03
10:1	1,19	0,30	0,07	0,09	0,15	0,29	0,21	0,14	0,27	0,25	0,14	0,17	--
11:0	tr	0,26	0,04	0,04	0,08	tr	tr	0,06	tr	tr	tr	--	--
12:0 lauric acid	1,55	4,16	4,37	4,04	3,01	2,55	2,57	2,21	3,31	3,32	4,01	5,94	2,70
12:1	0,07	0,45	0,95	0,17	0,39	tr	0,26	0,19	0,26	0,08	0,16	0,06	--
13:0 + 13:br ?	0,24	0,42	0,15	0,16	0,64	0,22	0,11	0,26	0,19	tr	0,09	0,09	0,26
13:1	0,10	0,09	tr	tr	0,50	tr	tr	tr	tr	tr	tr	--	--
14:0 myristic acid	5,83	10,55	4,52	5,01	7,63	7,77	7,79	5,81	10,84	9,21	6,26	8,11	11,28
14:1	0,31	0,57	0,09	0,13	0,40	0,22	0,47	0,21	0,61	0,53	0,06	0,43	--
15:br ?	0,09	0,08	0,06	0,07	0,16	0,22	0,07	0,08	0,12	0,09	0,06	0,06	0,24
15:0	0,67	0,65	0,12	0,17	0,75	0,80	0,62	0,39	1,05	0,76	0,45	0,55	--
16:br ?	0,14	0,39	0,08	0,09	0,26	0,20	0,20	0,13	0,30	0,21	0,15	0,16	--
16:0 palmitic acid	18,46	28,03	34,04	32,28	23,57	21,23	23,02	28,47	26,16	24,22	29,61	25,08	19,37
16:1	0,44	0,74	0,06	0,16	0,66	0,60	0,80	0,19	0,81	0,54	0,33	0,50	--
17:br ?	0,08	0,17	0,01	0,04	0,17	0,13	0,11	0,10	0,10	0,15	0,05	0,98	--
17:0	0,44	0,48	0,14	0,19	0,40	0,47	0,41	0,33	0,71	0,47	0,34	0,37	--
17:1	0,21	0,40	0,06	0,10	0,16	0,21	0,24	0,11	0,32	0,27	0,16	0,17	--
18:0 stearic acid	8,05	9,01	4,88	5,29	7,13	8,04	7,16	6,79	11,25	8,83	7,25	6,58	2,58
18:1 oleic acid	27,79	24,26	33,37	35,24	36,43	37,59	35,40	31,84	29,15	28,03	31,63	32,60	15,65
18:2 linoleic acid	27,71	8,94	11,33	11,48	10,21	10,69	12,86	15,86	3,67	13,07	11,85	11,34	21,58
18:3 linolenic acid	1,52	0,34	0,77	0,79	0,39	0,80	0,52	1,09	0,78	0,95	0,78	0,60	0,54
20:0	0,75	0,28	0,41	0,54	0,38	0,88	0,23	0,65	0,56	0,41	0,42	0,54	0,42
20:1	0,54	0,09	0,42	0,49	0,46	0,76	0,39	0,40	0,47	0,54	0,36	0,50	0,31

As for fatty acids (Tab. 4), it may be noted that the ratios of unsaturated total fatty acids, for the most part, found to be lower than the recommended figure to a greater or lesser extent (Tab. 5).
For unsaturated essential fatty acids (EFAs), linoleic acid content was assayed because the acid is present in markedly greater quantities than other EFAs. Linoleic acid-originating calories were seen to be lower than the recommended levels in 3 samples and higher in two (Tab. 6). This possibly harmful excess (6) may be attributed to an attempt to "humanise" powdered cows' milk, which is poorer in EFAs than human mothers' milk, by the addition of vegetal oils. In relation to this fact, the Vitamin E : linoleic acid ratio is important (Tab. 8) (7,8).

Tab. 5 - Percentage og unsaturated fatty acids
Recommended : 52-54 % as human milk

1	2	3	4	5	6	7	8	9	10	11	12	13
58	35	46	49	49	51	50	50	36	44	45	45	38

Tab. 6 - Calories from linoleic acid as percentage of total Calories
Recommended : 3-6 % as human milk

1	2	3	4	5	6	7	8	9	10	11	12	13
10,2	2,6	5,0	4,8	2,4	4,2	3,9	5,5	1,1	6,1	5,7	4,7	10,8

Tab. 7 - Linoleic / linolenic acid ratio
Recommended : 6 / 1 as human milk

1	2	3	4	5	6	7	8	9	10	11	12	13
18	26	15	15	26	13	25	15	5	14	15	19	40

Tab. 8 - Viamin E / linoleic acid ratio
Recommended : 0,8 sufficient
              0,6 minimum

1	2	3	4	5	6	7	8	9	10	11	12	13
0,9	3,8	2,1	2,1	7,9	4,5	2,7	1,8	--	0,6	1,5	1,9	0,6

All samples may, however, be seen to fall within minimum recommended limits. A balanced ratio between linoleic and linolenic acids also seems important because an excess of the former or a deficiency of the latter may lead to neural disturbances (9). All samples investigated, except one, exhibited an excessively high ratio of linoleic:linolenic acid (Tab. 7).

Evaluation of the Lipid Composition of Some Infant Formula

Tab. 9 - Sterol content of dried milk: mg % g

Sterols	1	2	3	4	5	6	7	8	9	10	11	12	13
Cholesterol	30,5	42,4	22,4	27,3	30,2	42,2	99,6	20,1	23,8	68,6	24,5	32,5	5,4
Campesterol	12,1	3,0	2,0	2,3	1,7	2,2	2,7	2,1	--	5,5	0,9	1,2	9,5
Stigmasterol	4,4	1,0	1,2	1,3	0,7	0,9	1,9	0,7	--	1,9	0,8	0,8	3,2
β - Sitosterol	34,1	11,0	5,2	6,6	4,7	7,9	21,5	7,8	--	20,4	2,9	4,1	25,8

Tab. 10 - Milligrams of cholesterol / 100 ml of reconstituted dried milk
Recommended : ? (30-40 mg / 100 ml in human milk)

	1	2	3	4	5	6	7	8	9	10	11	12	13
3000	3,1	--	2,6	--	3,4	--	3,3	--	--	8,1	--	4,1	--
3500	3,4	3,1	2,8	2,2	3,9	2,7	--	2,3	1,8	--	3,1	4,2	0,7
4000	3,7	3,2	2,9	2,7	4,0	3,3	--	2,8	2,6	6,7	2,5	4,3	0,7
4500	3,8	4,0	3,2	2,9	4,0	3,6	--	3,1	2,6	8,2	2,9	4,3	0,7
5000	3,8	4,3	3,1	3,9	3,8	3,1	--	3,1	2,5	8,9	3,2	4,4	--
5500	3,7	4,3	3,0	3,1	3,9	3,8	--	3,2	2,5	9,5	3,4	4,6	0,7
6000	3,8	--	3,0	--	4,3	--	3,9	--	3,1	--	--	--	--
6000 (last)	--	5,5	--	--	--	--	13,1	3,1	--	9,8	3,1	4,5	0,7

The underlined values indicate the reconstitutions recommended by pediatricians, others by manufacturers'

T:B. 11 - Percentage of fatty acids in the 2 position

Fatty acid	Human milk	1	2	3	4	5	6	7	8	9	10	11	12	13
12:0	5,6	--	1,5	1,3	0,5	0,9	0,5	1,2	--	1,8	4,5	3,4	4,2	--
14:0	14,8	3,2	1,6	11,8	2,3	7,2	5,5	11,3	12,9	11,9	8,3	15,0	8,6	--
16:0	61,9	14,6	16,8	34,6	14,6	19,2	16,7	28,9	67,2	29,5	22,3	58,9	3,8	--
18:0	1,6	2,6	0,9	4,6	1,6	2,3	3,5	1,2	11,0	4,2	3,6	9,3	0,9	--
18:1	10,4	3,52	6,53	32,3	6,50	40,5	53,3	47,9	9,1	83,7	44,6	14,7	42,5	--
18:2	5,7	44,3	13,9	15,3	15,9	29,8	20,6	9,5	--	19,0	16,7	0,8	--	--

Palmitic and stearic acids are among the fatty acids which influence adequate lipid absorbability. Palmitic and stearic acid contents should not, according to ESPGAN norms, be greter than 20 and 10%, respectively, of total fatty acids.

Lauric and myristic acids, which are considered to be atherogenic fatty acids, should be present to within at least 20% of their corresponding values in human milk. However, considering the variability in the extreme values of human milk composition reported by various authors (10, 11) (C12:0 3.0-17.1; C14:0 2.0-22.0), it may be seen that only one sample exhibited distinctly high lauric acid content.

The terol assay did not detect the addition of any oils other than those normally present in infant formula (Tab. 9). Particular attention was given to cholesterol in 100 ml of reconstituted milk (Tab. 10). Even if precise cholesterol demand values do not exist in the literature, it may be seen that powdered milk cholesterol contents are markedly lower than for human milk. This is understandable in view of cow milk's lower cholesterol content (10-15 mg/100 ml) (12) and the defattening processes to which these milks are subjected. The cholesterol content normally observed in infant formula (1-3 mg/100 ml) (12) was verified in only some of the samples assayed. The underlined values reported in Tab. 10 indicate the reconstitution figures recommended by paediatricians, other values are those based on manufacturers' recommendations.

As for triglyceride structure, it has been shown that a high percentage of palmitic acid in Position 2, as occurs in human milk, increases lipid absorbability. Table 11 reports the percentages of predominant Position 2 fatty acids. Values for human milk (13) have been manipulated so as to be compatible with the same parametres used for the samples assayed.

REFERENCES

1 G. Bellomonte et Al., Minerva Pediatrica 32 (1980), 1177
2 Ministero Agricoltura e Foreste, Metodi ufficiali di analisi per gli oli ed i grassi Suppl. 3 (1972)
4 Ministero Agricoltura e Foreste, Metodi ufficiali di analisi per gli oli ed i grassi Suppl. 2 (1971)
3 Manuel suisse des denrées alimentaires 1 (1969), 543
5 Ministero Agricoltura e Foreste, Metodi officiali di analisi per gli oli ed i grassi Suppl. 1 (1963), 12
6 P. Giorgy, Am. J. Clin. Nutr. 34 (1971), 970
7 P.L. Harris, N.D. Embree, Am. J. Clin. Nutr. 13 (1963), 385
8 M.K. Horwitt, Vitamins Hormones 20 (1962), 541
9 M.A. Crawford, 2° Congr. Int. Biol. Val. of olive oil Torremolinos May 1975, 35
10 Jensen et Al., Am. J. Clin. Nutr. 31 (1978), 990
11 Wilson et Al., Am. J. Clin. Nutr. 31 (1978), 1127
12 ESPGAN, Acta Paediatr. Scand. Suppl. 262 (1972)
13 R.M. Tomarelli et Al. J. Nutrition 95 (1968), 583

Immunological Detection of Meat from Turkey

A. SCHWEIGER, K. HANNIG, H.O. GÜNTHER[1] and S. BAUDNER[2]

Max-Planck-Institut für Biochemie, Martinsried bei München
[1] Landesuntersuchungsamt Gesundheitswesen Südbayern, München
[2] Behringwerke AG, Marburg-Lahn

SUMMARY

The identification of turkey meat in food products has been accomplished by immunological methods using antisera against the muscle protein troponin T. This protein (M_r=37 000) was isolated from fresh muscle in a procedure that included Free Flow Electrophoresis and was more than 95% pure.

INTRODUCTION

Turkey and chicken muscle are used in meat products, partly in mixtures replacing more expensive meat from beef or pork. These additions are, however, not sufficiently declared in many cases. On the other hand there are different rates of a customs reduction in the EG for imported meat products which contain 57-25% poultry meat. From these reasons it has become necessary to develop methods for detection and identification of poultry meat proteins.

The main difficulty in this work consists in the selection and isolation of typical protein species. Using polyacrylamide gel electrophoresis in the presence of SDS an identification of specific proteins may be achieved from the band pattern for pure meat or meat products but not for mixed preparations which contain different proportions of the various types of meat. In this situation the use of immunological techniques appeared more suitable. Previous attempts studying serum proteins like albumin, or the contractile proteins myosin or actin were not successful. Myosin becomes degraded during heating; actin which is heat-stable shows only very little differences in structure in different animal species (1) and therefore cannot be used for immunological identification. The interest then focused on another muscle protein, troponin, and it was possible to detect additions of poultry meat with the use of antibodies against crude preparations of troponin (2,3). Troponins C, I and T are involved in the myosin-actin system and thus regulate the contraction of muscle fibers. Troponin C is a Ca^{++}-binding protein, troponin I inhibits the dissociation of actomyosin specifically and troponin T interacts with tropomyosin to initiate muscle contraction. The three troponins have molecular weights of M_r = 18 000 (troponin C), 24 000 (troponin I) and 37 000 (troponin T). Tropomyosin does not induce specific antibodies.

RESULTS AND DISCUSSION

The most important step for a sensitive and more specific immunological identification of troponin was the preparation of this protein in a purified state. Fresh turkey muscle was minced and washed with cold water to remove actin. The dry powder was then obtained after treatment with ethanol, diethylether, acetone. The troponin was extracted with $LiCl-NaHCO_3$ buffer (4), precipitated with ammoniumsulfate and dialysed against dilute Tris-phosphate. The supernatant obtained after centrifugation was submitted to Free Flow Electrophoresis according to Hannig (5). There were three peaks in the effluent localized by the absorbances at 235 and 280 nm. The peaks of highest and lowest anionic migration contained minor protein components of various molecular weights (e.g. troponin I and actin). The third peak at fractions 45-55 was the largest and appeared rather symmetrical. This material was characterized by SDS polyacrylamide gel electrophoresis as one band with M_r = 37 000. This component, apparently troponin T, showed a purity of more than 95%.

Further evidence that the isolated protein was troponin T came from determination of the amino acid composition which is shown in Table 1. Comparing these data with the analysis given for rabbit troponin T (6) one can see that several amino acids are present in very similar amounts in both proteins while other values do not correspond. This partial identity has been previously described also for troponins from rabbit and chicken (7).

Table 1

Comparison of amino acid composition of troponin T isolated from turkey and rabbit muscle. Compositions are expressed in terms of moles per loo moles of recovered amino acids.

Amino acid	troponin T from turkey	rabbit (6)
Asp	9.6	8.o
Thr	2.5	2.2
Ser	4.7	2.6
Glu	28.3	23.7
Pro	-	3.2
Gly	2.o	2.o
Ala	12.8	6.8
Val	2.9	3.9
Met	1.9	2.o
Ile	2.4	3.1
Leu	lo.1	7.5
Tyr	3.o	2.2
Phe	1.o	2.5
Lys	12.1	15.5
His	o.6	2.6
Arg	5.7	12.1

Antisera against the purified turkey troponin T were produced by Behringwerke, Marburg-Lahn. Using these preparations it was possible to detect different amounts of turkey and chicken muscle added to beef and pork meat products. We have used the immunological techniques of double diffusion according to Ouchterlony, immunoelectrophoresis of Grabar-Williams and counterimmunoelectrophoresis described by Gocke and Howe (8). Immunoelectrophoresis with a "short trough" has been also applied. Counterimmunoelectrophoresis was the fastest method of identification since it allowed to perform the whole procedure within two days beginning from extraction of the product sample. In the presence of turkey muscle protein the immunoreaction pattern showed 1-2 precipitation lines which were confluent with those of control troponin T. In some cases additional precipitation lines have been observed. In particular, actin may lead to cross reaction and serious interferences in the test results. It was, therefore, essential to remove most of the actin and other contaminants from the troponin T used for antibody preparation by suitable methods e.g. Free Flow Electrophoresis. The antiserum obtained against highly purified turkey troponin T was then a valuable tool for the determination of specific proteins and in the quality control of meat products.

REFERENCES

1   M.E. Carsten and A.M. Katz, Biochem. Biophys. Acta 9o (1964), 534
2   A.R. Hayden, J. Food Sci. 42 (1977), 1189
3   I. Ohtsuki, J. Biochem., Tokio 75 (1974), 735
4   S. Ebashi, A. Kodama and F. Ebashi, J. Biochem., Tokio 64 (1968), 465
5   K. Hannig, in Modern Separation Methods of Macromolecules and Particles (T. Gerritsen, ed.) vol. 1 (1969) p. 45, John Wiley and Sons, Inc., New York, London, Sydney, Toronto
6   J.R. Pearlstone, M.R. Carpenter, P. Johnson and L.B. Smillie, Proc. Nat. Acad. Sci. (USA) 73 (1976), 19o2
7   T. Hirabayashi and S.V. Perry, Biochem. Biophys. Acta 351 (1974), 273
8   D.J. Gocke and C. Howe, J. Immunol. 1o4 (197o), 1o31

Determination of Glucose, Fructose and Small Amounts of Saccharose in Honey

J. JARÝ[a], M. MAREK[b] and J. BACÍLEK[c]

a) Laboratory of Monosaccharides, Prague Institute of Chemical Technology, 166 28 Praha 6, Czechoslovakia
b) Department of Biochemistry and Microbiology, Prague Institute of Chemical Technology, 166 28 Praha 6, Czechoslovakia
c) Bee Research Institute, Dol, 252 66 Libčice n/Vlt., Czechoslovakia

SUMMARY

The liquid chromatography of honey gives similar results as classical analytical methods in the case of glucose only. Polarographic determination of fructose and the suppression of oxygen peaks during the analysis was examined. In controlling the purity of honey samples it is advantageous, to examine the ratio between the height of the oxygen peak and the height of the reduction wave of fructose. This ratio is higher for artificial "honey" or adulterated natural honeys than the limit value for non-contamined honey.

INTRODUCTION

Polarimetric determination of sucrose may be influenced by other oligosaccharides, in honey particularly with erlose[1]. Enzymatic[2] and high-performance liquid chromatographic methods[3] have been devised for determining sucrose in honey. Chromatography on ion exchange resin in K^+ cycle was also described[4,5] for sugar-mixture separation.

Polarographic determination of fructose in honey and in invert sugar is a rapide and simple method[6]. During the analysis of natural honey, the oxygen peak is suppressed[6,7], whereas so-called "artificial honeys" (invert sugar) retain a high oxygen peak.

MATERIAL AND METHODS

Glucose content was examined iodometrically in a weakly alkaline medium[8]. Reducing sugars were estimated by the Meissl method using the reduction of Fehling's solution, the precipitated Cu_2O being determined manganometrically by the Bertrand method[8]. Sucrose[8] and invertase activity[9] was determined by double polarimetry of 20% stores before and after inversion.
Liquid chromatography of sugars was performed on a glass column (90 x 4 mm) containing Ostion LGKS ion exchanger 10-12 µm (Spolek pro chemickou a hutní výrobu, Ústí n/L., Czechoslovakia) in K^+ cycle, flow rate 6 ml water/h. The samples were diluted 1:5 with water, 3 µl of this solution were injected into the column. An RD-1 differential refractometer (Vývojové dílny Československé akademie věd, Praha, Czechoslovakia) was used as a detector. The upper pressure limit on the column was 2 MPa.
Samples were analysed on Polarograph LP 7 (Laboratorní přístroje, Praha) in 0.01 M lithium chloride (sensitivity 1/50). A 1% solution of each sample was gradually added to 10 ml of electrolyte. The heights of the polarographic waves were read on the polarograms at -1.7 V, corresponding to the reduction wave of fructose.
Various honeys and bee sugar stores were used in all experiments.

RESULTS AND DISCUSSION

Sugar bee stores were analysed by different methods and results are shown in Table I. By comparing classical methods of determination of each sugar with liquid chromatography, it is possible to trace some differences, especially in the sucrose and fructose contents. In the case of glucose no important differences were found. The quantity of fructose, determined as a difference between reducing sugars (Bertrand method) and glucose (iodometrically), was influenced by reducing oligosaccharides. Different results were also obtained for sucrose by the polarimetric method and liquid chromatography. Some natural oligosaccharides (like erlose) can disfigure the results of the polarimetry. The liquid chromatography is more precise for the estimation of

Table I. Comparison of liquid chromatography and classical methods in sugar bee stores analyse

		Time of storage after extraction (in days)	0	6			42		
		Temperature of storage(°C)		-30	0	+25	-30	0	+25
traditional methods		refractive index	1.4740						
		dry matter(%)	73.6						
		glucose (%)	23.4	23.1	23.3	35.5	24.3	23.8	30.7
		reducing sugars (%)	58.8	54.1	53.3	74.0	55.4	53.1	68.6
		fructose (%) +	35.4	31.0	30.0	38.5	31.1	29.3	37.9
		sucrose (%)	21.8	23.0	23.0	17.3	23.0	21.4	7.7
		activity of invertase (ukat/kg of stores)	59.96	56.93	57.82	42.23	59.12	52.30	33.78
ion-exchange chromatography		glucose (%)	23.2	22.1	22.1	24.9	23.2	27.3	33.4
		fructose (%)	26.3	24.8	24.8	27.3	27.1	21.3	28.9
		sucrose (%)	18.8	20.5	20.7	15.6	17.6	19.0	6.4
		oligosaccharides (%)	5.4	6.2	5.7	5.7	5.7	6.0	4.8
		fructose : glucose	1.133	1.119	1.119	1.094	1.161	0.779	0.865

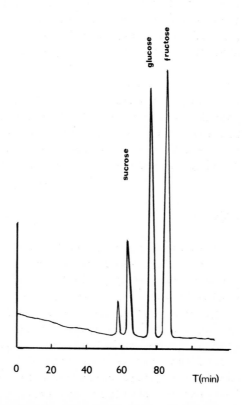

Fig. 1. Liquid-chromatography of sugar bee stores

Fig. 2. Dependence of the height of the fructose wave on the volume of 1% sample solution added

○ honeydew honey
● honeydew honey
◐ commercially processed honey
◑ nectar honey
◒ nectar honey from rape
◉ unsealed nectar-honey stores
□ winter stores
△ standard fructose
▲ invert sugar
■ adulterated honey (no.5+no.9, 1:1)

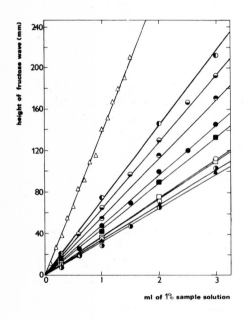

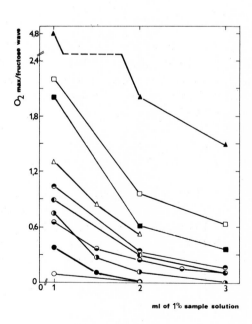

Fig. 3. Dependence of O_2 max./fructose wave on the volume of 1% sample solution added. See Fig. 2.

title sugars in honey (Fig.1). Polarographic determination of fructose was also performed (Fig.2). Markedly differing suppression of oxygen peaks was found in different kinds of natural honeys as well as in adulterated honeys[10]. The increase in fructose content in natural honeys is accompanied by the decrease of the ability to suppress the oxygen peaks. The ratio of the height of the oxygen peak to the height of the fructose wave can be suggested as the criterion of quality of honeys (Fig.3). If honey samples have a high sucrose content, the proposed method can be modified by repeated determination of fructose after inversion of the sucrose remaining in the sample. The parameter O_2 max./fructose wave is then transformed into:

$$O_2 \text{ max.} / \sqrt{F_1 (2F_1 - F_2 V_2/V_1)}$$

where O_2 max. is the value of the oxygen peak prior to inversion, F_1 and F_2 are heights of the reduction waves of fructose before and after inversion, and V_1 and V_2 are the volumes of samples in the polarographic cell before and after inversion, respectively.

REFERENCES

1 J.W. White, Jr., Composition of honey (1975). Pp. 157-206 from Honey: a comprehensive survey. E. Crane, ed., London: Heinemann
2 J.W. White, Jr., J. Ass. off. analyt. Chem. 60 (1977), 669
3 J.E. Thean, W.C. Funderburk, Jr., J. Ass. off. analyt. Chem. 60 (1977), 838
4 J.K. Palmer, W.B. Brandes, J. agric. Fd Chem. 22 (1974), 709
5 R.W. Goulding, J. Chromatogr. 103 (1975), 229
6 J. Heyrovský, I. Smoler, Collect. Czech. Chem. Commun. 4 (1932), 521
7 I. Vavruch, J. Wagner, Z. Hrudka, Listy cukrov. 63 (1946), 101
8 JAM, No. 21 (1950), Prague: Association Tchécoslovaque pour l´exploration et l´essai des matériaoux et construction techniques
9 H. Hadorn, K. Zürcher, Dt. Lebensmitt. Rdsch. (1966), 195
10 M. Marek, J. Bacílek, J. Jarý, J. Apic. Res. 19 (1980), 255

Gas-Chromatographic Analysis of Amino Acids in Fruit Drinks

J.P. ROOZEN and M.M.Th. JANSSEN

Department of Food Science, Agricultural University, De Dreijen 12,
6703 BC Wageningen, The Netherlands

Amino acid analysis is a useful method for determining the identity and juice content of fruit drinks. For routine analysis of several food constituents gaschromatography is available in almost every laboratory. Our investigation has been undertaken to study the special application of this technique for amino acid analysis in juices.

The amino acids had to be isolated from the samples before the derivatization procedure could be carried out. If present the cloud of the fruit juices was removed by centrifugation. As a standard 15 mg of 2-aminobutyric acid were added to 30 cm^3 of the juice serum acidified to pH < 2.5 which was then treated with amberlite CG-120 in order to adsorb the amino acids. After five times washing with demineralized water the amino acids were released from the cation exchanger with 7N NH_4OH and aliquots were freeze dried for derivatization (1). For gaschromatographic analysis the amino acids were volatilized by derivatization to their N-heptafluorobutyryl-isobutylesters (2,3). The procedure should be performed under stringent exclusion of water and oxygen in samples, reagents and reaction mixtures. One µl sample ($\approx$ 0,1 µMol amino acid derivatives) was injected on a 3% SE-30 Gaschrom Q or 6% SE-30 Chromosorb-WHP glass column in a Carlo Erba gaschromatograph, which was programmed for 6 min isotherm at 90°C and afterwards 3°C temperature increase per min (90°-240°C).

In Table 1 it can be seen from the selfpressed orange juice sample that there is a good agreement between the gaschromatographic method and the automatic amino acid analyser. Commercial fruit drinks have been employed for further evaluation. The data correspond very well with contents of amino acids published for orange (4), grape (5), grapefruit (6), apple (7), peach (8) and apricot juice (9). Obviously

Table 1:
Content of amino acids* in mg per 100 cm^3 fruit drink

Commercial sample	ALA	THR	SER	LEU	4-ABA	PRO	4-HYP	MET	ASP	GLU	ARG	Total
Orange juice	11	–	18	–	27	100	–	–	64	14	220	454
Grape juice	13	7	6	–	9	55	–	–	5.5	10	96	201.5
Grapefruit juice	10	–	19	–	6	44	–	–	186	55	67	387
Apple juice	2	–	2	–	5	1.5	3.5	2.5	62	–	–	78.5
Orange/peach nectar	12	1.5	8	–	16	25	6	9	208	–	59	344.5
Orange/apricot nectar	6	1.5	6	10	–	34	2	2	103	–	35	199.5
Self pressed orange juice	**20 / 21	– / –	24 / 27	– / –	32 / 35	210 / 242	– / –	– / –	124 / 104	38 / 44	80 / 106	528 / 579
Range variation coefficient (%)	3-13	–	5-20	–	21-30	7-15	–	–	3-25	20-60	14-35	–

* IUPAC abbreviations, except:
4-ABA = 4-aminobutyric acid
4-HYP = 4-hydroxy-methylprolin

** Automatic amino acid analyser

this is a good method for the determination of free amino acids in fruit juices. The standard deviation calculated, however, suggests that further optimalisation of the method should be investigated.

REFERENCES

1. F.E. Kaiser, C.W. Gehrke, R.W. Zumwalt and K.C. Kuo, J. Chromatogr. 94 (1974), 113
2. S.L. Mackenzie and D. Tenaschuk, J. Chromatogr. 97 (1974), 19
3. P. Felker, J. Chromatogr. 153 (1978), 259
4. S. Wallrauch, Flüss. Obst 47 (1980), 47
5. W.M. Kliewer, J. Food Sci. 34 (1969), 274
6. R.L. Clements and H.V. Leland, J. Food Sci. 27 (1962), 20
7. L.F. Burroughs, J. Sci. Food Agric. 8 (1957), 122
8. H. Kieninger and A. Eksi, Flüss. Obst 46 (1979), 124
9. L.F. Burroughs, in "The biochemistry of fruits and their products", A.C. Hulme, ed., Academic Press (1970), 133

Determination of Pyrethrins in Flour

I. SCHEIDL, W. PFANNHAUSER, H. WOIDICH

Forschungsinstitut der Ernährungswirtschaft
Blaasstraße 29, A-1190 Wien

SUMMARY

The determination of pyrethrins in flour is described. After having been extracted and cleaned up on a florisil column, the substances were quantified by gaschromatography using electron capture detector (1).

INTRODUCTION

Pyrethrin, the natural insecticide, results from a drying process or extraction at room temperature of the blossoms of some certain Chrysanthemum flowers. The active substances in Pyrethrin are Pyrethrin-I and Pyrethrin-II, Cinerin-I and Cinerin-II, Jasmolin-I and Jasmolin-II. They are used mainly as protection agents in the storage of cereals and vegetables. Some insect sprays contain also pyrethrins.

PRINCIPLE OF THE METHOD

Pyrethrins are extracted by dichlormethane together with chlorinated hydrocarbons and polychlorinated biphenyls (PCB). The chlorinated hydrocarbons and PCB disturb the gaschromatographic determination with the electron-capture detector (ECD). Therefore a clean up step by fractionation on a Florisil column using diethyl ether and petroleumbenzine as eluents is necessary. The first fraction contains chlorinated hydrocarbons and PCB. For elution of pyrethrins two fractions are to be taken. The first pyrethrin fraction contains Pyrethrin-I, Cinerin-I and Jasmolin-I, accompanied by some substances due to the flour, having the same retention time as pyrethrin-II. These substances could give rise to erroneous results, if pyrethrin-II were also in this fraction. The second fraction then contains only Pyrethrins-II, Jasmolin-II and Cinerin-II.

PROCEDURE

5 g of flour sample are extracted three times with dichlormethane. The combined extracts are evaporated by a rotary evaporator to a volume about 2 ml. The clean-up step is done by a Florisil column (30 g of Florisil desactivated with 5 % water). The eluents for the three fractions are: 150 ml of 20 % diethyl ether and petroleumbenzine (for chlorinated hydrocarbons and PCB's), followed by 70 ml of 50 % diethyl ether/petroleumbenzine (Pyrethrine-I, Cinerin-I, Jasmolin-I) and finally 60 ml 50 % diethyl ether/petroleumbenzine (Pyrethrin-II), Cinerin-II and Jasmolin-II). The second and third fractions are evaporated to exactly 0,5 ml and injected into the gaschromatograph.

GC-Conditions
GC: Varian 1400
detector: ECD (H^3)
column: 0,6 m glass (1/4" o.d.)
stationary phase: 2,5 % XE 60 on Chromosorb W/AW-DMCS (2)
carrier gas: N_2 : 10 ml min.
temperature: injector: $220°C$
              detecor : $250°C$
              column: program $160°-220°C$, $6°C/min.$
injection volume: 10 µl

Evaluation is done by the Lab. Automation System HP 3354 B.
The calibration curve is evaluated by summing peak areas of all
six pyrethrins due to the lack of single standards. This method
is possible because there is only a maximum value for this summary
reported in the Austrian regulation for pesticide residues. The
determination limit is 1 ng absolute, this means 0,01 mg/kg for a
0,5 ml extract of 5 g sample and an injection volume of 10 µl.

REFERENCES

1 I. Scheidl, W. Pfannhauser, H. Woidich: Deutsche Lebensmittel-
rundschau, 76(1980), 309

2 Y. Kawano, K.H. Yanagihara and A. Bevenue
J. Chromatogr. 90, 119-128 (1974)

Correlation of Isotopic Composition and Origin of Foods

E.R.SCHMID[1], H.GRUNDMANN[1], W.PAPESCH[2] and I. FOGY[1]

1) Institut für Analytische Chemie der Universität Wien, Waehringerstr.38, A-1o9o Wien, Österreich
2) Geotechnisches Institut der Bundesversuchs- und Forschungsanstalt Arsenal, Objekt 214, A-1o3o Wien, Österreich

SUMMARY

The content of radioactive and stable nuclides of foods can be used to determine their origin and authenticity. Methods and examples of such determinations are presented for vinegar, oil of bitter almonds and honey. They are based on the measurement of the specific ^{14}C and ^{3}H radioactivities and the $^{13}C/^{12}C$ isotope ratio.

INTRODUCTION

By measuring the isotopic composition valueable information on the origin and authenticity of foods can be obtained. For this purpose stable as well as radioactive nuclides can be determined. In this contribution the basis of methods to do this and some examples of their practical application are described.

RADIOACTIVE NUCLIDES

Specific ^{14}C radioactivity

All assimilating plants take up CO_2. The atmospheric CO_2 contains 10^{-14} g $^{14}CO_2$ per gram inactive CO_2. Therefore, all biogenic products have a ^{14}C radioactivity, even several thousand years after the plants died, since the half-life of ^{14}C is 57oo years. This is the reason why age determination is possible by measuring the specific ^{14}C radioactivity of biogenic organic material. For food analysis one can state that plants are used as food within a short period of time after harvesting. Therefore their specific ^{14}C radioactivity has to be the same as that of the atmospheric CO_2 during their growing period. On the other hand, substances obtained from fossil raw materials like coal and petroleum do not have ^{14}C radioactivity any more. Since coal and petroleum are millions of years old the ^{14}C radioactivity originally present has decayed completely. Therefore, the determination of the specific ^{14}C radioactivity is used as a basis for differentiating between biogenic material derived from plants and synthetic substances produced from coal or petroleum.

Specific ^{3}H radioactivity

Since plants also take up H_2O from the biosphere the specific ^{3}H radioactivity can be used in a similar way to determine the origin of food. By measuring the specific ^{3}H radioactivity one can also distinguish between biogenic substances obtained from plants and synthetic substances from coal and/or petroleum. ^{3}H has a half-life of 12 years. Measuring the specific ^{3}H radioactivity allows not only to decide on the origin of food but also on the age and the period for which a food product was stored. E.g. a wine or brandy is from a recent harvest, if the specific ^{3}H radioactivity is identical with that found in the atmosphere. In "old" products the specific ^{3}H radioactivity is decreased according to the half-life and the duration of the storage, e.g. to half of its value after a storage time of 12 years.

In reality the situation is more complicated, as both methods - the determination of the specific ^{3}H as well as the specific ^{14}C radioactivities - need a correction for the amount of artificial ^{3}H and ^{14}C brought into the atmosphere by nuclear explosions. A reliable and easy method of correction is the measurement of samples of authentic origin.

STABLE NUCLIDES

The elements H, C, N, O, and S are the main constituents of food and all of them possess more than one stable nuclide. Isotope effects occur during all physical processes and chemical reactions, of course also during the metabolism of plants. The isotope effects are in some cases so big that they can be measured and applied analytically. The following discussion is limited to carbon, which possesses the two stable nuclides ^{12}C and ^{13}C. A pronounced change of the $^{13}C/^{12}C$ isotope ratio takes place during the photosynthesis of plants. $^{13}CO_2$ is taken up slowlier than $^{12}CO_2$. Therefore, the carbon in plants is depleted in ^{13}C in comparison to the CO_2 in the atmosphere. The $^{13}C/^{12}C$ ratio is decreased in different extents by different plant types. Most of the plants growing in temperate zones are so-called C_3 plants. The name is derived from the fact that the first product of their photosynthesis is a substance containing three carbon atoms in the molecule, namely phosphoglycerinic acid. In further chemical reactions carbohydrates are built up from this substance.

A second group of plants are the so-called C_4 plants. This name results from the fact that the first product of the photosynthesis is a substance containing four carbon atoms in the molecule, namely oxalic acetate. Typical representatives of C_4 plants are maize and sugar-cane (Sugar-beets are C_3 plants).

Important for food control is the fact that the ^{13}C depletion is bigger in C_3 plants ($\delta^{13}C=-25‰$) than in C_4 plants ($\delta^{13}C=-12‰$). The $^{13}C/^{12}C$ ratio is usually expressed in the form of $\delta^{13}C$ values.

$$\delta^{13}C_{PDB} = \left(\frac{(^{13}C/^{12}C)^{sample}}{(^{13}C/^{12}C)^{PDB\ standard}} - 1 \right) \cdot 1000 \quad [‰]$$

On this basis it is possible to differentiate between foodstuff from C_3 plants and C_4 plants. Thus products obtained from sugar-beets can be clearly distinguished from those produced from sugar-cane.

A further possibility to determine the origin of foodstuffs is based on the carbon isotopic differentiation during the secondary metabolism in plants: carbohydrates have the least ^{13}C depletion (C_3 plants $\delta^{13}C=-25‰$, C_4 plants $\delta^{13}C=-11‰$), lipids have the biggest ($\delta^{13}C=-31‰$ and $-15‰$ respectively), and the depletion of proteins is lying in between ($\delta^{13}C=-27‰$ and $-13‰$ respectively). Petroleum originates from the lipid fraction of plants and possesses therefore less ^{13}C than the carbohydrates. Ethanol and acetic acid can be obtained either by fermentation of sugar-containing liquids or synthetically from petroleum. By measuring $\delta^{13}C$ of the total molecule one can find out the origin of the substances.

EXPERIMENTAL

The specific ^{14}C radioactivity is measured by liquid scintillation counting after isolation of the substances of interest - e.g. acetic acid from vinegar - by extraction and distillation. The counting efficiency is determined with ^{14}C standards. The results are presented in disintegrations per minute per gram carbon (dpm/gC).

The specific ^{3}H radioactivity is determined either again by liquid scintillation counting after combustion of the foodstuffs to water or by gas proportional counting after transformation of the hydrogen into ethane.

The $^{13}C/^{12}C$ ratio is determined with a mass spectrometer possessing a double or triple collector. The samples are combusted to CO_2. The intensities of mass 44, 45, 46 and 47 are measured, the unwanted contributions of ^{17}O and ^{18}O are corrected for. Afterwards the $\delta^{13}C$ values are calculated.

Details on the experimental procedure are described in three former publications (1-3).

RESULTS

The specific ^{14}C radioactivity of the acetic acid of 3o different vinegar samples was measured(1). Eight of them were of synthetic origin and the rest were biogenic samples. Among the latter were cider, wine, and different special vinegars. The mean value for the specific ^{14}C radioactivity for all the biogenic samples was as expected for the year 1977, namely 23,5 dpm/gC. For seven of the synthetic samples the specific ^{14}C radioactivity was - again as expected - much lower. The mean value was 6,o dpm/gC. Only one synthetic sample had - surprisingly enough - - a specific ^{14}C radioactivity of 22,o dpm/gC. It was the product of a big German company (glacial acetic acid, p.a. of E.Merck, Darmstadt, Federal Republic of Germany). Five batches of glacial acetic acid were analysed over a period of nearly two years. All of them gave the same result within the limits of the experimental error (3,1%, 1σ) of the method.

Also the specific ^{3}H radioactivity was determined (2). All the results were now as expected including the batches of glacial acetic acid of E.Merck, showing that also these samples were of synthetic origin.

To have a further check on these results, the $^{13}C/^{12}C$ isotope ratios were determined (3). Again, the results were as expected. They gave the final proof that the glacial acetic acid of E.Merck was really of syntheti origin and that its specific ^{14}C radioactivity had been adjusted to correspond to the natural amount. Six additional samples were analysed for their $\delta^{13}C$. The results are presented in Table 1.

Acetic acid from vinegar	$\delta^{13}C$ [‰]	
Sample 1	− 24,2	biogenic origin
Sample 2	− 23,9	biogenic origin
Sample 3	− 31,2	synthetic origin
Sample 4	− 30,2	synthetic origin
Sample 5	− 27,1	50% synthetic acid
Sample 6	− 27,7	60% synthetic acid
Oil of bitter almonds		
Biogenic oil of bitter almonds	− 22,0	
Synthetic benzaldehyde	− 26,5	
Honey		
Honey	− 24,0	
High fructose corn syrup (HFCS)	− 16,1	

Table 1: Results of $\delta^{13}C$ determinations for vinegar, oil of bitter almonds, and honey

Four samples were of known origin: Samples 1 and 2 were biogenic ones, samples 3 and 4 synthetic ones. Samples 5 and 6 were of unknown origin. They turned out to be mixtures containing 50% and 60% synthetic acetic acid.

The $\delta^{13}C$ values of acetic acid presented in Table 1 and in reference (3) differ. The values given in Table 1 were obtained by combustion of the complete acetic acid molecule, whereas in reference (3) $\delta^{13}C$ corresponds to the carboxyl group of the molecule. The $^{13}C/^{12}C$ ratio of the carboxyl group is less negative than the value of the complete molecule given in Table 1.

The biogenic oil of bitter almonds ($\delta^{13}C$=−22,0‰) can be distinguished from benzaldehyde of synthetic origin ($\delta^{13}C$=−26,5‰).

Honeys can be adulterated with maize syrup − so-called "high fructose corn syrup" − since the sugar composition of this product is very similar to that of honeys. The ^{13}C content of honeys collected from C_3 plants ($\delta^{13}C$=−24,0‰) and maize syrup ($\delta^{13}C$=−16,1‰) differ more than 7‰ since maize is a C_4 plant. Therefore, the determination of $\delta^{13}C$ allows to ascertain the genuineness of honey or its adulteration with high fructose corn syrup.

In summing up one can state that the determination of the isotopic composition can give valueable answers as to the authenticity and origin of foodstuffs. The content of radioactive isotopes and of stable nuclides can be measured for this purpose. The advantage of correlating the content of stable nuclides with the authenticity of foods is based on the fact that it is more complicated and expensive to manipulate isotope ratios of stable nuclides than to add radioactive nuclides.

REFERENCES

1 E.R.Schmid, I.Fogy and E.Kenndler, Z. Lebensm. Unters.-Forsch. 163 (1977), 121
2 E.R.Schmid, I.Fogy and E.Kenndler, Z. Lebensm. Unters.-Forsch. 166 (1978), 221
3 E.R.Schmid, I.Fogy and P.Schwarz, Z. Lebensm. Unters.-Forsch. 166 (1978), 89

Immobilization of Glucose Oxidase for Analytical Purposes

O. VALENTOVÁ, M. MAREK

Department of Biochemistry and Microbiology, Institute of Chemical Technology, 166 28 Prague 6, Czechoslovakia

SUMMARY

Glucose oxidase (GOD) was immobilized by means of covalent coupling on two basic types of polymers: glycidylmethacrylate copolymers and bead cellulose. These two supports were modified in many different ways and used for immobilization of native and oxidized GOD.
The obtained preparations of immobilized GOD were compared with some derivatives of GOD coupled to ordinary used commercial supports.
The best preparations of immobilized GOD with high absolute activity, good mechanical properties etc. were chosen for analytical application.

INTRODUCTION

Glucose oxidase (E.C.1.1.3.4) is the highly specific enzyme catalyzing the oxidation of β-D-glucose[1]. This selective reaction is very often used for the determination of D-glucose particularly in the presence of other sugars. Immobilized form of the enzyme has another advantages: higher stability in dependence on pH, ionic strength and temperature, lower sensitivity to various activators and inhibitors, possibility of repeated application of enzyme preparation, continual procedure etc. As the general rules of choice of the best immobilization technique for a given system has not yet been found, the determination of the optimum conditions consists in experimental verification of various binding methods and carriers.

MATERIAL AND METHODS

Glucose oxidase in commercial product Spoliase GCL-75 (Středočeská Fruta, yeast plant Kolín, Czechoslovakia) from Aspergillus niger was purified by dialysis, fractional precipitation with ammonium sulphate and chromatography on Sephadex G-200.
The determination of GOD activity together with binding methods have been described in previous paper[2].
Covalent coupling of GOD on nylon net was performed after partial hydrolysis with 3 M hydrochloric acid at 20°C, 20 min. The nylon net was then washed with water, ethanol and dried.
Cyclohexylisocyanide (0.1 - 0.2 μl), 2.5% water solution of glutardialdehyde (0.3 μl) and 15 μl of GOD solution (1 mg GOD/100 μl 0.1 M phosphate buffer, pH 6.8) was applied on partially hydrolyzed nylon net. After 24 h at 4°C was the nylon net with coupled enzyme thoroughly washed with 0.1 M phosphate buffer pH 7.1 and fixed on surface of the oxygen electrode.

RESULTS AND DISCUSSION

Glucose oxidase was immobilized by means of covalent coupling on glycidylmethacrylate copolymers[3] (G-60 and G-70) and bead cellulose[4] modified in different ways:

G-60	
G-70	$\vdash COOCH_2 CH\text{-}CH_2$ (epoxide)
G-60-NH_2	$\vdash COOCH_2 CH(OH)\text{-}CH_2 NH_2$
G-70-EDA	$\vdash COOCH_2 CH(OH)\text{-}CH_2 NH(CH_2)_2 NH_2$
G-70-HMDA	$\vdash COOCH_2 CH(OH)\text{-}CH_2 NH(CH_2)_6 NH_2$
G-70-EDApAB	$\vdash COOCH_2 CH(OH)\text{-}CH_2 NH(CH_2)_2 NHCO\text{-}C_6H_4\text{-}NH_2$

Bead cellulose + EDA $\vdash OCH_2\underset{OH}{CH}-CH_2NH(CH_2)_2NH_2$

Bead cellulose + HMDA $\vdash OCH_2\underset{OH}{CH}-CH_2NH(CH_2)_6NH_2$

The obtained preparations[2] of immobilized GOD were characterized and compared with some derivatives of the enzyme coupled to ordinary used commercial supports (Table I.). The determination of protein content, specific and relative activity, some kinetic data together with mechanical properties made possible to choose the most suitable preparation for analytical application of immobilized enzyme. Selected preparation, GOD bound to G-70-HMDA after glutardialdehyde activation, was used for determination of D-glucose in columnal arrangement with amperometric detection of H_2O_2 generated. Dependence of the height of the wave on the D-glucose content in the sample was linear in the range of concentrations $2.5 \times 10^{-4} - 2.8 \times 10^{-3}$ mol/l (Fig. 1.).

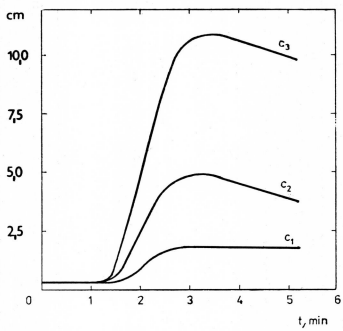

Fig. 1. Continual determination of D-glucose on the column 0.4x16 cm, 0.1M phosphate buffer + 10^{-4}M sodium azide, pH 5.7, flow rate 1 - 2 ml/min, injection 0.025 - 0.1 ml of D-glucose solution: $c_1 = 3 \times 10^{-4}$mol/l, $c_2 = 10^{-3}$mol/l, $c_3 = 2 \times 10^{-3}$mol/l.

Table I. Characterization of some selected samples of immobilized GOD

Carrier	Protein conc. mg/g dry carrier	Amount of bound prot. mg/g dry carrier	Specific activity μkat/mg protein	Relative activity[a] %	Immobilization efficiency[b]	$K_M^{glc}(app)$ (M)
G-60-NH$_2$	8.0	4.5	0.537	21.3	1198	1.68×10^{-3}
G-60-NH$_2$(ox.GOD)	8.0	5.3	0.777	32.5	2153	1.21×10^{-2}
G-70-EDA	8.0	6.4	0.511	19.9	1592	1.13×10^{-2}
G-70-HMDA	8.0	7.7	0.221	8.6	828	2.40×10^{-3}
G-70-EDApAB	30.0	15.5[c]	0.063	2.4	124	
Bead cellulose + EDA	8.0	2.5	1.168	45.6	1425	1.18×10^{-2}
Bead cellulose + HMDA	8.0	2.6	1.177	45.9	1492	1.29×10^{-2}
Sepharose 4B + EDA	30.0	5.6	2.110	49.6	926	
Sepharose 4B + HMDA	30.0	30.6	1.070	25.1	2510	

[c] Protein content determined by the ninhydrin and Lowry methods
[a] Ratio of activities of bound and free enzymes in the same amounts
[b] Product of relative activity and immobilization yield (ratio of amounts of adsorbed and employed enzymes x 100)

D-glucose was also determined by means of enzyme electrode. GOD was immobilized on the partially hydrolyzed nylon net through Ugi's reaction[5] and consequent formation of Schiff's base. The modified nylon net with bound GOD was fixed to the surface of oxygen electrode[6]. Velocity of the oxygen uptake was directly proportional to the concentration of D-glucose in the range from 6.25×10^{-5} to 7.5×10^{-4} mol/l (Fig. 2.).

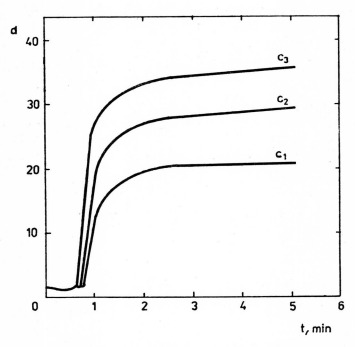

Fig. 2. Determination of D-glucose by enzyme electrode. Total volume of reaction mixture 2 ml, 0.05M citrate buffer pH 5.8, samples 0.05 - 0.2 ml of D-glucose solution: $c_1 = 10^{-4}$ mol/l, $c_2 = 2.5 \times 10^{-4}$ mol/l, $c_3 = 3.5 \times 10^{-4}$ mol/l.

The advantage of both described analytical methods consists mainly in high specificity, short time of analysis and possibility of manyfold application of the immobilized enzyme preparation.

REFERENCES

1 H.J. Bright, J.T. Porter, Flavoprotein oxidases (1975). P. 422 from The Enzymes, Vol.12, P.D. Boyer ed., New York: Acad. Press
2 O. Valentová, M. Marek, F. Švec, J. Štamberg, Z. Vodrážka, Biotechnol. Bioeng. (1981) in press
3 F. Švec, J. Hradil, J. Čoupek, J. Kálal, Angew. Makromol. Chem. 48 (1975) 135
4 J. Štamberg, J. Peška, B. Philipp, D. Paul, Acta Polymerica 30 (1979) 734
5 I. Ugi, Angew. Chem. 74 (1962) 9
6 L.C. Clark Jr., R. Wolf, D. Granger, Z. Taylor, J. Appl. Physiol. 6 (1953) 189

Investigation of Colloid Substances in Must and Wine

J.-C. VILLETTAZ[1], R. AMADO and H. NEUKOM

Swiss Federal Institute of Technology, Zürich, Dept. of Food Science
1) Present Address: Swiss Ferment Co. Ltd., Basle

SUMMARY

The colloid substances of the must and wine are mainly composed of neutral polysaccharides, proteins and pectic substances. The compounds have been investigated by use of gas chromatography, amino-acid analysis and carbazole test. These analytical methods showed, in most cases, a complete identification of the colloid compounds with a high reproductibility. Special attention was given to the investigation of the neutral polysaccharide fraction as well as the protein fraction.

INTRODUCTION

Grapes contain colloid substances which pass into the juice and finally to the wine. These colloids are of high molecular nature and contain neutral polysaccharides, proteins, pectic substances and polyphenols. The total amount of these substances as well as the amount of each compound depends on the stage of wine making (must or young wine), grape variety, climatic conditions, degree of maturity and the processing technology used (1,2,3). Due to their high molecular nature, the colloid substances are often responsible for clarification and filtration difficulties of the wines. The aim of this work was to characterize this high molecular fraction with further details in order to know their nature and to successfully solve the filtration and clarification problems. Until now, several investigations have been conducted (2,4,5,6), however this field still remains of high complexity.

GENERAL METHODS

Neutral Sugar Analysis

The neutral sugars were analysed and determined by gas chromatography as aldonitrile acetate after acid hydrolysis of the neutral polysaccharides (7,8). 5 mg polysaccharides were hydrolysed in a sealed tube with 1 ml 2M trifluoracetic acid for 1 h at 120°C. 0,5 mg D,L Xylose puriss. FLUKA, Buchs, (CH), used as internal standard, was added to the polysaccharide before hydrolysis. After hydrolysis, the trifluoracetic acid was removed by drying under reduced pressure at 50°C. by means of a rotary evaporator. The residue was stored under vacuum at 20°C. with phosphorpentoxyd for 12 h. Then 10 mg dry hydroxylamine hydrochloride and 0,5 ml pyridine were added and heated again for 30 minutes at 90°C. The aldonitrile acetate obtained was analysed by means of a gas chromatograph CARLO ERBA, Type 2300 (CARLO ERBA, RODANO, MILANO, I) under the following conditions:

A glass column (200 x 0,2 cm) packed with chromosorb G80/100 with 3% carbowax MT PA. The oven temperature was kept for 15 min. isotherm at 210°C., then programmed from 210 to 240°C. at 2°C./min. and finally 15 min. isotherm at 240°C. The injector and detector temperature was 250°C. Helium was used as carrier gas at 2,5 bar at a flow rate of 20 ml/min.

Left Column	:	2,08 bar	at 240°C.
Right Column	:	2,12 bar	
FID : Air	:	1,2 bar	
H_2	:	0,75 bar	

The quantitive evaluation from each peak was done by a SPECTRA PHYSICS INTEGRATOR SP 4000 which was directly coupled with a recorder.

Protein Analysis

The protein contents were determined as the sum of the various amino acids after acid hydrolysis fo the ethanol precipitated colloids. 20 mg colloids were hydrolysed with 2 ml 6M HCl during 16 h in sealed tubes under nitrogen. After cooling off, the hydrolysate was filtered and the hydrochloric acid evaporated until dry by a rotary evaporator. The residue was dissolved in a few ml water and concentrated again until dry. This operation was repeated 3 times. Finally, the residue was dissolved in 5,0 buffer pH 2.2 and filtered. 250 ml were injected in the Amino Acid Analyser (BIOCAL B C 201, BIOCAL, München BRD). The analysis was done with a single column procedure with 4 elution buffers as described by WERNER (9). The quantitive evaluation was done with the use of an AUTOLAB SYSTEM AA Integrator (SPECTRA PHYSICS, Darmstadt, BRD).

Determination of the Pectic Substances

The acidic polysaccharides were determined by the carbazole method (10) at 530 mm.

Isolation of the Colloids

The samples investigated were a grape juice and two red wines from the variety Pinot Noir issued by different technologies (11). The samples were supplied by the Swiss Federal Agricultural Research Station at Wädenswil, Switzerland. The samples (50 litres each) were first concentrated under reduced pressure at 40°C. in a rotary evaporator and then dialysed against distilled water during 3 days at 2°C. The non-dialysable material was centrifuged at 18000 g for 20 min. and 5 vol 95% ethanol was added to the supernatant. The precipatate formed was left overnight at 2°C. and then centrifuged (18000 g, 20 min.). The sediment was redissolved in water and centrifuged again. Finally, the supernatant was lyoplilsed. The powder obtained was used for the colloid characterisations.

Results

The total amount of neutral polysaccharides of the wines and must were determined by GC with xylose as internal standard. Table 1 shows the total amount of polysaccharides aswell as the distribution of the sugar monomers.

TABLE 1   Characterisation of the Neutral Polysaccharide Fraction

	Rha %	Fuc %	Ara %	Man %	Glu %	Gol %	Total in mg / l
HKZE-TS	4,1	2,9	12,9	19,9	12,6	48,5	181,5
HKZE-W	6,1	2,2	20,8	34,3	7,3	30,3	596,7
SD	2,1	-	9,0	65,4	6,6	16,9	755,4

HKZE-TS   Hot treated grape juice
HKZE-W   Wine from HKZE-TS
SD   Rosé wine

The protein level in must and wines as well as the distribution of the amino acids after acid hydrolysis of the proteins are show in Table 2.

TABLE 2   Characterisation of the Protein Fraction

Amino Acids	HKZE-TS	HKZE-W	SD
Hydroxyproline	13,37	1,96	3,20
Aspartic Acid	6,10	6,22	9,21
Threonine	11,56	10,15	13,35
Serine	20,52	15,19	11,53
Glutamic Acid	6,51	11,22	9,58
Proline	2,55	3,36	5,72
Glycine	1,59	9,76	5,17
Alanine	12,70	0,51	7,39
½ Cystine	-	-	2,53
Valine	1,55	1,91	5,11
Methionine	2,10	-	1,75
Isoleucine	2,69	3,00	3,29
Leucine	3,55	3,63	3,76
Tyrosine	1,79	2,17	3,18
Phenylalanine	1,20	1,57	3,22
Histidine	1,73	-	0,37
Lysine		5,10	6,09
Arginine	0,63	16,19	2,56
Protein Content mg/l	6,5	55,2	126,2

TABLE 3   Indicate the Amount of Pectic Substances found in the Must and Wine Colloids

	HKZE-TS	HKZE-W	SD
Pectic Substances in mg/l	46,6	31,6	12,5

These analytical data show that the neutral polysaccharide (Table 1) represents the greater part of the must and wine colloid fraction. The influence of the alcoholic fermentation on the composition of the sugar monomers is also clearly shown. Further investigations showed that during alcoholic fermentation a yeast mannane passes into the wine (11,12). The same observation was made for the protein fractions (Table 2). The amount of pectic substances decreases during the alcoholic fermentation and represents only a few percent of the total colloid amount of the wines. The amount of identified substances is compared with the total colloid amount obtained by ethanol precipitation (Table 4).

TABLE 4   Summarizing Table

	HKZE-TS	HKZE-W	SD
Total amounts of colloids after ethanol precipitation (mg/l)	252,0	721,6	896,0
- Polysaccharides	181,5	596,7	755,4
- Proteins	6,5	55,2	126,2
- Pectic Subst.	46,6	31,6	12,5
Total identified in mg	234,6	683,5	894,1
Total identified in %	93,1	94,7	99,8

These figures indicate clearly that the use of gas chromatography, the amino-acid analysis and the carbazole test permits a very good characterisation of the colloid substances of must and wine. In some cases, a quantitive determination of the polyphenols might be necessary in order to obtain a full clarification of the colloids.

REFERENCES

1. Danilatos, N., Bull. OIV, 52, Nr. 580, 456 (1979)
2. Doubourdieu, D., Thèse, Université Bordeaux II, (1978)
3. Boehringer, P. and Doelle, H., Z. Lebensm. Unters. U. Forsch., 111, 121, (1959)
4. Buechi, W., Diss ETH Nr. 2332 (1954)
5. Usseglio-Tomasset, L. and Castino, M., Riv. Vitic. Enol., 28, 328; ibidem 374; ibedem 401, (1975)
6. Mourgues J., Ann. Technol. agric., 28, (1), 121, (1979)
7. Albersheim, P. Nevins, D.J., English, P.D. and Karr A., Carbohydr. Res., 5, 340, (1967)
8. Mergentahler, E. und Scherz, H., Z. Lebensm. Unters. u. Forsch., 162, 25, (1976)
9. Werner, G.C., Diss ETH Nr. 5754 (1976)
10. Bitter, T. and Muir, H.M., Anal. Biochem., 4, 330-334 (1962)
11. Villettaz, J.C., Diss ETH Nr. 6503
12. Villettaz, J.C. Amado, R. Neukom, H., Horisberger, M., and Hormon, J., Carbohydr. Res., 81, 341, (1980)

Investigation of Proteins Regarding their Functionality for Baking

H. J. G. WUTZEL

Verein zur Förderung des Bäckergewerbes, A 1080 Wien

SUMMARY

The Brabender Extensograph makes possible to determine the functionality of proteins for baking. Many proteins deteriorate the loaf volume of bread when added to a given wheat flour. But besides wheat gluten also whey proteins can improve the loaf volume. This effect is increased by physical modification of whey proteins. Therefore investigations must be intensified to find a) other proteins b) physical and chemical methods c) additives to proteins improving the dough forming properties of wheat flour and other starch products.

Introduction

The development of methodologies for determination the functional properties of proteins may be of interest. The rheologic method using the Brabender Extensograph according ICC standard No. 114 is suited to determine the following dough-forming properties:

Area of Extensogramm (cm^2) = Expected bread volume

Ratio of dough resistance to dough extensibility = Aging condition of dough

RESULTS

TABLE I

Influence of adding 10 % protein to wheat flour

	Area of Extensogramm %
Wheat flour	100
Soya concentrate	94
Single Cell Protein (B. methylomonas)	90
Single Cell Protein (S. cerevisiae)	85
Soya isolate	84
Egg albumine	84
Casein	74
Potatoe protein	72
Gliadin	65

The addition of protein deteriorates in many cases the dough forming properties of a given wheat flour (depression of loaf volume resulting from dilution of gluten according Knorr and Betschard).

TABLE II

Influence of adding 10 % gluten or whey to wheat flour

	Area of Extensogramm %
Wheat flour	100
Gluten	111
Powdered sweet whey	105
Modified sweet whey	129
Whey protein	131

Dough forming properties (loaf volume) are increased by some proteins and protein products as gluten, whey and whey products.

TABLE III

Influence of physical and biochemical modifications of proteins on dough forming properties. 10 % protein added to wheat flour.

	Area of Extensogramm %
Wheat flour	100
Skimmed milk powder	90
Casein	74
Whey protein (denaturated)	131
Single Cell Protein (B. methylomonas)	90
Single Cell Protein (B. methylomonas – modified by proteolytic enzymes)	101

Dough forming properties can be improved by physical and biochemical methodes. So the fractionation of milk proteins leads to: 1) Casein with poor dough forming properties and 2) whey protein with increased baking quality. On the other side the deteorating effect of single cell protein can be altered by enzymatic modification.

TABLE IV

The influence of gluten on starch diluted wheat flour

% Wheat flour	% Potato starch	% Gluten	Area of Extensogramm %
100	–	–	100
80	20	–	94
70	30	–	68
60	30	10	108

Adding of gluten improves the dough forming properties of a starch diluted wheat flour.

DISCUSSION

Investigations must be intensified to improve the dough forming properties of wheat flour and other starch products by

proteins,
physical and chemical methods modifying such proteins,
additives to proteins.

The target should be to produce proteins and methods which act as gluten on starch enriched wheat flour. The Brabender Extensograph seems the given method for such a work.

REFERENCES

1   ICC Standard No 114., Intern. Ass for Cereal Chemistry,
    A 2320 Schwechat, Austria.
2   D. Knorr and A.A. Betschard, Lebensm.-Wiss. u. Technol. 11.
    198 - 201 (1978).
3   T.P. Labuza and al. Journ. Food Science 37., 103 - 107 (1972).
4   T.P. Labuza and K.A. Jones, Journ. Food Science 38., 177 (1973).
5   M. Lindblom, Lebensm.-Wiss. u. Technol. 10., 341 - 345 (1977).

Investigation about Lead and Cadmium in Wild Growing Edible Mushrooms from Differently Polluted Areas

J. DOLISCHKA, I. WAGNER

Institut für Technologie und Warenwirtschaftslehre der Wirtschaftsuniversität Wien, Austria

SUMMARY

Wild edible mushrooms and the corresponding substrate samples from differently polluted areas were examined as to their cadmium and lead content by means of flame-atomic absorption spectrophotometry. A very high degree of contamination was found in mushrooms and substrates from the surroundings of a lead and zinc plant. But even in unpolluted areas mushrooms (Macrolepiota procera) were found with an increased cadmium content due to their strong enrichment capacity. The different cadmium distribution in stalks, caps and capillary tubes/gills should be considered when consumed.

INTRODUCTION

The emission of more and more quantities of lead and cadmium through most diverse industrial processes, the combustion of coal and mineral oil products and also the use of phosphate fertilizers containing cadmium and sewage sludge in agriculture have led to the fact that lead and cadmium must be regarded as ubiquitious trace elements. This is why the control of foodstuffs as to these and other toxic elements and the study of the causes of contamination becomes more and more important.

As different investigations have shown already, higher mushrooms are especially suited to be bio-indicators for toxic heavy metal pollution of different regions.(1,2) Therefore, the objective of the present study was to examine lead and cadmium in wild growing edible mushrooms from differently polluted areas. Furthermore the content of both heavy metals were to be determined in the corresponding substrates of the different mushroom samples. The distribution of lead and cadmium in the different parts of the mushroom seemed to be interesting, too. Therefore, according to the available quantity of samples the different mushrooms were to be divided into stalks, caps and tubes/gills and each examined separately.

EXAMINATION MATERIAL AND ANALYTIC METHOD

The examined mushrooms and the corresponding substrate samples were collected in the following areas:
a) Unpolluted, montain areas without traffic in the province of Carinthia (Rubland, Wiederschwing, Sternberg).
   The following species of edible mushrooms were chosen for examination: Cantharellus cibarius (C.c.), Boletus edulis (B.e.) and (Macro)-Lepiota procera (L.p.).
b) Proximity zone of a bypass road in the periphery of Villach (St. Martin, St. Johann).
   Here also the species Cantharellus cibarius, Boletus edulis and Macrolepiota procera were chosen as samples.
c) Surroundings of a lead and zinc plant in Arnoldstein in Carinthia (Stossau, Klausen).
   Since in this area only boletuses were found, which is rather interesting, the following mushrooms were chosen: Boletus edulis, Boletus badius (B.b.), Boletus subtomentosus (B.s.) and Boletus chrysenteron (B.c.).

The different mushroom samples were cleaned, sliced, dried at a temperature of $80^{\circ}C$. Then the dried mushrooms were thoroughly homogenized in a blendor and put into ground glass bottles.
Determination of the cadmium content: According to the cadmium content to be expected up to 1 g of the dry mushroom powder was weighed and wet-ashed by means of concentrated sulfuric acid and concentrated nitric acid. Perhydrol was used for the final elimination of coal rests.(3)
Determination of the lead content: Up to 10 g of the mushroom powder was dry-ashed in a porcelain bowl with magnesium nitrate as incineration agent at $500^{\circ}C$ and from time to time wetted with concentrated nitric

acid if necessary. The pure white ash thus obtained was dissolved in little halfconcentrated hydrochlorid acid.(4)
As quantitative determination method, AAS was used for both elements. The decomposed sample solutions were adjusted to a certain pH-value (Cd...pH 3,9 Pb...pH 3,2) and the heavy metal ions complexed with ammoniumpyrrolidine-thiocarbamate and extracted by 4-methylpentanone(2). The heavy metal complex extract was sprayed into the flame of AAS.
The different soil samples were leached with a solution of ammonacetate and the heavy metal content of the substrate extract measured after filtration directly by AAS.(5)

RESULT AND DISCUSSION

The most important results of this investigation are shown in the following figures. Detailed values of the different analyses are shown in (6). The dependence of the heavy metal content on the locality but also on the mushroom species are shown in figure 1.

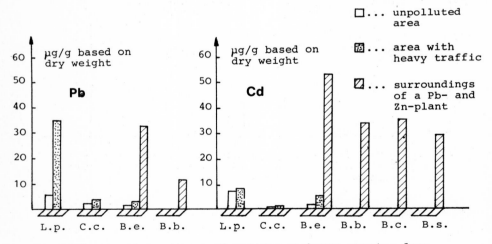

Fig.1: Heavy metal content of various mushroom-species from differently polluted areas

As mentioned above, certain mushroom species like the Macrolepiota procera and the Cantharellus cibarius could not be found in the heavy polluted surroundings of the lead and zinc plant. This is why only boletuses were available for analysis. As expected, mushroom samples from areas with heavy traffic showed both a higher lead and cadmium content than samples from unpolluted areas. A statistically significant difference in the heavy metal concentrations of mushrooms from these two areas could not be determined, however! The altogether low heavy metal content of the species Cantharellaceae should be particularly mentioned. The comparative data of the metallurgical region Arnoldstein, however, were significantly increased. (Probability of accuracy of the t-test: Pb...95%; Cd...99,5%). Peak values of 32,5 ppm of lead and 53 ppm of cadmium (both based on dry weight) were found in a big Boletus edulis. The average content of boletuses from this area amounted to 15 ppm Pb and 35,5 ppm Cd.

The high heavy metal content of the mushrooms from this region can be understood when looking at the lead and cadmium pollution of the corresponding substrate samples (figure 2). The normal values for unpolluted soils of 10 ppm Pb and 0,01-0,1 ppm Cd referred to in (7) are exceeded by 100 to 200 times (Pb) or rather by 200 to 2000 times (Cd).

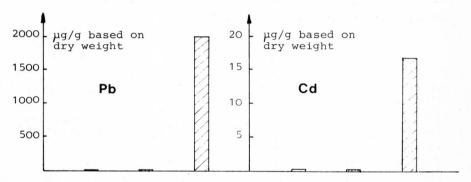

Fig.2: Average heavy metal concentration in substrates (unpolluted area, area with heavy traffic, surroundings of a lead- and zincplant)

Figure 3 shows the different concentrations of lead and cadmium in substrates and in mushrooms. While the lead concentration is on the average twice as high in the substrate than in the mushroom, cadmium indicated a significant enrichment in the mushroom. When the values in the substrate were extremely high, great quantities of cadmium and lead were absorbed, but the enrichment factor remained considerably lower. For example:

	Mushroom (ppm)	Substrate (ppm)	Enrichment Factor
Lead	B.e. 32,5	1.220	0,03
	B.b. 11,9	5.800	0,002
Cadmium	B.e. 53,4	11,4	4,7
	B.b. 46,3	23,2	2,0

A possible explanation for this increased mobility of the cadmium presents itself here over the formation of protein complexes, which play an important role in the cadmium metabolism of human being and animals. (8) Buth neither more recent detailed examinations could completely clarify this fact. (9)

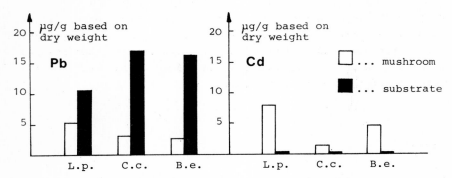

Fig.3: Comparison between heavy metal concentrations in mushrooms and substrates (unpolluted area and area with heavy traffic)

The distribution of the heavy metals lead and cadmium over different parts of the mushroom have repeatedly been examined. The results of our study widely confirms the statements thereof.(10,11) Figure 4 shows an almost equal lead distribution in the mushrooms, while cadmium concentrates in the ratio of approx. 1:2:4 for stalks, caps and tubes/gills, respectively. The calculation showed, that on the average 60% (in the Boletus edulis even 75-80%) of the total cadmium content of a mushroom are concentrated in the tubes or rather the gills.

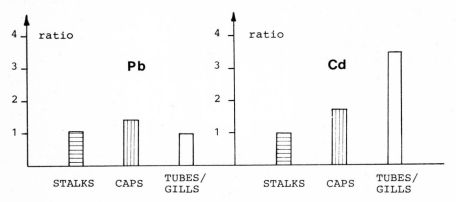

Fig.4: Ratio of the average heavy metal concentration in stalks, caps and tubes/gills

Thus it is recommended not to eat these parts of the mushroom, since a considerable reduction of the cadmium intake can be achieved in this way.

Nevertheless mushrooms coming from heavily polluted regions should not be consumed, since already 1 kg of fresh mushrooms exceeds the tolerance value set by the FAO/WHO of 1,6 - 2,0 mg of cadmium per man and month.

Finally it should be noted, that obtaining sufficient examination material constitutes a considerable problem. It is not always possible to make a representative examination of the different localities and/or mushroom species. We recommend further studies but also increased cooperation of the bodies concerned.

REFERENCES

1 W. Rauter, Z. Lebensm. Unters. Forsch. 159 (1975), 149
2 M. Enke, H. Matschiner and M.K. Achtzehn, Die Nahrung 21 (1977), 331
3 H. Woidich and W. Pfannhauser, Z. Lebensm. Unters. Forsch. 155 (1974) 72
4 H. Woidich and W. Pfannhauser, Beiträge - Umweltschutz, Lebensmittelangelegenheiten, Verterinärverwaltung 1977, 4,3
5 H. Kahn, F. Fernandez and S. Slavin, Atomic Absorption Newsletter 11 (1972), 42
6 J. Dolischka, Untersuchungen über die Schwermetallbelastung unterschiedlich stark exponierter Gebiete an Hand der Blei- und Cadmiumgehalte von Pilzen, Diplomarbeit, Wirtschaftsuniversität Wien 1980
7 ÖKO-Institut Freiburg, Wissenschaft aktuell 1980, 31
8 W. Mücke, Chem. Rd. 31 (1978), 1
9 H. Kruse and A. Lommel, Z. Lebensm. Unters. Forsch. 168 (1979), 444
10 P. Collet, Dtsch. Lebensm.-Rdsch. 73 (1977), 75
11 R. Seeger, Z. Lebensm. Unters. Forsch. 166 (1978), 23

Evaluation of Lead by Automated Anodic Stripping Voltammetry in Canned Juice in Presence of High Tin Concentration

A. CARISANO

Laboratorio Ricerche e Controllo Star, S.p.A. Agrate B. Milan, Italy

G.P. CELLERINO

Environmental Sciences Associates, Bedford, Massachusetts, USA

G.C. DELLATORRE

Laboratorio Ricerche e Controllo Star, S.p.A. Agrate B. Milan, Italy

SUMMARY

Anodic stripping voltammetry is a fast, practical, very versatile and accurate method for Pb evaluation in foodstuffs, which has been available for a few years. Nevertheless, when in the sample to be tested for Pb concentration the amount of Sn exceeds 50-100 ppm the analytical results can be greatly affected. In practice, this is a concrete possibility in the evaluation of Pb in fruit and vegetable juices canned in tins with an Sn coating. This interference has been overcome by us by the oxidation of Sn contained in the sample of 300 μl juice, using 50 μl 1.1 N $KMnO_4$. In this way the Sn present is totally transformed into SnO_2, a non active electrochemical species which does not interfere with Pb detection, and hence the fast, accurate measurement of Pb despite the presence of high Sn concentrations is made possible.

INTRODUCTION

The evaluation of the Pb concentration in foods is normally performed in laboratories of canned food industries. Several techniques are available for this evaluation. At present the most popular are: atomic absorption, polarography, spectrophotometry, and anodic stripping voltammetry (ASV). ASV in the automated version recently available using Metexchange reagent is very convenient because it is possible to use the sample directly, without chemical treatment or solvent extraction (1).

ASV is based on the preconcentration of the element to be measured into a thin mercury film supported by a special graphite electrode, to which a negative potential (plating potential) is applied. The ions originally present in the solution are transformed into neutral atoms and are dissolved in the mercury film forming an amalgam. At the end of the plating step the potential is changed in the anodic direction (positive); in this way the atoms are stripped from the electrode and return as ions into the starting solution.

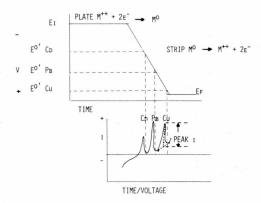

Figure 1   The electrochemical processes involved in anodic stripping voltammetry

When a certain metal begins to strip from the electrode it generates a sharp current peak, as shown in fig. 1. The peak area is proportional to the metal's concentration in the solution, while the peak position, measured in volts, indicates the metal species. The plating and the stripping steps are summarized in figs. 2 and 3.

Previously this technique could not be used for testing products containing Sn concentrations greater than 30-50 ppm (2). This was due

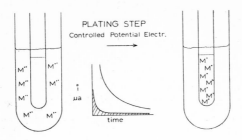

Figure 2   Plating of the metal (M)

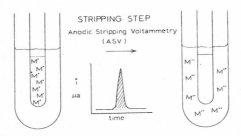

Figure 3   Stripping of the metal (M)

to the Sn peak either reinforcing or cancelling the Pb peak, giving apparent Pb variation. A similar phenomenon has also been observed using polarography (3). When the determination by ASV on a sample containing a high Sn concentration is repeated several times, the Sn electrochemical peak decreases steadily. This is because Sn oxidation generates a certain amount of SnO_2, an electrochemically non active species. Considering this, we decided to remove the interference of the Sn with the Pb evaluation by chemical oxidation of the Sn present.

MATERIALS AND METHOD

Materials:
solution of 1.1 N $KMnO_4$
Metexchange reagent
decontaminated plastic tubes
micropipettes

trace metal analyzer, ESA model 3010 A
standard solution containing 0.5 ppm Pb
standard solution containing 10 000 ppm Sn

Method:

0.3 ml homogeneous juice to be tested were introduced into a 5 ml decontaminated plastic tube. 50 µl 1.1 N $KMnO_4$ were added to the above tube and mixed. 2.9 ml Metexchange reagent (ion exchange reagent) were then added to the above mixture, which was analyzed directly on the trace metal analyzer.

Instrument's Parameters:

recorder set point	-0.560 V
integration set point	-0.500 V
blank correction	8.0
calibration	21.5
time	2 minutes
scale expansion	×1 or ×10

RESULTS AND DISCUSSION

To check the Pb recovery obtained by this method we added variable amounts of Pb, spiking an apple juice containing no Pb. Table 1 shows the results.

Table 1 Apple juice samples to which variable amounts of Pb had been added

sample no.	theoretical Pb conc. (ppm)	measured Pb conc. (ppm)
1	0.35	0.37
2	0.15	0.18
3	0.07	0.08
4	0.04	0.04

It was observed that in some old tins of canned juice the Sn content exceeded some hundred ppm. To verify the effect of Sn interference, we spiked pear juice containing no Pb with variable amounts of Sn. The data obtained are shown in table 2. The instrument readout of

Table 2 Pear juice samples (no Pb) with different Sn additions

Sn conc. (ppm)	instrument readout
0	1
50	1
100	1
150	1
175	1
200	35
225	203
250	139
275	74
300	169
330	13
600	-52

Table 3 Pear juice samples (no Pb) with various Pb, Sn and $KMnO_4$ additions

Pb (ppm)	Sn (ppm)	$KMnO_4$ (µl)	readout (ppm)
0.0	0	0	0.03
0.5	0	0	0.47
0.0	300	0	-0.50
0.5	300	0	-0.22
0.0	300	50	0.05
0.5	300	50	0.45
0.5	600	50	0.49

table 2 divided by 100 indicates the apparent Pb concentration in ppm due to Sn interference; this is because the instrument had already been calibrated with Pb standards. As shown in table 2 the Sn interference can be either positive or negative; it is not reproducible, depending upon the actual concentration during the plating step of Sn electrochemically active in the solution.

With the same pear juice we checked the effect of $KMnO_4$ addition to remove the Sn interference. Table 3 shows the sample treatment and the result obtained. The above data indicate that the addition of 50 μl 1.1 N $KMnO_4$ completely removes the Sn interference at least up to 600 ppm Sn. Therefore, it is possible to measure by ASV the actual Pb concentration in canned juice, even with Sn concentrations higher than the normally accepted values (300 ppm Sn). We also tested other juices with similar results (table 4).

Table 4   Additions of Sn, Pb and $KMnO_4$ to different fruit and vegetable juices

juice	Pb (ppm)	Sn (ppm)	$KMnO_4$ (μl)	readout (ppm)
apple	0.5	0	0	0.49
apple	0.5	300	0	-0.31
apple	0.5	300	50	0.53
peach	0.5	0	0	0.56
peach	0.5	300	0	-0.13
peach	0.5	300	50	0.56
apricot	0.5	0	0	0.65
apricot	0.5	300	0	-0.01
apricot	0.5	300	50	0.67
carrot	0.5	0	0	0.62
carrot	0.5	300	0	-0.19
carrot	0.5	300	50	0.61
paprika	0.5	0	0	0.61
paprika	0.5	300	0	1.91
paprika	0.5	300	50	0.62

CONCLUSION

By a simple spike of 50 μl 1.1 N $KMnO_4$ to the sample of juice, it is possible to remove the interference of Sn in Pb determinations, up to at least a concentration of 600 ppm Sn. Therefore, in samples with a possibly high Sn concentration, we suggest that the Sn concentration should be evaluated before the Pb determination is carried out. If the Sn concentration exceeds 100 ppm it is necessary to add $KMnO_4$. Obviously, in such a case the instrument's parameters (especially the integration set point) must be set in the presence of $KMnO_4$.

REFERENCES

1 E. Zink, W. Matson, S. Pfeiffer and A. Pietrzk, J. Ass. Offic. Anal. Chem. 61 (1978), 653
2 Trace metal analyzer ESA 3010 A manual
3 E. Desimoni, F. Palmisano and L. Sabbatini, Anal. Chem. 52 (1980), 1889

Index

AAS (Atomic Absorption Spectroscopy)   372
Amines   234
Amino Acid Analyser   97
Amino Acids   190, 458
Antioxidative Activity   409
Arsenic   367
Automated Anodic Stripping Voltammetry   492

Baking   482
Beans   292
Bioassays   205
Brain Function   234

Cadmium   372, 389, 486
Caffeate, Ethyl   41
Caffeate, Methyl   41
Caramelisation   397, 409
Carbamate Pesticides   437
Carbonisation   397
Carboxylic Acids, Short Chain   47
Cereals   383
Chlorine Containing Organic Compounds   322
Chromatography   306
Chromium   389
CID-MIKE Mass Spectrometry   190
Colloid Substances   476
Copper   389
Cruciferae Oils   174

Dairy Products   377
Dyes   64

Electrochemical Detection   41
Electrophoresis   264
Elements, Analysis   335
Enzymatic Activities   300
Enzymatic Determination of Starch   286
Enzymes   270, 306

Fish   114, 367

Flavours, Artificial   168
Flavour Mixtures   162
Flour   403, 461
Food Additives   59, 149, 428
Fructose   453
Fruit Drinks   458
Fruit Juice   97, 229
Fruits   105
Fusarium Mycotoxins   143

Galacturonic Acid   53
Gas Chromatography   120, 174, 183, 458
GC/MS   183
Gel Chromatography   105, 108
Gelling and Thickening Agents   327
Glass Capillary Columns   143
Glucose   453
Glucose Oxidase   470

Heating of Food Proteins   416
Heavy Metals   383
Honey   137, 453
HPLC (High Performance Liquid Chromatography)   3, 14, 27, 33, 41, 47, 53, 64
HPTLC (High Performance Thin-Layer Chromatography)   114

Immobilization of Glucose Oxidase   470
Immunochemistry   215
Immunology   449
Infant Formula   443
Inorganic Contaminants   356
Ionisation, Chemical   183
Ion Monitoring, Selected   196
Ion-Pair Liquid Chromatography   59
Isoelectric Focusing   247, 264
Isotopic Composition   464

Juice, Canned   492

Lead   372, 377, 389, 486, 492

Lipids   443
Lipoproteins   403
Lysine   422

Maillard Reaction Products   409
Manganese   389
Mapping   264
Mass Spectrometry   155, 174, 183, 190
Meat   70, 300
Mercury   389
Microbial Contamination   229
Milk   377
Morphology in Meat   300
Mushrooms   486
Mussel   372
Must   264, 476
Mycotoxins, Fusarium   143

Near Infrared Reflectance   397

Oxygen Probe   286

Pesticides   83, 183, 437
Phaseolus vulgaris   292
Phenols, in Smoke   70
Polycyclic Aromatic Hydrocarbons   76, 149
Post Mortem Changes   300
Propylene Carbonate   149
Protein Analysis   215, 247
Protein Hydrolysates   322
Protein Mapping   264
Proteins   264, 306, 416, 482
Proteolytic Activities   240
Pyrethrins   461

Reagent Gases   162
Reverse Phase - HPTLC   114

Saccharose   453
Sensory Evaluation   422
Sensory Profiles   108
Smoke   70
Smoked Meat Products   70
Soap - TLC   114
Starch   286
Statistics   377
Sugars, Rare   105
Sugars, Reducing   422
Sweeteners, Artificial   428

Thermal Processing   416
Tin   492
Trace Contaminants   196
Trace Elements   389
Trypsin Inhibitor   292
Tuna Fish   114
Turkey   449

Ultrathin-Layer Isoelectric Focusing   264

Vegetables, Determination of Methyl- and Ethylcaffeate in   41
Vitamins   27, 33
Vitamin B_6 132
Volatile Components, Honey   137
Voltammetry   372, 492

Wheat Flour   403
Wine   264, 476

Zinc   389